THE GENUS

CYMBIDIUM

A BOTANICAL MAGAZINE MONOGRAPH

THE GENUS

CYMBIDIUM

David Du Puy and Phillip Cribb

With notes on Culture and Breeding by
Michael Tibbs

Kew Publishing
Royal Botanic Gardens, Kew

PLANTS PEOPLE
POSSIBILITIES

First published in 2007 by
Royal Botanic Gardens, Kew
Richmond, Surrey, TW9 3AB, UK
www.kew.org

ISBN 978-1-84246-147-1

Also published in South East Asia by: Natural History Publications (Borneo) Sdn. Bhd. (216807-X)
A913, 9th Floor, Wisma Merdeka Phase 1, P.O. Box 15566, 88864 Kota Kinabalu, Sabah, Malaysia.
Tel: 088-233098   Fax: 088-240768
info@nhpborneo.com
www.nhpborneo.com

British Library Cataloguing in Publication Data
A catalogue record for this book is available from the British Library

Cover design by Kew Publishing,
Royal Botanic Gardens, Kew

Plates by Claire Smith, with additional plates by George Bond, J. Eyre, W.H. Fitch, Mary Grierson, W. Griffith, C. Mukerjai, Hugh Low, G. Mann, Charles Parish, Stella Ross-Craig, Matilda Smith, Lilian Snelling, Margaret Stones and H. Weddell

Line drawings by Claire Smith, with additional figures by Andrew Brown, Deborah Lambkin, Mutsuko Nakajima, Susanna Stuart-Smith and Hazel Wilkes

Additonal photographs by L. Averyanov, K. Barrett, C.L. Chan, Chen Sing Chi, C. Clarke, M. Clements, the late J.B. Comber, N. Cruttwell, S. Dorji, S. Gale, T. Harwood, F. Hoeck, T. Hurrell, B. Klein, Luo Yi Bo, D. Menzies, H. Perner, the late G. Seifenfaden and N. Trudel

Printed in Italy by Printer Trento

For information or to purchase all Kew titles please visit
www.kewbooks.com or email publishing@kew.org

All proceeds go to support Kew's work in saving the world's plants for life

CONTENTS

LIST OF COLOUR PLATES

1. *Cymbidium elegans*. Unknown Indian artist for J. Cathcart. Kew collection.
2. *Cymbidium erythraeum* [as *C. longifolium*]. Unknown Indian artist for J. Cathcart. Kew collection.
3. *Cymbidium iridioides* [as *C. giganteum*]. Unknown Indian artist for J. Cathcart. Kew collection.
4. *Cymbidium chloranthum*. Curtis's Botanical Magazine t. 4907, del. W.H. Fitch.
5. *Cymbidium floribundum*. China, *Andrew* s.n., cult. Kew, del. Claire Smith.
6. *Cymbidium canaliculatum*. Curtis's Botanical Magazine t. 5851, del. W.H. Fitch.
7. *Cymbidium suave*. Australia, George Bond for A. Cunningham, 1823, Kew collection.
8. *Cymbidium aloifolium*. *Seth* 154, cult. Kew, del. Claire Smith.
9. *Cymbidium atropurpureum*. Curtis's Botanical Magazine t. 5710, del. W.H. Fitch.
10. *Cymbidium atropurpureum*. Sarawak, *Giles & Woolliams* s.n., cult. Kew, del. Claire Smith.
11. *Cymbidium bicolor* subsp. *pubescens*. Sarawak, del. Hugh Low, Kew collection.
12. *Cymbidium finlaysonianum*. Sarawak, *Giles* 600, cult. Kew, del. Claire Smith.
13. *Cymbidium devonianum*. Curtis's Botanical Magazine t. 9327, del. Lilian Snelling.
14. *Cymbidium devonianum*. NE India, cult. Kew, del. Claire Smith.
15. *Cymbidium dayanum*. Thailand, *Menzies & Du Puy* 72, cult. Kew, del. Claire Smith.
16. *Cymbidium erythrostylum*. Curtis's Botanical Magazine t. 8131, del. Matilda Smith.
17. *Cymbidium eburneum*. Curtis's Botanical Magazine t. 5126, del. W.H. Fitch.
18. *Cymbidium elegans*. N India, *Seth* 109, cult. Kew, del. Claire Smith.
19. *Cymbidium hookerianum*. Curtis's Botanical Magazine t. 5574, del. W.H. Fitch.
20. *Cymbidium hookerianum*. *Young* s.n., cult. Kew, del. Claire Smith.
21. *Cymbidium insigne* subsp. *insigne*. Curtis's Botanical Magazine t. 8312, del. Matilda Smith.
22. *Cymbidium insigne* subsp. *seidenfadenii*. Thailand, *Menzies & Du Puy* 500 (right) and 501 (left), del. Claire Smith.
23. *Cymbidium iridioides*. Curtis's Botanical Magazine t. 4844, del. W.H. Fitch.
24. *Cymbidium lowianum* var. *lowianum*. Cult. Kew, del. Claire Smith.
25. *Cymbidium mastersii*. NE India, Khasia Hills, del. W. Griffith, used by Sarah Drake as model for Edward's Botanical Register 31: t. 50 (1845).
26. *Cymbidium parishii*. Myanmar, Megala Chyoung. Illustration of the type by Charles Parish, 1867, Kew collection.
27. *Cymbidium sanderae*. *Seth* 87, cult. Kew, del. Claire Smith.
28. *Cymbidium schroederi*. Curtis's Botanical Magazine t. 9637, del. Lilian Snelling.
29. *Cymbidium tracyanum*. Cult. Kew, del. Mary Grierson, Kew collection.
30. *Cymbidium tracyanum*. Cult. Kew, del. Stella Ross-Craig, Kew collection.
31. *Cymbidium whiteae*. Sikkim, del. C. Mukerjai from Annals of the Royal Botanic Garden, Calcutta t. 258 (1898).
32. *Cymbidium tigrinum*. NE India, *Rittershausen* s.n., cult. Kew, del. Claire Smith.
33. *Cymbidium ensifolium* subsp. *ensifolium*. China, del. H. Weddell, Curtis's Botanical Magazine: t. 1751 (1815).
34. *Cymbidium goeringii*. Japan, cult. Kew, del. Margaret Stones.
35. *Cymbidium sinense*. Cult. Kew, del. Claire Smith.
36. *Cymbidium sinense*. Hong Kong, del. J.Eyre.
37. *Cymbidium lancifolium* var. *aspidistrifolium*. *Andrew* s.n., cult. Kew, del. Claire Smith.
38. *Cymbidium macrorhizon* var. *macrorhizon*. NE India, Khasia Hills, Shillong, del. G. Mann.

Front cover: *Cymbidium tracyanum* by Mary Grierson, Kew collection

ACKNOWLEDGEMENTS

Facilities and materials were kindly provided by Professor Simon Owens, Keeper of the Herbarium of the Royal Botanic Gardens at Kew. Much of the living material has been grown in the Living Collections at Kew by Chris Bailes, Dave Menzies, Sandra Bell and Kathy King, under the curatorships of John Simmonds and Nigel Taylor. Visits have been made to herbaria at Bangkok, Bogor, Beijing, Brisbane, Edinburgh, Guangdong, Kunming, Harvard, Leiden, the Natural History Museum, London, Melbourne, New York, Paris, Singapore, Sydney, Taipei, Tokyo and Vienna. We would like to thank the Directors and staff of these institutions for their assistance. Fieldwork has been undertaken in Thailand, China, Vietnam, Bhutan, India, Malaysia, Indonesia, New Guinea and Australia. Work in Thailand was made possible through funding generously donated by the Bentham-Moxon Trust, the Studley College Trust, D. Clulow, R. Bilton, the Orchid Review and B. Rittershausen. We would like to thank TOBU Department Store, Tokyo for supporting fieldwork in China and Vietnam.

We would like to thank the Orchid Digest Foundation, inspired by the late Walter Bertsch, for their support in publishing *The Genus Cymbidium* (1988), our first monograph of the genus. The Orchid Digest were supported in this venture by the following sponsors: the San Diego Orchid Society, the Cymbidium Society of America, the South Bay Orchid Society and the San Gabriel Orchid Society, Helen Congleton, the Orchid House, Mr & Mrs Don Bradish and Bill Bailey. The authors are particularly grateful for the support of Ernest Hetherington, Harold Koopowitz, the late Trudi and Fordyce Marsh, Lance Birk, the late Don Herman and other members of the Foundation.

Many friends and colleagues have helped us with advice, material and inspiration over the past 25 years, notably Keith Andrew, Leonid Averyanov, Gloria Barretto, Chris Bailes, Andrew Bacon, Ray Bilton, Chan Chew Lun, Lawrence Chau, Chen Sing Chi, Mark Clements, David Clulow, Julian Coker, the late Jim Comber, Andy Easton, Brian Ford-Lloyd, Anna Haigh, Meta and Fritz Held, Nguyen Tien Hiep, Gosta Kjellsson, Arnold Klehm, Bert Klein, Tony Lamb, P.S. (Bill) Lavarack, Gwen Lee, Phan Ke Loc, Luo Yi Bo, Dave Martin, David Menzies, Justin Moat, Alan Moon, Henry Oakeley, Holger Perner, Qin Hai Ning, Finn Rasmussen, Tom Reeve, Brian Rittershausen, Lady Lisa Sainsbury, the late Gunnar Seidenfaden, Kit Seth, P.S. Shim, Gloria Siu, the late G. Hermon Slade, the late Tem Smitinand, Peter Taylor, Mike Tibbs, Niklaus Trudel, Z.H. Tsi, Jeffrey Wood, and the late Eric Young.

Claire Smith, now of Zürich, painted a number of the exquisite watercolours that grace this book during a three-year visit to Kew. They were justly awarded a Gold Medal at the Royal Horticultural Society in spring 1985. The large composite line figures are also her work. Other watercolours come from the archives of the Royal Botanic Gardens, Kew and many were featured in *Curtis's Botanical Magazine*. We also thank Andrew Brown, Deborah Lambkin, Chan Chew Lun, Mutsuko Nakajima, Peter O'Byrne, P.S. Shim, Claire Smith, Susanna Stuart-Smith, and Hazel Wilkes for their black and white line drawings. The diagram indicating the pollination strategy of *Cymbidium insigne* subsp. *seidenfadenii* and *Dendrobium infundibulum* was kindly lent by the late Ben Johnson (Denmark).

Photographs have been kindly loaned for use by Martin Ahring, Leonid Averyanov, Gloria Barretto, Chris Bailes, Sheila Collenette, F. Hoeck, T. Hurrell, T. Illenseer, Chan Chew Lun, Lawrence Chau, Chen Sing Chi, Mark Clements, the late Jim Comber, Andy Easton, Bert Klein, Tony Lamb, Luo Yi Bo, David Menzies, Alan Moon, Peter O'Byrne, Holger Perner, Royal Botanic Gardens, Edinburgh and Kew, the late Gunnar Seidenfaden, Gloria Siu and Niklaus Trudel.

The chapters on *Cymbidium* breeding and cultivation have been kindly contributed by Michael Tibbs, expanding on the ideas of Andy Easton, Ernest Hetherington and Alan Moon who contributed similar chapters to the previous monograph.

Alec Pridgeon and Kay Lyons have kindly edited the text and have made a number of valuable suggestions. Chan Chew Lun of Natural History Publications (Borneo), Kota Kinabalu, Sabah, has designed the book to his usual high standard. Gina Fullerlove, John Harris and Lloyd Kirton have overseen the publication at Kew.

Finally, we would like to thank Professor Sir Peter Crane, former Director of the Royal Botanic Gardens, Kew, for access to the facilities of Kew and to his staff for seeing it through to publication.

PREFACE

Cymbidium has been arguably the most important genus of orchid in horticulture during the past century. The reasons are several: they are easy to grow, the diversity in the genus is great, several species have showy large long-lasting flowers, they are easily hybridised and the resulting seedlings are easy to grow and flower, especially in temperate regions. The variety of form and colour resulting from hybridisation has provided a steady stream of novelties to the trade over the years. Cymbidiums are the quintessential buttonhole and flower-arrangers' orchids and, nowadays, are also popular for sprays and as pot plants.

Cymbidium growing is a phenomenon with a venerable history. Terrestrial species with graceful foliage and fragrant flowers have been cultivated in China for at least 2500 years, the earliest written records dating from the time of Confucius (about 500 BC). The same species and their varieties continue to be of importance in cultivation in the Orient to the present day. The first *Cymbidium* species were introduced to Europe from China at the beginning of the 18th century, and Linnaeus described two species, as *Epidendrum aloifolium* and *E. ensifolium*, in 1753 in his *Species Plantarum*. Relatively few species were seen in cultivation in Britain until the time of the Industrial Revolution, which provided both the leisure time and the money for an explosion of interest in orchid-growing. From the mid-19th Century onwards, extensive exploration and collection of new species took place, and many were introduced into cultivation. The genus *Cymbidium* is now considered to comprise some 52 species, and is distributed from Sri Lanka, Nepal and India to China and Japan, south through the Malay Archipelago, to the Philippines, New Guinea and north and east Australia.

The advent of orchid hybridisation in the latter half of the 19th century led to the development of a new interest in *Cymbidium* cultivation, based on the large-flowered Himalayan species. For over a hundred years they have been hybridised to produce plants with flowers of rich texture and colouring, and large size. These have formed the basis of a cut-flower and, more recently, a pot-plant industry which is now world-wide. *Cymbidium* hybrids are probably the most commercially important orchids in cultivation at the present time.

The modern hybrids usually have a complex origin, involving several species in their ancestry, but the species themselves have continued to be grown for their own merits and still play a role in hybridisation. In last quarter of the 20th century many of the smaller-flowered species have been introduced into *Cymbidium* hybridisation. Hybrids involving species such as *C. devonianum, C. floribundum, C. tigrinum, C. ensifolium* and *C. madidum*, with large-flowered hybrids, have led to the production of smaller plants with smaller flowers which are suitable for the pot-plant trade. The resurgence of interest in *Cymbidium* species has highlighted the taxonomic confusion that is still to be found in the genus.

In our first monograph of the genus (Du Puy & Cribb, 1988), we outlined a revised classification of the genus and an assessment of specific delimitation and nomenclature within the genus that has been widely followed. We also suggested that the delimitation of *Cymbidium* needed to be reassessed. Since then a considerable number of new species have been described. We have been able to examine a number of species in cultivation and in the wild, and have a better appreciation of their natural variation. Finally, we and others have undertaken phylogenetic analyses based upon DNA data that clarify the circumscription and classification of the genus.

As explained in our original monograph, horticultural involvement in the genus, in both the Western world and in China and Japan, has itself created confusion. Minor differences continue to be given taxonomic recognition, particularly in species which are naturally rather variable and especially in eastern Asia, where extreme natural variants are selected and highly prized in cultivation. Further confusion has arisen in widely distributed species, through the naming of separate species in geographically restricted regional floras. In some cases, the complete range of variation within a species has not been examined by authors describing new taxa, and extremes of the variation have been recognised as distinct. These confusions are resolved in this work: synonymies are identified, and the literature concerning individual species is listed. Full accounts of all species, sections and subgenera are provided here.

We have been able to examine a wide range of herbarium collections principally from Beijing, Bogor, Edinburgh, Geneva, Guangdong, Harvard, Kew, St Petersburg, Leiden, the Linnean Society and the Natural History Museum in London, Missouri, Paris, Singapore and Vienna. Many individuals around the world have given us advice during the course of preparing this work but we are responsible for all the taxonomic opinions in the book.

Fieldwork in Bhutan, India, China, Thailand, Vietnam, Malaysia, Indonesia, New Guinea and Australia has allowed the examination of many species in their wild habitats, contributing valuable information concerning the ecology of the species, natural variation of wild populations and conservation status.

David Du Puy
Phillip Cribb
December 2006

CHAPTER 1

HISTORY

Orchids have probably been cultivated in China since before the time of Confucius (551–479 BC). *Cymbidium* species were amongst the earliest to be cultivated (Chen & Tang, 1982), their attraction being the simple but elegant form of the plant, and the shape and delicate fragrance of the flowers. Since then, selection of desirable and unusual variants has taken place, and some clones in cultivation are now far removed from their more typical wild ancestors. Historically, artificial hybridisation was not used to produce new variation, but naturally occurring hybrids were selected and brought into cultivation (Chow, 1979). Selected plants having monstrous flowers or variegated foliage have also been particularly prized. These two factors have made the classification of cultivated specimens difficult, especially in the complexes related to *C. ensifolium*, *C. kanran* and *C. goeringii*, and have led to the publication of many specific and varietal names for variants which have little or no relevance to the biology of wild populations.

Cymbidium cultivation has been significant in the culture and ethnic history of China for 2500 or more years. Linked with the elite, cymbidiums have come to epitomise such human qualities as elegance, refinement and nobility.

The Chinese word *lan* means 'orchid', whereas *lan hua* refers specifically to *Cymbidium*. However, in writings before the time of Confucius this word referred to several aromatic plants that were used in religious ceremonies and were said to ward off evil spirits. Indeed the Chinese verb *lan* can mean to ward off, hinder or enclose. Perhaps the first use of *lan* in connection with orchids in Chinese literature is in the *Shih Chi* (The Classic Songs), a collection of poetry and songs that predates Confucius. In this, the plant was used to signify love and the courtship of a young couple (Hu, 1971). This theme is continued in a story reputed to be from the Shin (Ch'in) Dynasty (249–207 BC). Yohki-hi, a woman of legendary beauty, and the wife of the Emperor Shi-kotei, was unable to have any children. However, the Emperor acquired a *Cymbidium* plant for his wife who placed it in her room where it flowered in the following autumn, emitting a sweet fragrance that perhaps had an intoxicating effect on the lady. In due course she bore a son, and this was repeated each year until they had 13 wise and brave sons. This orchid is reputed to have been an albino variant of *C. ensifolium,* which is said to have about 13 flowers on each scape (Nagano, 1960).

Confucius was influential in attaching the name *lan* solely to orchids. He knew of the wild habitat of orchids as is evident in the line:

> 'The *chih-lan* that grows in the deep gorges does not withhold its fragrance because of lack of appreciation.'

This saying continues:

> 'The superb person strives for self-discipline, maintenance of principle and establishment of virtue. He does not alter his integrity because of poverty and distress.'

That he was familiar with cultivated *Cymbidium* species is also evident through such sayings as:

> 'The association of a superior person is like entering a hall of *chih-lan* [*Cymbidium*]. In the course of time one becomes accustomed to the superior ways of life, and gets used to the fragrance.'

(Opposite). **Plate 1.** *Cymbidium elegans.* Drawn by an unknown Indian artist for J. Cathcart. Kew collection.

Considering the importance of his teaching to Chinese philosophy, it is easy to see how such comparisons could lead to the high esteem in which *Cymbidium* species were, and still are, held. For example, Confucius commented that *lan* produced the fragrance of a King. Later scholars developed this theme, making *lan* a symbol of royalty and the loyalty of the King's subjects (Hu, 1971).

During the Eastern Chin Dynasty, the ruling Wang family built an orchid pavilion (AD 354), to serve as a gathering place for the leading scholars of the time. It is still on the original site at Nanjing.

The earliest book on orchids was published in AD 1233, by Chao Shih-keng, of Fukien Province, the area which was at that time the centre of orchid cultivation. It was called *Chin Chang Lan Pu*, and includes descriptions of 22 orchids, mostly *Cymbidium* species. In AD 1247, Wang Kuei-Lsueh, also from Fukien, published a second treatise, *Lan Pu*, in which 37 orchids are described and full cultural details are given. He wrote:

'It is a symbol of perfect personality, the quality of a superb person.'

These two works illustrate the popularity of orchid cultivation, and especially of cymbidiums, at that time (Nagano, 1960; Hu, 1971). Their popularity and, to a large extent, their mystique remain today. Two ink paintings of *C. goeringii*, reproduced by Hu (1971), make the same point. One was painted by Chao Meng-chien (1199–1264) during the Southern Sung Dynasty, the second by Cheng Sze-shiao in 1306. Chinese orchid paintings are usually in black and white, reflecting the Chinese appreciation of the simplicity and refinement of the *Cymbidium* orchids, which they express most effectively using this technique (Hu, 1971). Two examples of ink paintings from the Ming Dynasty are given in Du Puy & Cribb (1988).

In the literature of the Ming and Ch'ing dynasties (*ca.* AD 1400–1800), *lan* was used in front of a noun to modify it in such a way that it took on the sentiment of good, fine, elegant or refined. Examples of this are *lan-chang* (= orchid writing) meaning 'fine manuscript', *lan-i* (= orchid manner) meaning 'handsome person', *lan-hui* (= orchid instruction) meaning 'good teaching' and *lan-hsin* (= orchid heart) meaning 'refined lady' (Hu, 1971).

Orchid cultivation in Japan does not appear to be so ancient. The first Japanese orchid book was published in the early 18th Century, written by Matsuoka. He described several orchid genera and illustrated two cymbidiums. Orchid cultivation was popular at that time in Japan, with each class preferring to grow a different orchid. Thus, the Imperial family and its circle grew *Dendrobium monile* Sw., and the Samurai and even some Shoguns grew *Neofinetia falcata* (Thunb.) H.H. Hu. The merchants and other wealthy people, including immigrants from China and other parts of Asia (Nagano, 1953, 1960), grew *Cymbidium*.

Orchids are grown in Japan primarily for their scent, and for the gracefulness of their leaves, and although the form and beauty of their flowers is also appreciated, they are of less significance. Specimens with variegated leaves are highly prized. The scents of orchids such as *C. kanran* were said to be transmitted to the clothes of lovers (Nagano, 1952, 1953, 1955).

(Opposite) **Plate 2.** *Cymbidium erythraeum* [as *C. longifolium*]. Drawn by an unknown Indian artist for J. Cathcart. Kew collection.

Cymb. longifolium Dr.

Cymbidium giganteum, Wall

EARLY EUROPEAN DESCRIPTIONS OF *CYMBIDIUM*

The first descriptions of cymbidiums by European botanists to reach the West were those of Hendrik Adriaan van Rheede tot Drakenstein (1637–1691) and Olof Rudbeck (1630–1702), who both described the plant now known as *Cymbidium aloifolium*. Rheede's description and illustration of '*Kansijram-maravara*' were published posthumously in his monumental *Hortus Malabaricus* in 1693. Rudbeck's '*orchid abortiva, flore majore rubra, folio aloes*' appeared in the second volume of his *Camporum Elysium*, a work that was almost lost in 1702 when a fire destroyed almost all the copies. The source of both appears to have been the Malabar coast in southern India, and it seems likely that Rudbeck's source was *Hortus Malabaricus*. Rheede's black and white etching is the first published illustration of a *Cymbidium*.

Linnaeus (1753) described two *Cymbidium* species, as *Epidendrum aloifolium* (= *Cymbidium aloifolium*) and *Epidendrum ensifolium* (= *C. ensifolium*) in his *Species Plantarum*. The former was based on van Rheede's and Rudbeck's accounts, the latter on a collection from China made by Pehr Osbeck (1723–1805), one of Linnaeus' students, who visited Canton (now Guangzhou) in 1751.

A black-and-white illustration of *Cymbidium goeringii* was published, as '*Fokouri*', in M. Shimada's *Kwa wi*, an early Japanese plant book published in 1754, but it remained un-noticed in Europe until translated by L. Savatier in 1875.

Olof Swartz established the genus *Cymbidium* in 1799. It was broadly circumscribed, and he cited 44 species, most of which find no place in the genus today. Indeed, of the species he listed only *C. aloifolium*, *C. ensifolium* and *C. pendulum* (= *C. aloifolium*) remain in the genus.

CYMBIDIUM IN CULTIVATION IN THE WEST

The earliest record of the genus in cultivation in Europe is a note by Joseph Banks (1791) in Kaempfer's *Icones selectae plantarum* that Fothergill flowered *C. ensifolium* (as *Limodorum ensatum*) in England in August 1780. Kaempfer's drawing of an inflorescence, drawn in Japan, appears as plate 3 of the same work. A cultivated specimen of the same species was illustrated in 1815 in *Curtis's Botanical Magazine*. The plant originated in China, probably from Canton (Guangzhou). Henry Andrews (1802) illustrated a cultivated specimen of *C. sinense,* another Chinese species, in his *Botanical Repository* and noted that it had been in cultivation since 1793 by J. Slater of Leytonstone, Essex, but had only just flowered. In 1818, Loddiges also published a coloured illustration of *C. sinense*, in the *Botanical Cabinet*. Two further Chinese species, *C. xiphiifolium* (= *C. ensifolium*) and *C. aloifolium*, followed in 1818 and 1824 respectively, the former in Loddiges' *Botanical Cabinet* and the latter in Curtis's *Botanical Magazine*.

EXTENSION OF THE RANGE OF *CYMBIDIUM*

The distribution of *Cymbidium* was rapidly shown to range from mainland Asia into the adjacent archipelagos. William Hooker (1823) described and illustrated *C. lancifolium* from the "East Indies" in his *Exotic Flora*, a plant that Dr Wallich had sent to Shepherd at the Liverpool Botanic Garden which flowered in May 1822. In his *Bijdragen*, Carl Blume (1825) illustrated *C. bicolor* from Gunong Salak, and described *C. javanicum* (= *C. lancifolium*) from G. Seribu, both in Java. Walter Fitch's watercolour illustration of *C. chloranthum*, distributed in Java, Sumatra and Borneo, appeared in *Curtis's Botanical Magazine* in 1856 and is probably the first published illustration of a *Cymbidium* species from the region.

(Opposite) **Plate 3.** *Cymbidium iridioides* [as *C. giganteum*]. Unknown Indian artist for J. Cathcart. Kew collection.

Lindley (1833) revised the genus and included 40 species, but only 17 of these are now included in the genus, the remainder falling in genera as diverse as *Eulophia*, *Cymbidiella*, *Luisia* and *Maxillaria*. He had access to the herbarium of Nathaniel Wallich which included collections from India, Sri Lanka, Nepal and peninsular Malaya. He described five of the Himalayan species as new to science, unfortunately ignoring David Don's earlier descriptions based on the same material.

Indian species began to appear in cultivation in the 1830s. The publication of the first coloured illustration of an Indian species was that by William Roxburgh of *Epidendrum pendulum* (= *C. aloifolium*) in his *Plants of the Coast of Coromandel* (t. 44: 1796). Coloured illustrations of *C. giganteum* (now *C. iridioides*) and *C. elegans* appeared in 1837 and 1838, respectively, in Lindley's *Sertum Orchidaceum*. Most of the other Himalayan species followed over the next decade. Three unpublished watercolour paintings of Sikkim plants, drawn by an unknown Indian artist for John Cathcart in the 1820s and 1830s, are reproduced here (Plates 1–3). The first Burmese species were brought into cultivation in the 1860s, mainly by Charles Parish and Captain Benson, collecting for Messrs Hugh Low and Company. *Cymbidium tigrinum* was illustrated in *Curtis's Botanical Magazine* in 1864.

In Australia during Matthew Flinders' circumnavigation of the continent, Robert Brown discovered *C. canaliculatum* in Queensland and *C. suave* in Sydney [Port Jackson] and described them in 1810 in his *Prodromus* of Australian plants. However, the first illustration of an Australian species to be published was of *C. canaliculatum*, which was figured in *Curtis's Botanical Magazine* of 1870 from a plant collected by John Veitch in Queensland and flowered in cultivation in England at the Veitch nursery in Chelsea.

Both *Cymbidium atropurpureum* and *C. finlaysonianum* had been discovered in the Philippines by 1882 (Ames in Merrill, 1925). Vietnamese species, which have had such an impact on cymbidium breeding, were introduced in the early years of the 20th Century by professional collectors, notably Wilhelm Micholitz, travelling for Messrs F. Sander and Sons. He discovered *C. insigne*, *C. erythrostylum*, *C. sanderae* and *C. schroederi*, all on the Langbian plateau of southern Vietnam. The first plants of *C. insigne* and *C. erythrostylum* flowered in cultivation in 1906 and illustrations were published in the same year in Cogniaux's *Dictionnaire* and the *Gardeners' Chronicle* respectively.

Rudolf Schlechter recorded the first *Cymbidium* from New Guinea in 1913, describing *C. papuanum*. Few additional species were described from the end of the First World War until the late 1970s, and those were mostly relegated to synonymy. Comber and Nasution (1977) described *C. hartinahianum* from Sumatra and three species, the Bornean *C. borneense* and *C. elongatum* and the Vietnamese *C. sanderae*, were recognised during the 1980s (Wood, 1983; Du Puy & Cribb, 1988).

RECENT ADDITIONS TO THE GENUS

The discovery and description of numbers of new species of *Cymbidium* from China in recent years is remarkable. Of several novelties that have been described in recent years, the most notable is *C. wenshanense* from south-eastern Yunnan, a large, white-flowered species that somewhat resembles *C. erythrostylum*. It has yet to appear in cultivation in the West and be used in hybridisation. Most of the other Chinese novelties are closely allied to well-known and widespread species, whereas others are undoubtedly conspecific with known species. Sorting out the identity of the many new taxa from China has been one of the most difficult tasks for this new monograph.

CHAPTER 2

MORPHOLOGY

Morphology remains the basis for identification of orchids, including *Cymbidium*. The terminology applied to the various organs in the descriptions is summarised in Figure 1.

VEGETATIVE MORPHOLOGY

The species of *Cymbidium* are usually autotrophic, terrestrial, epiphytic or lithophytic **herbs**. In many instances, species that are usually epiphytic are also encountered as lithophytes. The species in sections *Jensoa* and *Pachyrhizanthe* are entirely terrestrial, and include *C. macrorhizon,* which is heteromycotrophic (or mycoparasitic) and grows entirely below the soil surface, except when the flower spike is produced. Plant size is variable, but some species form large clumps with the leaves reaching over a metre (39 in) in length.

Most species have thick **roots** up to about 8 mm (0.3 in) in diameter that are covered in thick, spongy, white or greyish velamen, and have a core of vascular tissue. Only the root apex lacks velamen and is usually green, containing chlorophyll. Erect, acute roots (aegotropic roots), which provide a means for the orchid to gather leaf litter about its base, are found in a number of species including *C. aloifolium, C. bicolor, C. finlaysonianum, C. atropurpureum, C. dayanum* and *C. tracyanum*. They are also found in allied genera such as *Grammangis, Grammatophyllum, Graphorkis, Ansellia* and *Acriopsis*.

The erect stems are usually swollen to form prominent **pseudobulbs**, which are often slightly flattened, but which may be reduced to a slight swelling of the base of the stem. The usual growth habit is sympodial, with new pseudobulbs being produced annually on a short rhizome, resulting in the pseudobulbs being tightly clustered in the mature plant. The pseudobulbs of *C. eburneum* and its allies grow and flower continuously for 2–3 years before a new shoot is produced. In *C. elongatum* and *C. suave*, each shoot grows continuously for many years, forming a long stem rather than a pseudobulb and giving the plant an apparently monopodial growth habit. In *C. ensifolium* the pseudobulbs are sometimes subterranean and corm-like. *Cymbidium macrorhizon* lacks pseudobulbs and has a subterranean rhizome covered with small tubercles, sites where mycorrhizal infection is concentrated.

Cataphylls enclose the young shoot before the true leaves emerge, and can be distinguished from the true leaves by their lack of a distinct lamina and abscission zone. The cataphylls quickly become scarious and eventually disintegrate, leaving numerous fibres around the base of the pseudobulb.

Each growth has from 3–12, distichous, green **leaves**. The lamina of the leaf is articulated to the leaf base by an abscission zone, and although the lamina is eventually deciduous, it usually persists on the pseudobulb for 2–4 years. The broad, sheathing leaf bases often completely enclose the pseudobulb, and persist on it after the lamina has been shed. In *C. tigrinum* the pseudobulbs have 2–4 apical leaves, and are almost completely exposed. In the heteromycotroph *C. macrorhizon,* leaves are reduced to scales.

Cymbidium leaves are always duplicate (folded along the mid-vein) in development, but may be conduplicate (V-shaped in transverse section) or plicate (ribbed) in appearance when mature. No species has leaves that can be categorised as truly plicate in the sense of Withner *et al.* (1975), as plicate leaves have convolute (folded like a fan) development. Other differences from the true plicate leaf category are evident in the leaf anatomy of *Cymbidium* (see Chapter 4).

Most *Cymbidium* species have relatively thin-textured and flexible leaves of a similar ribbed appearance. However, some species have leaves that are duplicate in development, and conduplicate

7

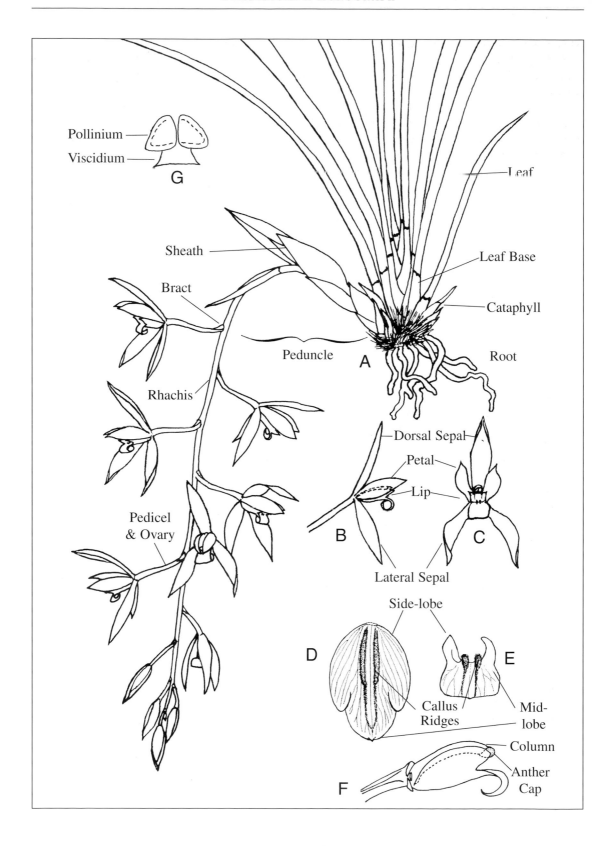

when mature, consistent with the leathery leaf category of Withner *et al.* (1975). The species in the *C. aloifolium* alliance (section *Cymbidium*) and *C. canaliculatum* (section *Austrocymbidium*), are epiphytic or lithophytic species that are often exposed to arid conditions: they have hard, leathery leaves, with no protruding veins apart from the mid-vein, and many sclerenchyma fibres near the leaf surface, making the leaf rigid. The leaves are also thick (about 3 mm (0.12 in) or more), and have a thick cuticle, and a large volume of spongy mesophyll. The evidence therefore supports the suggestion of Withner *et al.* that leathery leaves are a xerophytic modification of the more primitive, thin leaves with plicate appearance. The leaves of these species appear to assume the function of water storage organs instead of the normal swollen pseudobulbs. It is possible that these adaptations may be linked to crassulacean acid metabolism (Winter *et al.*, 1983). Several other anatomical and micromorphological characters can also be interpreted as xerophytic modifications in these species, and are discussed in Chapter 4.

Other modifications in leaf shape that can be linked to the ecology of the species include the broad, elliptic leaves of the shade-loving species *C. lancifolium* and *C. devonianum*, and the long, thin, relatively narrow leaves of most of the forest epiphytes that grow in shaded, humid conditions, including many of the species in section *Cyperorchis*.

The leaf apex is usually acute and slightly unequal, but in section *Cymbidium* it is obtuse and often unequally bilobed. In *C. mastersii* and *C. eburneum* it is conspicuously forked, with a small mucro in the sinus formed as an extension of the mid-vein.

FLORAL MORPHOLOGY

The **inflorescence** in *Cymbidium* is unbranched and may be erect, arching or pendulous. The peduncle (scape) is usually covered by about eight sheaths. One or two inflorescences are usually produced from the base of the mature pseudobulb. The species allied to *C. eburneum* produce inflorescences in the axils of the leaves towards the apex of the pseudobulb, and *C. elongatum* and *C. suave* also produce them in the axils of the upper leaves. In *C. lancifolium* they arise from one of the central nodes of the pseudobulb.

The **flowers** are usually about 3–10 cm (1.2–4 in) in diameter. Flower number varies from one in *C. goeringii* var. *goeringii* and *C. eburneum*, to about 50 in *C. canaliculatum* and *C. suavissimum*. Each flower has an inferior **ovary** and short **pedicel** that are difficult to delimit from each other, and is subtended by a small scarious **bract**. The flower comprises a dorsal and two lateral **sepals**, two free **petals** that may be spreading or covering the column, a lip, and a central column. The **lip,** which is a modified petal, is hinged at the base of the column, or at the base of a short column-foot in *C. devonianum* and some species in section *Cymbidium*. The base of the lip is fused with the base of the column for 3–6 mm (0.15–0.25 in) in all species in section *Cyperorchis*. The three-lobed lip may be weakly saccate at the base in some species of section *Cymbidium*, but it is never spurred. It is usually three-lobed, the two side-lobes being erect and loosely clasping the column, and the mid-lobe is usually recurved. Two parallel **callus** ridges are commonly borne on the upper surface of the lip. However, in *C. eburneum* and *C. erythrostylum* a wedge-shaped callus is found, whereas in *C. suave* and *C. madidum* the callus ridges are replaced by a glistening depression. Sections *Jensoa* and *Pachyrhizanthe* are characterised by a callus of two ridges that converge towards their apices, forming a short tube at the base of the mid-lobe. In *C. devonianum* and *C. borneense*, the callus is reduced to two small swellings at the base of the mid-lobe, whereas in *C. aloifolium* and *C. bicolor* the callus ridges are often strongly sigmoid and may be broken in the middle.

(Opposite). **Fig. 1.** The morphology of *Cymbidium*. **A.** Plant of *C. dayanum* (×0.5). **B.** Flower: side view (×0.7). **C.** Flower: front view (×0.7). **D.** Lip: flattened to allow the measurement of constituent parts (×1.5). **E.** Lip: front view (×1.5). The mid-lobe is recurved. **F.** Column and lip: side view with petals and sepals removed (×2). **G.** Pollinarium (×10). Enclosed by the anther-cap. All drawn by D. Du Puy.

The **column** is weakly winged, more strongly so towards the apex. The anther usually contains two **pollinia**, each of which is deeply cleft behind. Each pollinium may be visualised as a pair of unequal, flattened pollinia fused along the inner margin. In sections *Jensoa* and *Pachyrhizanthe* and in *C. borneense*, this fusion is absent, and there are two pairs (four) of unequal pollinia in the anther. The pollinia are usually triangular, but in *C. eburneum* and its allies they are usually quadrangular, and in *C. elegans* and its allies they are club-shaped. The pollinia are placed almost directly onto a sticky **viscidium**, and are held there by extremely elastic but short caudicles. The caudicles are derived from the anther, whereas the viscidium is derived from the rostellum, which is part of one of the stigma lobes. The viscidium, caudicles and pollinia constitute the **pollinarium** The **stigma** is a sticky concavity located on the underside of the column, directly behind the anther.

PROTOCORM STRUCTURE

Little has been written about the structure of protocorms in *Cymbidium*, although horticulturists raising cymbidiums *in vitro* from seed have been aware that the Oriental small-flowered species have a different protocorm development from the other species. It is apparent that two quite distinct types of protocorm are found in the genus. The hard-leaved species and those of sections *Annamaea*, *Bigibbarium*, *Cyperorchis*, *Himantophyllum* and *Parishiella* have typically ovoid to subspherical protocorms, whereas those of sections *Jensoa* and *Pachyrhizanthe* are elongate and termed mycorhizomes (Hisanchi, 1958). Further work is needed to fully establish the development of these structures. Tahara (2001) reported that primary hybrids of Oriental *Cymbidium* species took between six and twelve years to develop from germination to flowering.

CHAPTER 3

SEED MORPHOLOGY

INTRODUCTION

A range of *Cymbidium* species were surveyed by Du Puy (1986) using a scanning electron microscope to allow high magnification. Several taxonomically useful conclusions have been drawn from the results. The pollinium micro-structure was also examined, but the results were taxonomically inconclusive.

The seed of *Cymbidium*, quite typical of the majority of orchid seeds, is minute and dust-like, 520–1900 μm (*c.* 0.5–1.9 mm, 0.02–0.07 in) long, and fusiform or filiform in outline. A spherical embryo without endosperm is enclosed in a thin testa, one cell layer thick. The testa cells are largest towards the centre of the seed. The anticlinal walls of the testa cells are thick, appearing as a net on the surface of the testa (Fig. 2A). The junctions of this net may protrude somewhat, sometimes giving the seed a spiny appearance. The surface walls between this net have fine secondary thickening which appears as minute striations which are either longitudinal, running parallel with the long axis of the seed (Figs. 2B & C) or transverse, running at right angles to the long axis of the seed (Figs. 3C & F).

VARIATION IN *CYMBIDIUM* SEEDS

Most *Cymbidium* species have fusiform seeds between 520–1200(1400) μm long (Fig. 2). Seeds of *C. devonianum*, *C. elegans*, and of *C. canaliculatum* (section *Austrocymbidium*) were shortest in the genus, only 520–650 μm long, while those of *C. eburneum*, *C. lowianum* and their allies were the largest with fusiform shape, usually 800–1200 μm long. Seed shape and size varied somewhat, both within and between samples.

A second, distinct type of seed has been found only in *Cymbidium* sections *Jensoa* and *Pachyrhizanthe* (Fig. 3), and in some species of *Eulophia*. The seeds of these species are filiform, 1400–1900 μm long and very slender.

The surface cell walls of the seed have a striated appearance. Longitudinal striations were found in seeds with a fusiform shape (Fig. 2), while lateral striations were restricted to filiform seeds (Fig. 3).

In many of the species examined, the junctions of the anticlinal walls were marked by an extension of the walls into a small, protruding, pyramidal structure. In a few species this was very pronounced, giving the seeds a spiny appearance.

THE IMPLICATIONS FOR
THE CLASSIFICATION OF CYMBIDIUM

The fusiform seed shape found in most epiphytic and lithophytic *Cymbidium* species is similar to the related genera *Grammatophyllum*, *Ansellia* (Fig. 4), *Porphyroglottis*, *Eulophiella*, and *Cymbidiella*. These seeds also have in common the longitudinal striations of the walls of the testa cells, and the somewhat protruding junctions of the net of cell walls which are often capped with waxy deposits. This type of seed is therefore characteristic of the epiphytic species of *Cymbidium* and the related genera in the Cyrtopodiinae.

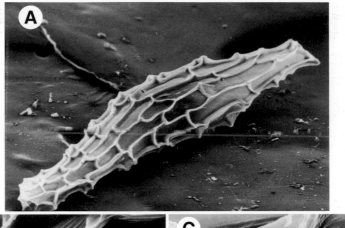

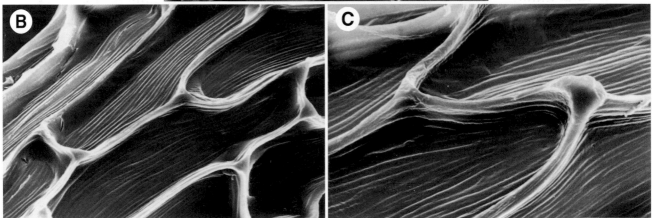

Fig. 2. Section *Cyperorchis*—*C. lowianum* (S. 879). **A.** Whole seed, showing the fusiform shape and network formed by the anticlinal walls of the testa cells (×75). **B.** Testa cells, showing the slightly raised junction of the net of cell walls (×380). **C.** Testa cells, showing the longitudinal secondary striations of the surface walls (×765).

Within the epiphytic *Cymbidium* species, the seeds vary in length, in their comparative breadth and in the degree to which the junctions of the anticlinal walls of the testa cells protrude. Seed length and comparative breadth vary both within and between species, so that only the most general taxonomic conclusions can be drawn. Species in sections *Cymbidium* and *Bigibbarium* usually have the most broadly fusiform seeds. The shortest seeds are found in *C. elegans* and its allies, *C. devonianum* and *C. canaliculatum*. Section *Austrocymbidium* shows great variation in seed morphology, in size, relative breadth and in the protrusions at the junctions of the net of cell walls. Seeds of *C. canaliculatum* are easily distinguished by the presence of long, distinct protrusions with a distinct waxy cap, giving the seeds a spiny appearance. This type of seed is also found in the related genus *Ansellia* and, to a lesser extent, in *Grammatophyllum* (Fig. 4). This is probably an adaptation to facilitate seed dispersal, by increasing the surface area to volume ratio of the seed, making it more buoyant in the air. This may be important in increasing the chance of seed finding a suitable niche in which to germinate, as for example in *C. canaliculatum* which grows in rotten wood in trees in the arid zones of northern Australia. The trees are often well-spaced, and the probability of the seed being deposited on a suitable site might be greater if the buoyancy of the seed were increased. It is unlikely that this type of seed indicates close taxonomic affinity between these taxa.

The narrow, filiform seeds found in the species of sections *Jensoa* and *Pachyrhizanthe* are very distinct from those in the other *Cymbidium* species. Furthermore, they have transverse secondary

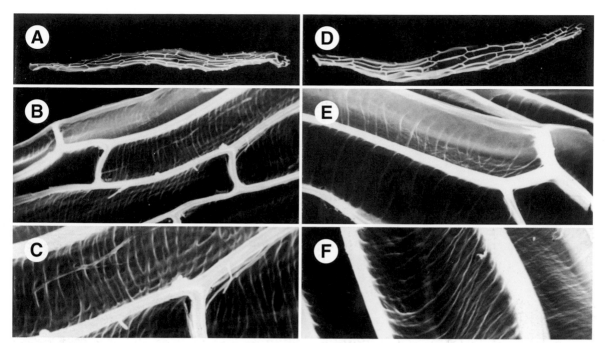

Fig. 3. Section *Jensoa—C. cyperifolium* (S. 1051). **A.** Whole seed, showing the long, filiform shape (×45). **B.** Testa cells, showing the weakly raised junctions of the net of cell walls (×450). **C.** Testa cells, showing the transverse secondary striations of the surface walls (×900). Section *Jensoa—C. ensifolium (Streimann and Kairo* in NGF 39369). **D.** Whole seed (×50). **E.** Testa cells (×900). **F.** Testa cells (×1010). The species in subgenus Jensoa all exhibit a strong similarity to each other in seed shape and direction of the surface wall striations.

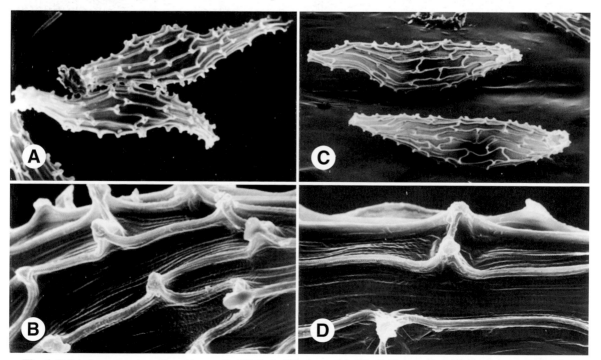

Fig. 4. Related genera. *Ansellia africana* (S. 1014). **A.** Whole seed (×85). **B.** Testa cells (×400). *Grammatophyllum* sp. (S. 1018). **C.** Whole seed (×65). **D.** Testa cells (×650). Both genera show strong similarity to the subgenera *Cymbidium* and *Cyperochis*, having fusiform seeds and longitudinal striations of the periclinal walls. *Ansellia* particularly resembles *C. canaliculatum* (section *Austrocymbidium*) with strongly raised, wax-capped junctions of the anticlinal walls. This may create a greater buoyancy in the air, increasing the chances of the seed lodging in a suitable branch. This could be especially important in the more open, arid environments, with a sparse tree cover, in which these species grow.

striations on the walls of the testa cells, and the junctions of the net of cell walls are raised only slightly if at all. This combination of characters is also found in some species of the genus *Eulophia*, the Asiatic species *E. burkei* and *E. keithii* both having seeds with strikingly similar morphology to those of species in section *Jensoa*.

Other workers (Clifford & Smith, 1969; Arditti *et al.*, 1979, 1980; Barthlott & Zeigler, 1981) found very little or no variation within genera, but noted occasional differences between related genera, and expressed the opinion that seed morphology was most useful above the generic level. The similarity between seeds of the epiphytic *Cymbidium* species and the related genera such as *Grammatophyllum*, *Ansellia*, and *Porphyroglottis*, would seem to confirm this view. It is therefore highly unusual to find that the seeds of sections *Jensoa* and *Pachyrhizanthe* are so distinct. Clifford & Smith (1969) found that variation in seed shape, seed length and testa wall secondary thickening were important at the tribal level. The last character was also shown by Barthlott & Zeigler (1981) to be variable at the tribal level. They also found, however, that the genus *Habenaria* included species with both inflated and thread-like seeds. *Cymbidium* is therefore unusual, but not unique, in this variation. However, the variation in secondary thickening of the surface walls is highly unusual within a genus, and usually occurs at the tribal level.

The large differences between sections *Jensoa* and *Pachyrhizanthe* and the rest of *Cymbidium* are reflected in their leaf micromorphology and anatomy (see chapter 4). However, the relationship between the sections cannot be disputed as hybrids have been made, and flowered, between species in section *Jensoa* and hybrids derived from species of large-flowered *Cymbidium*.

Furthermore, the genus most similar in its seed morphology to section *Jensoa* is *Eulophia*, which is usually regarded as being rather distantly related to *Cymbidium* (Dressler, 1981). Within *Eulophia* itself, seed morphology varies, some species being distinct with a clavate shape and a different distribution of cell sizes in the testa (small cells only towards one end of the seed). This type of seed is not found in *Cymbidium*.

It is possible that certain species of *Eulophia*, such as *E. keithii* and *E. burkei*, are more closely related to *Cymbidium* section *Jensoa* than are the rest of *Eulophia*. Macromorphological studies suggest that *Eulophia* should not be split in this way, while DNA studies indicate that *Cymbidium* is monophyletic. It is more likely that this is a case of parallel evolution of similar seed types in the two genera. It is suggested, however, that further studies of seed morphology in the Cyrtopodiinae could provide valuable data to add to the DNA data with which to elucidate the relationships within this subtribe. The combined data from studies of seed morphology and leaf morphology and anatomy provide evidence that sections *Jensoa* and *Pachyrhizanthe* together form a natural group which is significantly distinct, and more distantly related, to the other sections of *Cymbidium*.

CHAPTER 4

ANATOMY

Detailed accounts of the vegetative anatomy of *Cymbidium* have been given by Du Puy (1986), Du Puy & Cribb (1988) and Yukawa & Stern (2002).

LEAF

The bulk of the *Cymbidium* leaf is made up of a mass of loosely packed mesophyll cells containing chloroplasts which provide sugars for the plant through photosynthesis. Running through this mesophyll tissue are several vascular bundles, which transport water, sugars and micronutrients. These bundles are strengthened by caps of thick-walled cells (sclerenchyma). Further strengthening is provided by a series of fibre bundles (again of thick-walled sclerenchyma cells), which run along the length of the leaf, just below the epidermis (Figs. 5 & 9). The epidermis itself is surrounded by a thick, waterproof cuticle, which covers the entire leaf, preventing desiccation. A thin cross-section of a typical *Cymbidium* leaf is shown in Fig. 9A. The red dye indicates where the cell walls have been strengthened with lignin, whereas the blue dye has stained the thin cellulose walls of the other cells.

The stomata are mainly confined to the undersurface of the leaf. They are openings in the epidermis and cuticle, through which the exchange of photosynthetic and respiratory gases and water vapour between the leaf and the surrounding air can be controlled. Each stoma is formed by a pair of cells and is covered by a dome of waterproof cuticle, with a small pore in the top. These can be seen in Fig. 6, a high-magnification photograph of a typical leaf surface.

VARIATION IN THE ANATOMY AND SURFACE MORPHOLOGY OF *CYMBIDIUM* LEAVES

Subepidermal fibre bundles, composed of lignified sclerenchyma cells, are a common feature of this genus, and are characteristic of members of the Cymbidiinae (Kaushik, 1983). In most *Cymbidium* species, fibre bundles are located in a row below the epidermis, on both sides of the leaf (Du Puy, 1986; Du Puy & Cribb, 1988). They provide some taxonomically informative characters:

- the species in section *Cymbidium* are characterised by the presence of a continuous uniseriate row of sclerenchyma cells directly adjacent to the abaxial and adaxial epidermis, linking the densely spaced fibre bundles together, broken only below the stomata (Figs. 5A & 9B). This has a strengthening function, surrounding the soft mesophyll cells in a lignified protective layer. It may also be a xerophytic adaptation, helping to control water loss from the leaf.
- an alternating series of subepidermal fibre bundles (beneath both abaxial and adaxial surfaces) that are widely separated by spongy mesophyll cells (Figs. 5B & 9A) is characteristic of many *Cymbidium* species (including those in sections *Cyperorchis, Annamaea, Parishiella, Borneense, Bigibbarium, Floribundum* and *Austrocymbidium*). Yukawa & Stern (2002) noted that a variation on this theme is found in section *Himantophyllum* (*C. dayanum*), in that the fibre bundles beneath the adaxial surface are separated by several chlorenchyma (rather than spongy mesophyll) cells.
- occasional widely separated subepidermal fibre bundles (separated by chlorenchyma cells) occur in sections *Jensoa* and *Pachyrhizanthe*, although they are absent in *C. faberi* and *C. goeringii* (Table 1), and present only beneath the upper surface in *C. cyperifolium*. The strengthening function is taken over by exceptionally developed sclerenchyma caps on the vascular bundles.

15

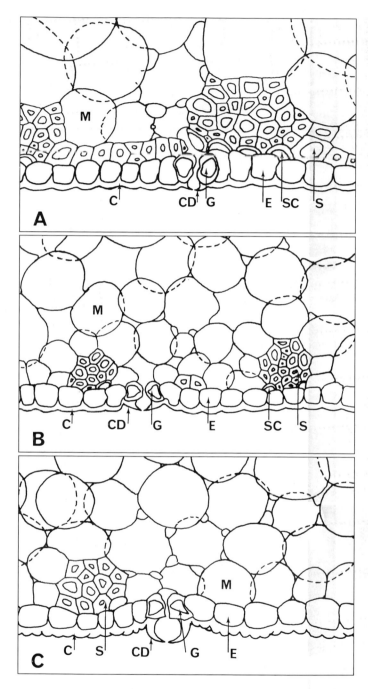

Fig. 5. Transverse section of a leaf. **A.** Section *Cymbidium*, *C. aloifolium* (Kew no. 180-34, 11201). The cuticular dome covering the stoma is not raised above the surrounding cuticle; the epidermal cells are not papillose; there is a subepidermal layer of sclerenchyma fibres linking the fibre bundles and broken only below the stoma. **B.** Section *Himantophyllum*, *C. dayanum* (Kew no. 120-83.01291). The cuticular dome is not raised; the epidermal cells are not papillose; the subepidermal fibre bundles are isolated from each other. This arrangement also occurs throughout section *Cyperorchis*. **C.** Section *Jensoa*, *C. ensifolium* (Kew no. 040-83.00337). The cuticular dome covering the stoma is raised well above the surrounding cuticle; the epidermal cells are strongly papillose; the subepidermal fibre bundles are isolated from each other. **KEY: C** — cuticle, **CD** — cuticular dome covering the stoma, **G** — guard cell, **E** — epidermal cell, **S** — sclerenchyma cell, usually confined to subepidermal fibre bundles, **SC** — silica cell, **M** — mesophyll cell.

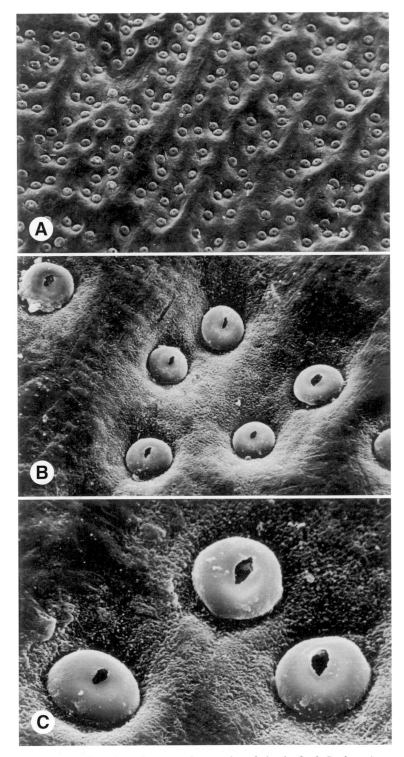

Fig. 6. Section *Bigibbarium*—Scanning electron micrographs of the leaf of *C. devonianum* (Kew no. 102-83.01272). **A.** Abaxial epidermis (×75). **B.** Stomata (×375). **C.** Stomatal cover (×750). The stomatal covers are circular, not raised above the surrounding epidermis, and the pores are fusiform. The absence of troughs marking the anticlinal walls of the epidermal cells is unusual and somewhat characteristic.

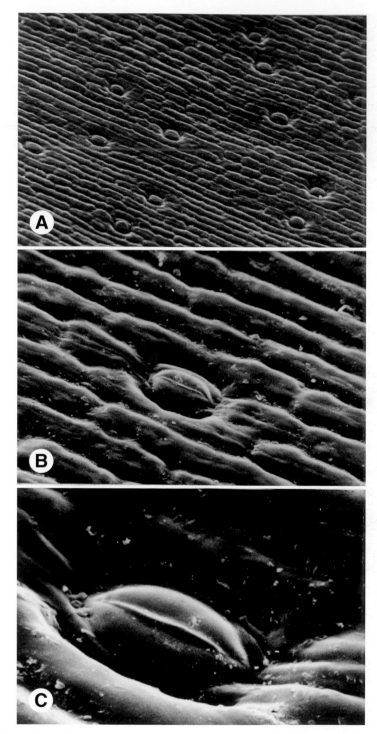

Fig. 7. Section *Cymbidium*—Scanning electron micrographs of the leaf of *C. atropurpureum* (Kew no. 417-63.41701). **A.** Lower epidermis (×165), showing the scattered stomata. **B.** Single stoma with the surrounding epidermal cells (×750), showing the general shape of the cuticular stomatal cover and the related epidermal cells. **C.** Single stoma (×2065), showing the shape of the pore in the stomatal cover. Scanning electron micrographs show that the species in this section have characteristically elliptic stomatal covers, with a slit-shaped pore which extends almost the full length of the cover. The covers are not raised above the level of the surrounding epidermal cells.

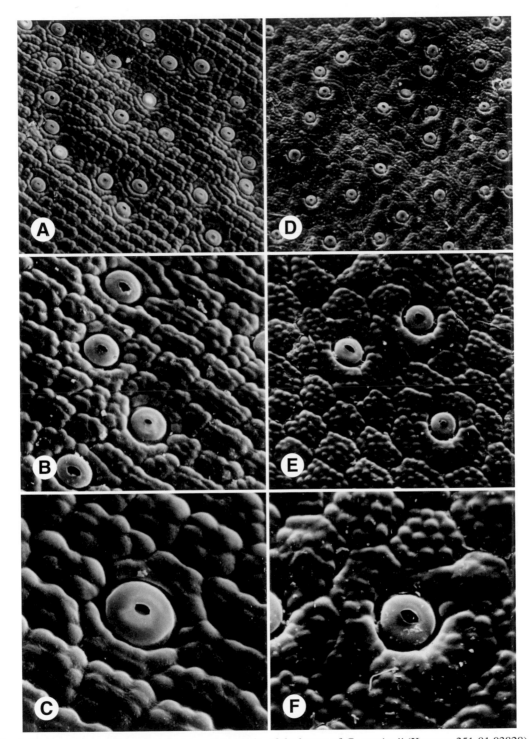

Fig. 8. Section *Jensoa*—Scanning electron micrographs of the leaves of *C. goeringii* (Kew no. 351-81.03828). **A.** Abaxial epidermis (×140). **B.** Stomata (×350). **C.** Stomatal cover (×700). Section *Pachyrhizanthe*, *C. lancifolium* (Kew no. 102-83.01227). **D.** Abaxial epidermis (×125). **E.** Stomata (×325). **F.** Stomatal cover (×625). These species show the characteristic features of sections *Jensoa* and *Pachyrhizanthe*, including the papillose epidermal cells, the raised stomatal covers and the rather circular pores. The epidermal cells of *C. lancifolium* are broader, with correspondingly more numerous papillae, than the species in section *Jensoa*.

Yukawa and Stern (2002), in their review of the anatomy of *Cymbidium*, examined in particular the correlation between anatomy and habitat preferences. They confirmed that in the forest floor species, such as *C. goeringii*, subepidermal sclerenchyma was absent, and they postulated that development of sclerenchyma in other species was correlated with degree of epiphytism.

Section *Cymbidium* is further characterised by leaves with several layers of loosely packed palisade (elongated mesophyll) cells (Fig. 9B, Table 1). Yukawa and Stern suggested that development of palisade mesophyll is another adaptation correlated with migration to exposed habitats with high light intensity.

They also noted that stegmata (flat cells containing a mass of silica) are found in the leaves of almost all *Cymbidium* species, and also in the roots of a few epiphytic species which is an unusual feature in vascular plants. The exception, *C. macrorhizon*, lacks green leaves and has present only much-reduced non-photosynthetic scales, a feature associated with its heteromycotrophic existence: it also exhibits degeneration of stomata (anomocytic stomata).

Sections *Cyperorchis*, *Parishiella* and *Annamaea* have a characteristic leaf margin in transverse section (Table 1). Whereas the species in other sections have obtuse to rounded margins, these sections along with *C. devonianum* have slender, acuminate, decurved margins, appearing as narrow, recurved, hyaline margins on living leaves, suggesting a close relationship between them. However, *C. devonianum* lacks the fusion of the bases of the lip and column that characterises sections *Cyperorchis*, *Parishiella* and *Annamaea*, and there do not appear to be any other morphological characters that reiterate this connection.

The scanning electron microscope (SEM) enables the micromorphology of the leaf surface to be seen, and does not destroy the cuticular cover over the leaf surface. The lower surface was examined because in most species the stomata are restricted to that surface. The only exceptions appear to be the distantly related species *C. canaliculatum* (section *Austrocymbidium*) and *C. borneense* (section *Borneense*). Several useful characters have been obtained from the morphology of the dome of cuticular material formed by the outer ledges of the guard cells, and from the shape of the pore in this dome. In the majority of species, these stomatal covers are level with the rest of the epidermis, round in outline, and have a central, spindle-shaped pore (Fig. 6). The elliptic stomatal covers with narrow, slit-shaped pores found in section *Cymbidium* are distinctive (Fig. 7), which may be another xerophytic adaptation in this section.

The stomatal morphology of *C. borneense* is almost identical to that of the species in section *Cymbidium*. Although the leaves lack the other anatomical characters of that section (complete subepidermal sclerenchymatous layer and palisade-like mesophyll cells), the similarity of the distinctive stomata indicates that they may be closely related.

The presence of about 6–16 rounded papillae on each of the epidermal cells in sections *Jensoa* and *Pachyrhizanthe*, giving them a raspberry-like appearance (Fig. 8), is unique in the genus. The domed covers over the stomata are also raised above the rest of the leaf surface (Fig. 5C). This unusual and highly characteristic seed morphology and the presence of four rather than two pollinia (the latter also found in section *Pachyrhizanthe*), characterise this group of species.

These anatomical and leaf surface characters are summarised in Tables 1 and 2.

RELATED GENERA

Related genera have a variable epidermal and stomatal morphology (Fig. 10; Table 2). Of the more closely related genera (Dressler, 1981), *Ansellia* appears to be the most similar to those *Cymbidium* species in sections *Austrocymbidium*, *Floribundum*, *Bigibbarium*, *Annamaea*, *Cyperorchis* and *Parishiella*. The epidermal cells are smooth, and the stomatal covers are circular with a central fusiform pore (Fig. 10A), possibly indicating a close relationship between these taxa. *Grammatophyllum* has a

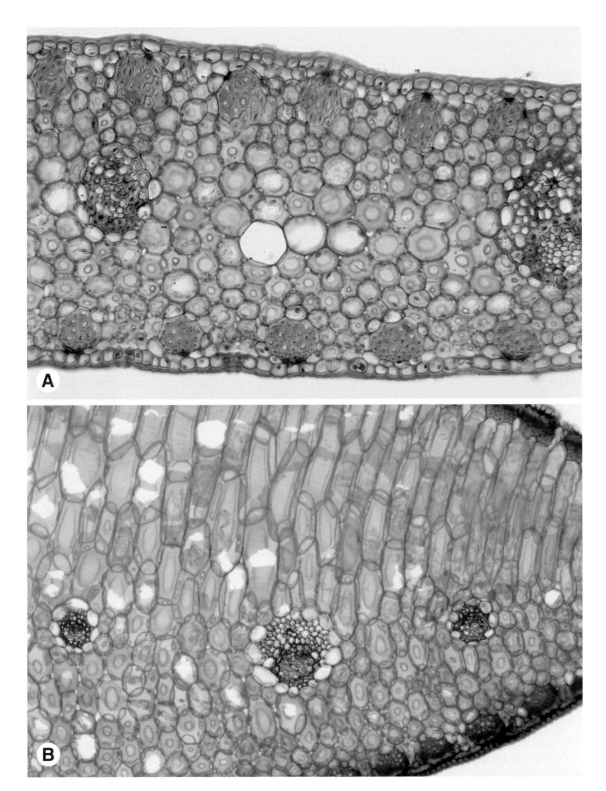

Fig. 9A. Transverse section of leaf of *Cymbidium floribundum*, typical of many *Cymbidium* species. **B.** Transverse section of leaf of *Cymbidium rectum*, characteristic of species in section *Cymbidium*.

TABLE 1

TAXONOMICALLY USEFUL CHARACTERS OF THE LEAF ANATOMY (DU PUY, 1986)

Section	Species	Kew Accession No.	Presence of the sub-epidermal fibre strands	Complete layer of sub-epidermal sclerenchyma cells	Palisade-like cells in mesophyll	Leaf margin acuminate in T.S.
Cymbidium	C. aloifolium	120-34.11201	+	+	+	−
	C. bicolor	558-65.55825	+	+	+	−
	C. atropurpureum	417-63.41701	+	+	+	−
	C. rectum	214-83.02467	+	+	+	−
Floribundum	C. chloranthum	214-83.02460	+	−	+	−
	C. floribundum	340-82.03508	+	−	−	−
	C. floribundum	120-81.01688	+	−	−	−
	C. floribundum	120-81.01664	+	−	−	−
	C. suavissimum	120-81.01681	+	−	−	−
	C. elongatum	229-83.02923	+	−	−	−
Austrocymbidium	C. canaliculatum	040-83.00318	+	−	−	−
	C. canaliculatum	481-81.06520	+	−	−	−
	C. madidum	189-78.01942	+	−	−	−
Borneense	C. borneense	248-82.02403	+	−	−	−
Bigibbarium	C. devonianum	102-83.01272	+	−	−	+
	C. devonianum	120-81.01670	+	−	−	+
Himantophyllum	C. dayanum	324-78.03481	+	−	−	−
	C. dayanum	120-81.01695	+	−	−	−
	C. dayanum	102-83.00863	+	−	−	−
	C. dayanum	151-83.01893	+	−	−	−
	C. dayanum	102-83.01291	+	−	−	−
Annamaea	C. erythrostylum	430-81.04913	+	−	−	+
Cyperorchis	C. tracyanum	120-81.01676	+	−	−	+
	C. longifolium	120-81.01684	+	−	−	+
	C. hookerianum	323-81.03514	+	−	−	+
	C. lowianum	269-80.02506	+	−	−	+

	C. insigne	120-81.01683	+	−	−	+
	C. sanderae	120-81.01678	+	−	−	+
	C. eburneum	120-81.01673	+	−	−	+
	C. mastersii	102-83.01182	+	−	−	+
	C. mastersii	120-81.01662	+	−	−	+
	C. elegans	120-81.01685	+	−	−	+
	C. whiteae	551-82.05677	+	−	−	+
Parishiella	C. tigrinum	323-81.03531	+	−	−	+
Jensoa	C. ensifolium	102-83.01053	+	−	−	−
	C. ensifolium	481-81.06516	+	−	−	−
	C. ensifolium	474-82.05087	+	−	−	−
	C. ensifolium	040-83.00324	+	−	−	−
	C. ensifolium	040-83.00336	+	−	−	−
	C. ensifolium	040-83.00338	+	−	−	−
	C. ensifolium	040-83.00337	+	−	−	−
	C. ensifolium	102-83.01111	+	−	−	−
	C. sinense	325-81.03538	+	−	−	−
	C. sinense	077-83.00594	+	−	−	−
	C. sinense	424-82.05088	+	−	−	−
	C. cyperifolium	439-84.04707	Below adaxial surface only	−	−	−
	C. faberi	439-84.04748	−	−	−	−
	C. faberi	077-83.00593	−	−	−	−
	C. goeringii	351-81.03828	−	−	−	−
	C. goeringii	077-83.00592	−	−	−	−
Pachyrhizanthe	C. lancifolium	340-82.03514	+	−	−	−

Note: + = present.

TABLE 2

TAXONOMICALLY USEFUL CHARACTERS OF EPIDERMAL AND STOMATAL MORPHOLOGY (DU PUY, 1986)

Section	Species	Kew Accession No.	Stomatal cover shape	Stomatal cover raised above the surrounding epidermis or not	Pore shape	Epidermal cell surface
Sections						
Cymbidium	C. bicolor	032-75.00438	elliptical	level	slit-shaped	smooth
	C. rectum	214-83.02467	elliptical	level	slit-shaped	smooth
	C. atropurpureum	417-63.41701	elliptical	level	slit-shaped	smooth
Floribundum	C. chloranthum	731-60.73101	circular	level	fusiform	smooth
	C. chloranthum	214-83.02460	circular	level	fusiform	smooth
	C. floribundum	120-81.01688	circular	level	fusiform	smooth
	C. suavissimum	120-81.01681	circular	level	fusiform	smooth
Austrocymbidium	C. canaliculatum	040-83.00318	circular	level	fusiform	smooth
	C. madidum	336-82.03312	circular	level	fusiform	smooth
	C. suave	340-82.03504	circular	level	fusiform	smooth
Borneense	C. borneense	214-83.02513	elliptical	level	slit-shaped	smooth
	C. borneense	248-82.02403	elliptical	level	slit-shaped	smooth
Bigibbarium	C. devonianum	102-83.01272	circular	level	fusiform	smooth
Himantophyllum	C. dayanum	102-83.01291	circular	level	fusiform	smooth
	C. dayanum	151-83.01893	circular	level	fusiform	smooth
Annamaea	C. erythrostylum	040-83.00316	circular	level	fusiform	smooth
	C. erythrostylum	551-82.05676	circular	level	fusiform	smooth
Cyperorchis	C. lowianum	120-81.01685	circular	level	fusiform	smooth
	C. sanderae	120-81.01678	circular	level	fusiform	smooth
	C. insigne	102-83.01280	circular	level	fusiform	smooth
	C. eburneum	120-81.01673	circular	level	fusiform	smooth
	C. mastersii	102-83.01182	circular	level	fusiform	smooth
	C. elegans	120-81.01685	circular	level	fusiform	smooth

Genus	Species	Number				
	C. cochleare	167-84.01164	circular	level	fusiform	smooth
	C. whiteae	551-82.05677	circular	level	fusiform	smooth
Parishiella	*C. tigrinum*	323-81.03513	circular	level	fusiform	smooth
Jensoa	*C. ensifolium*	214-83.02491	circular	raised	circular	papillose
	C. ensifolium	474-82.05087	circular	raised	circular	papillose
	C. ensifolium	340-82.03501	circular	raised	circular	papillose
	C. ensifolium	102-83.01053	circular	raised	circular	papillose
	C. kanran	429-83.05475	circular	raised	circular	papillose
	C. sinense	325-81.03538	circular	raised	circular	papillose
	C. sinense	474-82.05088	circular	raised	circular	papillose
	C. faberi	077-83.00593	circular	raised	circular	papillose
	C. goeringii	195-79.01884	circular	raised	circular	papillose
	C. goeringii	351-81.03828	circular	raised	circular	papillose
	C. goeringii	077-83.00592	circular	raised	circular	papillose
Pachyrhizanthe	*C. lancifolium*	340-82.03514	circular	raised	circular	papillose
	C. lancifolium	102-83.01277	circular	raised	circular	papillose

Other genera

Species	Number				
Grammatophyllum speciosum	032-75.00440	circular	level	slit-shaped	smooth
Ansellia africana	052-77.00323	circular	level	fusiform	smooth
Cymbidiella humblottii	340-82.03517	circular	level	slit-shaped	papillose
Eulophia keithii	400-65.40007	circular	level	fusiform	smooth
Eulophia stenophylla	040-81.01284	circular	level	fusiform	smooth

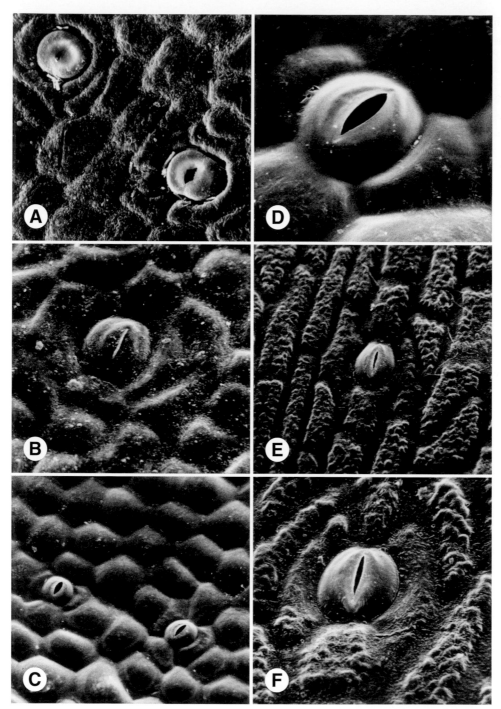

Fig. 10. Related genera. Scanning electron micrographs of the abaxial leaf surface of: *Ansellia africana* (Kew no. 052-77.00323). **A.** Stomata (×325). *Grammatophyllum speciosum* (Kew no. 032-75.00440). **B.** Stomata (×825). *Eulophia stenophylla* (Kew no. 040-81.01284). **C.** Epidermis with stomata (×230). **D.** Stomatal cover (×800). *Cymbidiella humblotii* (Kew no. 340-82.03517). **E.** Epidermis (×420). **F.** Stomatal cover (×850). These related species all differ from the *Cymbidium* species examined in their epidermal morphology. The most similar are the leaves of *Ansellia* and species in section *Cymbidium* and *Cyperorchis*, and *Grammatophyllum* resembles section *Cymbidium* with its slit-shaped pores in the stomatal covers. A more detailed survey may help to indicate the relationships between the genera in the subtribe Cyrtopodiinae.

slit-shaped pore similar to section *Cymbidium*, but the cover is circular, not elongated, and the arrangement of epidermal cells is different (Fig. 10B). *Cymbidiella* has papillose epidermal cells similar to those of section *Jensoa*, but the stomatal cover is not raised, is elongated, and has a slit-shaped rather than a circular pore, indicating that these two taxa are not closely related, and that the papillae on the epidermal cells have evolved in two different lineages (Figs. 10E & F). *Eulophia* is distinct (Figs. 10C & D) and the epidermal morphology does not echo the affinity to section *Jensoa* suggested by seed micromorphology. Characters of the epidermis and stomata would appear to be useful in the elucidation of the relationships among genera in the Cyrtopodiinae.

PSEUDOBULB ANATOMY

Yukawa and Stearn (2002) examined the pseudobulb anatomy of a number of species in the genus. The cuticle varies from 2 μm in *C. aliciae* to 42 μm in *C. tigrinum*. Stomata and trichomes are absent. The epidermis comprises square, rectangular, rounded or variously shaped cells depending upon the species. Epidermal cell wall thickening is often U-shaped with the base of the U adjacent to the cuticle and thick in most species. *Cymbidium aliciae* is an exception with thin cell walls. The ground tissue consists of large, more or less circular, thin-walled water-storage cells, sometimes with pleated walls surrounded by smaller, thin-walled cells amongst which are small triangular intercellular spaces. Vascular bundles with xylem and phloem occupy the centre of the ground tissue. Raphides sometimes appear in the ground tissue. The vascular bundles are collateral, numerous and usually scattered through the ground tissue. Stegmata bearing conical, rough-surfaced silica bodies are associated with phloem sclerenchyma and with fibre bundles in *C. bicolor*.

ROOT ANATOMY

Root anatomy has been examined in a number of species by Yukawa & Stearn (2002). The roots of *Cymbidium* species are covered with a multi-seriate velamen, ranging from 3–5 cells thick in *C. kanran* to 10–16 cells thick in *C. ensifolium*. Cells of the epivelamen, the outermost layer, are polygonal and isodiametric. Hairs occur sporadically. Baculate tilosomes are present in all species. Exodermal cells are thin-walled or slightly thickened in some species, e.g. *C. kanran*. The cortex varies from 7–8 cells wide in *C. dayanum* to 18–25 cells wide in *C. ensifolium*. Most cells of the middle layer are thin-walled, circular to oval or angular. Pelotons and dead masses of fungal hyphae are common in the roots. Cruciate starch grains, silica, and raphides occur in the the cells of the middle layer. The endodermis and pericycle are unseriate. The vascular cylinder is variously polyarch with elliptic to oval or circular phloem clusters alternating with the xylem rays around the circumference of the vascular cylinder. In all species there is a central pith of thin-walled cells with prominent intercellular spaces.

AGEOTROPIC ROOTS

These roots grow erect from the normal roots and are found in various species, including *C. aloifolium*, *C. bicolor*, *C. finlaysonianum*, *C. dayanum,* and *C. tracyanum*. They are similar in anatomical structure to normal roots but differ in the clustering and small diameters of their xylem bundles (Yukawa & Stearn, 2002). Such roots are also found in *Ansellia*, *Grammangis*, *Grammatophyllum,* and *Graphorkis*.

TABLE 3

CHROMOSOME NUMBERS IN *CYMBIDIUM*

Section	Species	Chromosome number 2n=	Authority
Floribundum	*C. chloranthum*	40	Aoyama (1989)
	C. floribundum	40, 60, 80	Leonhardt (1979); Tanaka & Kamemoto (1974)
Austrocymbidium	*C. canaliculatum*	40	Aoyama (1989)
	C. madidum	40	Aoyama (1989)
Cymbidium	*C. aloifolium*	40	Aoyama (1989)
	C. finlaysonianum	40	Aoyama (1989)
Borneense	*C. aliciae*	40	Aoyama (1989)
Bigibbarium	*C. devonianum*	40	Aoyama (1989)
Himantophyllum	*C. dayanum*	40	Aoyama (1989)
Annamaea	*C. erythrostylum*	40	Aoyama (1989)
Cyperorchis	*C. eburneum*	40	Aoyama (1989)
	C. elegans	40	Aoyama (1989)
	C. hookerianum	40	Aoyama (1989)
	C. insigne	40, 60	Leonhardt (1979)
	C. iridioides	40	Aoyama (1989)
	C. erythraeum (as *C. longifolium*)	40	Aoyama (1989)
	C. lowianum	40	Aoyama (1989)
	C. mastersii	40	Aoyama (1989)
	C. tracyanum	40	Aoyama (1989)
	C. sanderae (as *C. parishii*)	40	Aoyama (1989)
	C. whiteae	40	Aoyama (1989)
Parishiella	*C. tigrinum*	40	Aoyama (1989)
Jensoa	*C. ensifolium*	40	Aoyama (1989)
	C. faberi	40	Aoyama (1989)
	C. goeringii	40	Aoyama (1989)
	C. kanran	40, 41	Aoyama (1989)
	C. sinense	40	Aoyama (1989)
Pachyrhizanthe	*C. lancifolium* (incl. *C. javanicum*)	38, 39, 43, 57	Aoyama (1989)
	C. macrorhizon	38	Aoyama (1989)

CHAPTER 5

CYTOLOGY

CHROMOSOME NUMBERS

T he diploid chromosome number of all species of *Cymbidium*, except in section *Pachyrhizanthe*, is 40 (Wimber, 1957a; Tanaka & Kamemoto, 1974; Leonhardt, 1979; Du Puy, 1986; Aoyama, 1989). The only other exceptions are a few triploid or tetraploid cultivars of some species: *C. insigne* 'Bieri' ($2n = 60$) and *C. floribundum* 'Yoshina' ($2n = 60$) in Leonhardt (1979), *C. floribundum* 'Geshohen' ($2n = 80$) in Tanaka & Kamemoto (1974) (Table 3). There have been occasional counts other than $2n = 40$ for species, but these are normally contradicted by other counts of the same species, and are either miscounts or variants of little importance to the taxonomy of the genus.

MEIOSIS AND SPORAD DEVELOPMENT

Meiosis and pollen formation in the genus *Cymbidium* have been studied by Wimber (1957b, 1957c). Several species from section *Cyperorchis* and *C. floribundum* (section *Floribundum*) were investigated. The normal cycle of meiotic and mitotic divisions produces a tetrad that does not separate into individual pollen grains even when mature. This was found in most cases, but a small percentage of abnormalities in individual sporads was noted in many of the species. These abnormalities included the formation of univalents and fragments that later formed micronuclei, creating polyads instead of tetrads. Bridges were also noted at telophase 1. These abnormalities were common in two species only. In *C. lowianum* 'Concolor' the higher percentage of abnormal sporad development was explained by Wimber as resulting from this cultivar being an extreme variant of the species. In *C. insigne* the high percentage of abnormality could only be a result of some lack of homologies inherent in the genomes of the specimens examined, perhaps due to some past hybridisation of *C. insigne* (Wimber, 1957b). It was also shown that the percentages of abnormal sporad development varied within a single plant, probably in response to minor environmental variations. Du Puy (1986) noted occasional disruption of the meiotic cycle in *C. lowianum* and *C. tracyanum*. This manifested itself in the presence of univalents or fragments at telophase 1.

In primary hybrids, abnormal meiosis and sporad development were found to be more common. The percentage varied, but as all of the crosses were between species within section *Cyperorchis*, the parental species were closely related, and the genomes (although divergent to some extent) were still close enough to allow production of some normal, fertile tetrads. In general, the irregularities were not common enough to reduce noticeably the fertility of the pollinium. For this reason there has been little problem in interbreeding to form the modern complex hybrids (Wimber, 1957c).

Several *C. floribundum* hybrids were shown to have a meiotic cycle that was almost entirely disrupted. These hybrids are almost completely sterile and few crosses with them have been successful. This is explained by the large difference between the genome of *C. floribundum* and the genomes of the hybrids involving the species in section *Cyperorchis* that were used as the second parent (Wimber, 1957c). Du Puy (1986) also showed that the meiotic cycle of *C.* Pumilow (*C. floribundum* × *C. lowianum*) was badly disrupted. Many abnormalities of the reduction division cycle and sporad development were observed. Univalents, fragments, bridges and unequal divisions were common. This disruption resulted in the production of many polyads instead of tetrads. There were also some normal-looking tetrads produced but in reduced numbers, resulting in high sterility. *Cymbidium ensifolium* has been hybridised with some of the large-flowered hybrids, but sterility is very high in resulting off-spring. Colchicine has been used to create a teraploid of *C.* Peter Pan 'Greensleeves' thereby restoring fertility, allowing hybrid lines to be developed using this species (see Chapter 8, p. 50). Examination of primary hybrids of species in different sections may indicate the degree of divergence between the taxa involved.

KARYOMORPHOLOGY

Aoyama (1989) examined the chromosome morphology of 30 taxa in *Cymbidium,* recognising two distinct types within the genus. Three taxa from section *Pachyrhizanthe* had a basic chromosome number of 2n = 38 and chromosomes of intermediate type between the simple and complex chromocentre types with an average chromosome length of over 2µm; in contrast, the remaining taxa, representing sections *Annamaea, Austrocymbidium, Bigibbarium, Borneense, Cymbidium, Cyperorchis, Himantophyllum, Jensoa* and *Parishiella,* had a complex chromosome type, averaging less than 2µm in length.

Fig. 11. *Cymbidium insigne* subsp. *seidenfadenii*. The albino form (*Menzies & Du Puy* 500) found in N. Thailand. Cult. Kew.

CHAPTER 6

POLLINATION BIOLOGY AND FLORAL FRAGRANCES

POLLINATION

Few observations of the natural pollination of *Cymbidium* species have been made. Two reports on the pollination of *C. madidum* indicate that the native Australian honey-bee (*Trigona* sp.) is the pollinating insect. Macpherson & Rupp (1935) studied a population of *C. madidum* near Proserpine, Queensland, in September 1934 and suggested that the bees were initially attracted by the sweet perfume of the flowers and later appeared to collect the viscid secretion from the disc of the lip. The bees were observed to gnaw at this exudation, which was transferred from the front legs to the second pairs of legs, and finally it was fastened to the pollen sacs on the rear legs. It was suggested that this substance is used as 'bee glue', to repair small cracks and gaps in the nest. The lip of the flower was observed to be 'irritable', in that it closed up against the column in 'several jerky movements' as the bee moved into the flower, imprisoning the bee and bringing its back into contact with the sticky stigmatic surface, on which pollinia were deposited. As the bee struggled backwards out of the flower, the anther was ruptured, and new pollinia were deposited on the hairs on the back of the bee. The bee was later (1936) identified as *Trigona kockingsi*. Smythe (1970) reported that another species, *Trigona caponaria*, was collected carrying pollinia of *C. madidum*.

Anthony Lamb (pers. comm.) observed the pollination of *C. finlaysonianum* by an unidentified bee species in Sabah. Liu and Nakayama (2005) published photographs of a small solitary bee entering the flower of *C. goeringii* in Japan and removing the pollinia on the back part of its abdomen. They postulate that the bees are attracted by the sweet fragrance of the flowers and scratch and possibly also bite the back of the callus. They also noted that dead bees are often found trapped between the lip callus and column in flowers that proceed to develop fruits. The fruits develop through the summer months and mature in autumn.

In northern Thailand in April 1983, Kjellsson *et al.* (1985) conducted field observations on the pollination of *C. insigne* subsp. *seidenfadenii*. The bumble-bee, *Bombus eximius* Smith, which is a regular pollinator of *Rhododendron lyi* Leveille in the area of the Phu Luang Wildlife Sanctuary, was observed to pollinate the *Cymbidium* and *Dendrobium infundibulum* Lindl. Observations indicated that flowers of the two orchid species are mimics of the flowers of *R. lyi* and that these species constitute a three-way floral mimicry system operated by *B. eximius*.

The flowers of the three species appear similar in shape and size (Kjellson *et al.,* 1985; Du Puy & Cribb, 1988). *Rhododendron lyi* has large, white flowers with a pale yellow patch at the base of one petal. The petals are fused basally into a slightly zygomorphic, conical corolla about 6 cm (2.4 in) across. The flowers of *D. infundibulum* are white with a light orange patch at the base of the lip. *Cymbidium insigne* subsp. *seidenfadenii* occurs in two colour variants in the area. Both have creamy-white flowers with a pale yellow patch on the mid-lobe of the lip. However, one form has red stripes on the lip (Fig. 17), whereas the other is unstriped (Fig. 11). The unstriped form has so far been recorded only from this locality. Flowers of both colour forms turn dull red when pollinated or when the pollinarium has been removed.

Dendrobium infundibulum flowers from the beginning of March until about the middle of April, and *C. insigne* subsp. *seidenfadenii* from early February until the middle of April. The *Rhododendron* flowers mainly in March and April. Its flowering season thus coincides to a great extent with that of the two orchids. Observations indicate that flowers of the two orchid species can last for about a month if

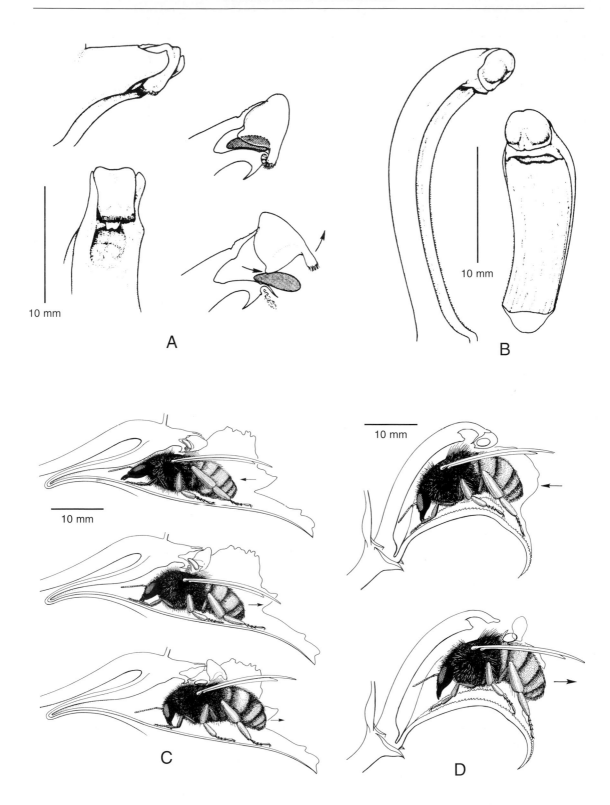

Fig. 12. A. The column of *Dendrobium infundibulum.* **B.** The column of *Cymbidium insigne.* **C.** Pollination of *D. infundibulum.* **D.** Pollination of *C. insigne.*

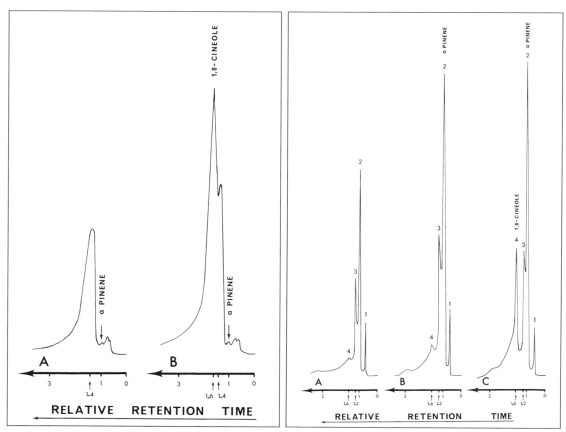

Fig. 13a (left). Analysis of the fragrance of *C. hookerianum*. **A.** Pure sample of *C. hookerianum*. **B.** *C. hookerianum* enriched with 1, 8-cineole; **Fig. 13b** (right). Analysis of the fragrance of *C. tracyanum*. **A.** Pure sample of *C. tracyanum*. **B.** *C. tracyanum* enriched with alpha-piene. **C.** *C. tracyanum* enriched with alpha-pinene and 1, 8-cineole.

the pollinia are not removed. *Bombus eximius* also shows seasonality. The worker females are found only during early spring (Frison, 1934). The condition of the specimens caught indicates that they had been flying for some time (O. Lomholdt & P. Williams, pers. comm.). Both orchid species were found growing in the immediate vicinity of *Rhododendron* scrub which occurs on a soil of almost pure sand broken by outcrops of sandstone and sandstone boulders (Fig. 112, p. 202). *Dendrobium infundibulum* usually grows on rocks and boulders (Fig. 15), and occasionally epiphytically, but *C. insigne* subsp. *seidenfadenii* is always terrestrial and appears to require at least a thin layer of soil. The cymbidium often grows in the shelter of the base of a shrub, and the flowering spikes can be seen protruding through the foliage of low-growing bushes (Fig. 111).

The bumble-bee *Bombus eximius* was found to be the main pollinator of the two orchids (Figs. 14 & 16). Most of the bees were worker females and, less frequently, queen bees. A second species, *B. rotundiceps* Friese, was seen only a few times carrying *D. infundibulum* pollinaria. Some other insects (*Anthophora himalayensis* Rad. and *Apis dorsata* Fabr. (bees), *Clinteria* sp. (a Cetoninae beetle) and two species of hawkmoth) sometimes visited the flowers, but were never seen carrying pollinia.

The flowers were visited most regularly during the morning between 5 a.m. and 10 a.m., except for the period from 6 a.m. to 7 a.m., when visits to flowers nearly doubled. It may be that the early foraging bees are most likely to make mistakes in their search for a food source. In the hot hours, from about 10 a.m. to 3 p.m., bumble-bee activity was much reduced, but it increased somewhat in the late afternoon. However, afternoon visits to orchid flowers were only observed twice, to *C. insigne* in both cases. The

33

Fig. 14 (above). Bumble bee pollinating *Cymbidium insigne* subsp. *seidenfadenii* flower. **Fig. 15** (bottom). *Dendrobium infundibulum* in flower, Phu Luang, Thailand.

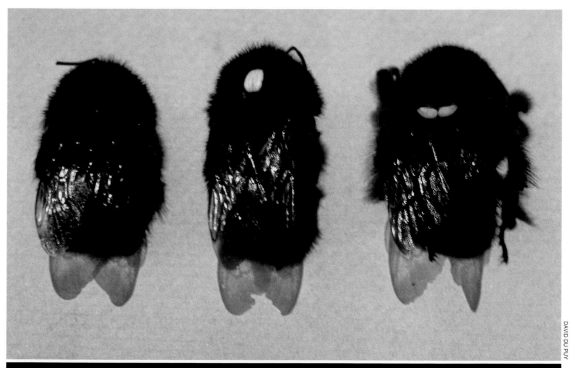

Fig. 16 (above). Bumble bees carrying pollinia of *Dendrobium infundibulum* (centre) and *Cymbidium insigne* subsp. *seidenfadenii* (right). **Fig. 17** (bottom). *Cymbidium insigne* subsp. *seidenfadenii* in flower.

visits that resulted in removal and/or deposition of pollinaria lasted 4 to 12 seconds in *C. insigne* but only 2 to 3 seconds in *D. infundibulum*.

Examination of the flowers of *D. infundibulum* and *C. insigne* subsp. *seidenfadenii* revealed neither nectar, nor any other reward for the pollinators, and no odour was detected.

It was possible for both orchid species to use the same bee as pollinator because they utilised different parts of the bee (Fig. 16). The cymbidium had a long column, and its pollinia were caught on the back of the main body of the bee figure (Figs. 11B & D), whereas the dendrobium had a shorter column, and placed its pollinia nearer to the head of the bee (Figs. 11A & C), precisely on a small bald, shiny patch on the bee's otherwise hairy back. The two orchids in this way ensured that neither collected the wrong pollen, their respective columns being the correct length to collect the appropriate pollinia when the bee entered the next flower.

This specialised morphology and the precise positioning of the pollinaria on the bees indicate a high degree of adaptation to the pollinator. The spatial separation of the different pollinaria on the bee ensures that there is no interference between the reproductive systems of the two orchids.

About 6% of all individuals of *B. eximius* seen were carrying pollinaria of *D. infundibulum*, whereas only a few carried pollinaria or remnants of pollinaria from the cymbidium. However, the frequency of flower visits was much lower in *D. infundibulum* than in the cymbidium. A possible explanation could be that the total number of flowers of *D. infundibulum* was much higher, so that a high number of *D. infundibulum* pollinaria were accumulated. Furthermore, the *Dendrobium* pollinaria are perhaps affixed more firmly to the bee than the *Cymbidium* pollinaria. The flowering season of these orchid species is extended over a period of nearly two months, thereby increasing the chances of pollination.

Observations and experiments indicate pollination by floral deceit, which has previously been reported in several orchids (Ackerman, 1983a; Bierzychudek, 1981; Boyden, 1980; Dafni, 1983; Dafni & Ivri, 1981a, 1981b; Nilsson, 1980, 1983). The flowers are similar enough in colour, size and shape to trick the bees into visiting the orchids, even though they offer no food reward, by mimicking the *Rhododendron* flowers with their abundant nectar and pollen. The similarity between the *Rhododendron* and *C. insigne* subsp. *seidenfadenii* appears to be enhanced in a large number of individuals in this population by the loss of the red markings on the lip.

FLORAL FRAGRANCES

Significant biological and taxonomic information has been obtained from the analysis of floral fragrances in the study of several genera of tropical American orchids, e.g. in subtribes Catasetinae and Stanhopeinae (Dodson & Hills, 1966; Hills, Williams & Dodson, 1968, 1972; Dodson *et al.*, 1969; Ackerman, 1983b) and several European species (Nilsson, 1978a, 1978b, 1979a, 1979b, 1981). In some of these cases the composition of the floral fragrance has been shown to be species-specific.

The fragrant species of *Cymbidium* include the strongly scented *C. eburneum*, the less fragrant *C. mastersii*, *C. tracyanum*, *C. iridioides* and *C. hookerianum* (section *Cyperorchis*), the coconut-scented *C. atropurpureum* (section *Cymbidium*) and *C. borneense* (section *Borneense*), and the sweetly fruit-scented *C. sinense*, *C. ensifolium*, *C. kanran*, *C. faberi*, some *C. goeringii* and *C. cyperifolium* (section *Jensoa*). Others, including *C. madidum* and *C. suave* (section *Austrocymbidium*), are less strongly fragrant.

The floral fragrances of two of these species were analysed by Du Puy (1986). That of *C. tracyanum* showed four peaks. Peak 2 was identified, by enrichment with pure chemicals, as alpha-pinene, and peak 4 as 1,8-cineole (Fig. 13b). Two peaks were observed in the analysis of *C. hookerianum*, the minor one was alpha-pinene but the major one did not correspond with 1,8 cineole (Fig. 13a).

These preliminary results indicate that floral fragrance analysis might provide useful taxonomic evidence when more species are surveyed.

CHAPTER 7

CULTIVATION

Cymbidiums are amongst the easiest of all orchids to grow well, provided that some simple instructions are followed. Their ease of culture has helped to make them as popular to grow as they are today. Cymbidiums make perfect pot plants with flowers that can last from eight to ten weeks, and they are also renowned as excellent producers of long-lasting cut flowers. The literature on their cultivation is extensive (e.g. Veitch, 1887–1894; Williams, 1894; Rittershausen & Rittershausen, 2000, 2001).

GENERAL CULTIVATION AND GROWING CONDITIONS

Cymbidiums, including the large-flowered hybrids which are most frequently cultivated, are mostly 'cool growers' preferring cool greenhouse conditions, but many will also grow outside in warm climates, or in heated or air-conditioned homes. In warmer temperatures the flowers will generally not last as long as they would if they were kept in cooler conditions. They enjoy a temperature range of between 10 and 30°C (50–85°F). These are ideal temperatures in which the plants will thrive, and temperatures below or above this will cause stress. However, in late summer and early autumn, they require lower night temperatures (max. 12°C (54°F)) to initiate the formation of flower spikes. This drop in night-time temperature is essential to ensure optimal flowering. Maturing buds can turn yellow and fall from the stems if the plants are kept too warm at night at this stage, or if they are subjected to extremely warm temperatures during the day. Therefore, ensure that the plants are kept in cool conditions until the flowers have opened.

Cymbidiums are found in a broad range of environments. The large-flowered Himalayan, Indo-Chinese and south-west Chinese species (section *Cyperorchis*), from which the large-flowered hybrids are derived, grow at higher elevations in shaded montane forests and they prefer cooler conditions. The smaller-flowered tropical and sub-tropical hard-leaved species (section *Cymbidium*) grow at low altitudes or even on the coast, often in exposed situations, and therefore need slightly drier and warmer conditions and require more light. The small-flowered oriental and Himalayan species (section *Jensoa*), some of which also occur in the tropics, are mainly terrestrial species, and although some enjoy cool conditions many can tolerate warm or even hot and wet summers followed by cold and dry winters. In temperate regions, the main group of species and hybrids which we grow as pot plants are ideally grown in shaded glasshouses, but in sub-tropical climates they can be cultivated out-of-doors. Cymbidiums are vigorous plants and they enjoy good light; indoors they should be placed out of direct sunlight but near a good source of natural light, while in a greenhouse they need to be slightly shaded in summer to avoid scorching of the leaves. Bright light will also ensure that the plants are hardened and do not produce soft, lanky growth that is reaching for the light. They will then flower more successfully and produce stronger stalks to carry the heavy spikes of flowers.

Air movement is as vital to the good health of cymbidiums as it is to all orchids: it maintains an even overall temperature and ensures that there are no pockets of cold or hot air. In glasshouses, adequate ventilation is vitally important in the late summer and early autumn, when cymbidiums require a cooler temperature to encourage the production of flower spikes. Outdoors, the temperature drops naturally in autumn, but be sure that plants are fully protected where frosts are likely to occur. It is a myth that allowing plants to be frosted will encourage more flower spikes. The plants must be brought into a frost-free, well-lit environment, removing any dead leaves and weeds.

Flower spikes should be staked as they emerge, with as thin a bamboo stake as possible, but one which is strong enough to ensure enough support for the fully open spike of flowers. Undoubtedly, the flowers become much heavier as they open. The young flowering stalks emerge from the base of a mature pseudobulb and, when they have reached 15–20 cm, should be trained upwards to avoid the stem spiralling and becoming crooked. Initially a stake should be inserted into the compost just behind the emerging stalk. Within a few weeks the first tie should be applied and later, as the stem hardens, it will become firm and preferably only one other tie need be applied just below the forming buds; too many ties will give the appearance of the stem being trussed. The ties can be rearranged as the stem elongates and matures. The stems should all be encouraged to stand erect and the flowers to emerge above the foliage for the best appearance.

After flowering, cymbidiums will survive quite happily if they are placed outdoors in a semi-shaded spot or in the shade of a tree, once all danger of frost has passed. However, a careful watering, misting and feeding programme should be followed, and also the plants will be more vulnerable to pests and diseases. In the controlled environment of a greenhouse or home, pests and diseases can be more easily controlled and are much less likely to appear.

PESTS AND DISEASES

Orchids are as susceptible as any other plant when it comes to pests, and bringing new plants into a collection from outside sources can also bring pests and diseases. Unless you have a quarantine area it is difficult to protect your plants from every problem.

Cymbidiums do not suffer from too many fungal diseases but they are prone to attack by cymbidium mosaic virus, which is incurable and is spread from plant to plant by using infected and un-sterilised tools, particularly knives and secateurs. The transfer of sap from an infected plant to a clean plant is the most common way of transmitting viruses. When dividing plants, ensure cutting instruments and blades are sterilised in a flame or sterilising agent before using them on another plant. Sadly, the only way to rid any collection of viruses is to destroy all infected plants. Virus infections will discolour the leaves and lead to deformed flowers and irregular colouration. Virus infection can be recognised by distinct yellow or white mottling or black streaking on the foliage, or colour breaks on flowers which may also discolour soon after the blooms have opened. However, be sure not to confuse this with insect damage.

The worst pests from which cymbidiums suffer are red spider mite, mealy bug and scale insects. They can often become infested with red or false spider mite that will cause yellowing of their leaves. These pests are minute and can be difficult to see, but they can be discouraged by maintaining a buoyant humid atmosphere. If a heavy infestation is found, a predatory mite *Phytocelius persimilis* can be introduced to control it. However, it is often necessary to resort to pesticides and arachnicides to control these pests. Aphids often attack flower spikes, buds and flowers, causing distortion and damage to the flowers. They can, in most instances, be removed with soapy water. Mealy bug and scale insects are best wiped off with cotton wool soaked in methylated spirits. Bad attacks of mealy bug can be controlled using the ladybird *Cryptolaemus montrouzieri*. If in doubt, a full strength programme of malathion could be applied with due care.

Slugs and snails are notorious for causing damage—especially to show plants just before they are to be exhibited! They are very prevalent on plants grown outside but are easily dealt with using a liquid or pelleted slug killer.

Vermin, especially mice, are encouraged into the warmth of the greenhouse in winter to avoid the cold, just as the flower buds are being formed and beginning to open on the cymbidiums. The pollen makes a tasty morsel, high in protein, for these hungry rodents who scurry up the stems damaging the buds and devouring the pollinia. On open flowers, once the pollen is removed the lip will quickly discolour and the flowers will soon fade. Therefore, be sure to block up any holes which are large enough to allow vermin to enter the greenhouse, and small traps may also be necessary as a precaution. Outdoors, this problem is more difficult to deal with.

REPOTTING AND COMPOSTS

Cymbidium plants are best repotted in the spring, after flowering, and must be repotted, if not annually, then every second year, especially if the compost is decomposed and there is damage to the root structure. They will not tolerate being over-potted and enjoy being root-bound; they like to be planted in a pot just large enough to accommodate one or two years' growth. The plants can be divided if they have become too large and bulky. It is simple to divide a *Cymbidium* by finding a natural break in the plant and simply cutting between the back bulbs and splitting the plant into divisions, each ideally comprised of four or more firm pseudobulbs. These latter contain sustenance which the plant can draw on until it re-establishes itself. Leafless pseudobulbs can also be removed individually and placed into a damp medium where they will often sprout and eventually produce a new plant. Although spring is the ideal time for repotting, if the plants suffer from being over-watered they can be repotted at almost any time. Repotting a healthy plant is much easier than repotting a sick or poorly one. The pot should be removed, any excess compost shaken off, and the plant placed into a slightly larger pot filled with fresh damp compost. Do not water the plant for at least a week as this will encourage the new roots to enter the fresh compost. Then continue with your standard watering regime and begin the fertilising programme as normal.

To repot a sick plant, it must be taken from the pot, have all the compost removed, any damaged or decaying root and leaf material eliminated, and the plant should be put into the smallest possible pot so that it is comfortable for at least one year. Remember that whatever is happening to the roots is reflected at the tips of the leaves; die-back on the leaf tip is usually a sure sign that there is a problem with the roots, either from over-watering or too much fertiliser. Always ensure that any knives or cutting utensils that are used are sterilised in between each plant, because this will limit the spread of viruses to which cymbidiums are susceptible.

Various potting media are available to choose from, the most popular option being a mixture of equal parts of medium grade bark and perlite or polystyrene. Variations of this recipe and additions to the base mix may be used. Peat or coconut fibre can be added as long as some dolomitic lime is used in conjunction with either of these additives to neutralise their high acidic content. Rockwool is the medium of choice for *Cymbidium* culture by professional growers, especially for plants that are used to produce cut flowers, although some large-scale production units prefer to use a coconut fibre and bark mix for pot plants as it offers a more natural and far less artificial appearance. Plants intended for sale as pot plants need to have a neat and acceptable appearance for the customer, whereas those cultivated for cut-flower production are cultivated in the way which requires least maintenance. These latter usually have their leaves cropped for ease of flower-stem maintenance and are often unsightly, and they are divided into individual backbulbs when the plants become too large, cumbersome and unkempt.

Rockwool is unsightly and can cause irritations if your skin is not protected because the fibres can penetrate the skin and cause a rash. Diatomaceous earth can be added to various rockwool mixes and has proved to be more stable than just using rockwool on its own. A horticultural grade of foam chips is another alternative as an additive to rockwool, and can assist in keeping the mixture aerated and reducing compaction. Rockwool mixes must never be allowed to dry out completely, as they are difficult to re-wet. If they do, then a wetting agent or a few drops of liquid detergent will be needed in order to re-wet them and allow them to hold water again. Large-scale producers and nurserymen have the advantage of highly sophisticated irrigation and irrigation/fertilisation ("fertigation") units that replenish the moisture regularly and ensure that the plants do not dry out too much, rinse the compost when necessary to avoid a build-up of excessive nutrients, and feed at just the right time. Simple forms of irrigation units are now available at a reasonable price, as are simple "fertigation" equipment kits for the hobbyist.

Thorogood (1992) recommended cultivation of the Australian species in plastic pots with a bottom mixture of four parts of polystyrene chips which will go through an 18-mm sieve to one part sphagnum pushed through a 6-mm sieve, and a top mix of equal parts of sphagnum and polystyrene chips passed through a 12-mm sieve. To both he added a quarter-cup of 9-month 'Osmocote' to a 10-litre bucket of mix. He also covered the top 10 mm with charcoal to stop algal build-up.

WATERING AND FEEDING

Watering cymbidiums is not a science and there are no exact values or amounts which they require. Generally, they enjoy having moisture at their roots and, as a rule, must be allowed to become almost dry before re-watering. They should be watered once a week in the winter and probably twice a week in the summer. The best way to tell if your plant requires water is to learn from experience how much the pot weighs, and as soon as it weighs approximately half of its wet weight it needs to be re-watered. Watering is best done as early in the day as possible so that the leaves have enough time to dry before the temperature drops in the evening, preventing bacterial rots from establishing in the new growths.

Ideally, rainwater should be used, or if only a few plants are grown, bottled spring water which is low in lime is also suitable. Tapwater can be used in areas with soft water, although it is better to allow it to stand before use to allow dissolved sterilisation gases to evaporate. Hard water, high in dissolved lime, will not only cause a build-up of lime in the compost, but will also leave unsightly white deposits on the foliage. During the flowering season the plants must be regularly watered to the point of being damp only—they do not like being saturated and should never be left standing in water. If they are too wet the roots will eventually rot and the decay will spread through the plant and could ultimately cause its death. Indoors, stand the plant in a planter which is deep enough to hold at least 2–3 cm of pebbles or gravel. The gravel should never be fully submerged in water, but some water in the gravel will create humidity around the plant, which is essential for the health of the plant and especially its foliage. The humidity will prevent crinkled leaves from being produced, as they are a sure sign that the plant is dehydrated. In a greenhouse, the plants can be watered overhead or with a lance, which will also wash dust from the leaves and keep the plants in their best condition. Obviously, the larger the area in which the plants are situated, the more air movement there will be, and therefore the quicker the plants will dry out.

Cymbidiums are hungry feeders: they must be fed, and will not flower well if starved. If they do flower without being fed properly, the flowers will be pale and the stems will be few-flowered and weak. A high-nitrogen fertiliser, such as in a 30:10:10 N:P:K formula (nitrogen: phosphorus: potassium (= potash)), should be used in spring to encourage new growth and this can be changed to more general feed (20:20:20) for the summer. During autumn or fall, a high potash fertiliser (10:10:30), combined with cooler nights, will induce flowering. This can then be changed back to a half-strength general feed throughout the winter. Feed cymbidiums at least once a week, ensuring that they are also watered well enough to rinse out any excess salts from the fertilisers: these salts could lead to a build-up of undissolved fertilisers in the potting media which in turn can burn the roots and eventually cause rotting.

CHAPTER 8

HYBRIDISATION
AND BREEDING

NATURAL HYBRIDISATION

Natural hybridisation has been reported a number of times in *Cymbidium*, although it has mainly been postulated on the basis of the intermediate morphological characteristics of the plants between the presumed parental species, rather than by direct observation in the field.

Five natural hybrids of *Cymbidium* are listed in *Sanders' List of Orchid Hybrids* (1946): *C.* × *ballianum*, *C.* × *bennettpoi*, *C.* × *cooperi*, *C.* × *gammieanum* and *C.* × *zaleskianum*. All are hybrids involving parents in section *Cyperorchis*.

The earliest of these to be identified was *Cymbidium* × *gammieanum,* described by King and Pantling in 1895. They suggested a parentage of the sympatric Himalayan species *C. erythraeum* (as *C. longifolium*) and *C. elegans*. Linden described *C.* × *zaleskianum* in 1902 and suggested a parentage of *C. grandiflorum* (= *hookerianum*) and *C. giganteum* (= *iridioides*). For reasons given under the account of *C. tracyanum*, we suspect that this is merely a well-marked form of *C. tracyanum*. Rolfe described *C.* × *cooperi* in 1904 as a natural hybrid of *C. insigne* with *C. schroederi*. Recent observations of the parental species in the Dalat region of Vietnam by Averyanov (pers. comm.) have confirmed this parentage. In the same year, Rolfe also described *C.* × *ballianum* with the parentage of *C. eburneum* and *C. mastersii*. The parentage of *C.* × *bennettpoi* was given as *C. grandiflorum* (= *hookerianum*) and *C. tracyanum* by Sanders (1946). We have seen no natural source material of this putative origin. Another natural hybrid reported amongst the larger-flowered species was *C.* × *pseudoballianum*, a name coined by Averyanov (1990) for a plant collected by Sigaldi (no. 212) at Dalat in Vietnam, which Guillaumin had referred to as *C.* × *ballianum* but with the comment that it was a hybrid of *C. eburneum* and *C. erythrostylum*.

Two recently described Chinese cymbidiums are possibly also of hybrid origin. *Cymbidium maguanense*, described by Liu in 1996 from Maguan in South-east Yunnan, is possibly a hybrid of *C. eburneum* and *C. mastersii*, both of which occur in the region. Its habit is close to that of *C. eburneum* but it has more and smaller flowers like *C. mastersii*. Likewise, *C. baoshanense*, described from western Yunnan by Liu and Perner in 2001, is almost certainly a natural hybrid of *C. lowianum*, probably with *C. tigrinum*. The progeny we have seen were rather varied in flower colour, suggesting that some degree of backcrossing had occurred in nature. The Chinese *C. gongshanense* is, almost certainly, an artificial hybrid with *C. floribundum* as one parent, and a complex hybrid as the other.

Du Puy *et al.* (1984, 1988) postulated that *C. schroederi* might be a stabilised hybrid between *C. iridioides* and *C. lowianum,* originating in a time when these three may have been sympatric. Neither putative parent is found in the range of *C. schroederi* today, although these species have both been recorded from the same locality in Yunnan.

Natural hybridisation has also been reported elsewhere in the genus, especially between species in sections *Jensoa* and *Pachyrhizanthe*. Makino named *C.* × *nishiuchianum* (*C. goeringii* × *kanran*), based upon a plant collected in Japan, a name recently validated by Shaw (2002). Maekawa (1971) and Masamune (1984) both provide excellent descriptions and illustrations of this hybrid. Cheng (1975, 1981) provided illustrations of several Taiwanese natural hybrids, none being formally named and described: *C. ensifolium* × *kanran*, *C. ensifolium* × *sinense*, *C. kanran* × *sinense*, *C. kanran* × *tortisepalum*, *C. goeringii* × *ensifolium*, *C. goeringii* × *kanran*, *C. goeringii* × *sinense* and *C. sinense* ×

tortisepalum. He also reported hybrids of *C. lancifolium* with *C. ensifolium* and *C. sinense,* and of *C. ensifolium* with *C. floribundum.* All of these purported hybrids are illustrated by Cheng but not all are convincing. Du Puy & Cribb (1988) indicated a possible hybrid origin for the taxon then recognised as *C. goeringii* var. *tortisepalum,* and now recognised as *C. tortisepalum* var. *tortisepalum* (q.v.), from Taiwan.

CYMBIDIUM BREEDING

Hybrid cymbidiums have been grown since the earliest times, probably since the time of Confucius (551–479 BC). The smaller-flowered Chinese species hybridise occasionally in the wild and any unusual plants that differed from the typical ones were collected and cultivated not only for their flowers but also for their fragrance and leaf variegation. A selected specimen would often accompany a Mandarin in his carriage or palanquin when he was being transported from place to place. Thus, the state official could relax and appreciate the beauty of the plants, keeping his mind free from the cares of state. Artificial hybridisation, however, is altogether more recent and started in the late 19th and early 20th centuries when large-flowered Himalayan species were crossed in the large European orchid nurseries and collections, particularly in England. The 50 or so species of *Cymbidium* offer a wide variety of flower size, shape and type, but less than a quarter are responsible for the breeding lines of modern hybrid cymbidiums (see Table 4).

A better understanding of the cultural needs of the large-flowered cymbidiums, which are generally orchids that prefer cooler growing conditions, accelerated their popularity and eventually their breeding. Initially, cymbidiums were cultivated in Europe in Victorian stove houses or heated glasshouses. The plants suffered from the stifling heat, despite other orchids from more tropical climes thriving in such conditions. Victorian glasshouses and stove houses were normally heated with hot water passing through iron pipes or in open troughs giving off steam and were usually heavily shaded. The iron pipes were sometimes laid beneath the floor or were passed through beds of bark, producing conditions akin to the warm tropics and far too warm for most *Cymbidium* plants. The glasshouses were often glazed with Hartley's Patent Glass which, while allowing the entry of some light, was too opaque for cymbidiums.

Progress in the quality of *Cymbidium* hybrids has, in many cases, been through the sudden appearance of a new species or superior plant of an individual clone of a hybrid, and subsequent line breeding. The discovery of the white- and pink-flowered *C. insigne* in Vietnam in 1905 signified the biggest breakthrough in the history of *Cymbidium* hybridising. Prior to that the hybrids that had been bred had not particularly excited growers, and the prospect of major advances in the production of *Cymbidium* hybrids that we know today had not been considered. Nowadays, the popularity of the genus is the result of the ease with which they can be cultivated, the kaleidoscope of flower colours offered by the hybrids, and the long-lasting qualities of both cut flowers and pot plants. Future developments will probably centre on the introduction of the oriental small-flowered *Cymbidium* species into modern breeding programmes, thereby bringing fragrance back into the progeny and breeding from newly discovered species such as *C. wenshanense.*

Cymbidium hybridisation began in a desultory manner. The nursery of Messrs James Veitch & Sons of King's Road, Chelsea, flowered the first artificial hybrid, *C. eburneum* × *lowianum* in 1889, using the latter as the pollen donor. It is now known that *C. lowianum* var. *concolor* was used in this crossing to produce *C.* Eburneo-lowianum. In this instance the seed from the capsule was sprinkled around the base of the mother plant and, whilst it was slow to germinate and the seedlings just as slow to grow, there was jubilation when the first new *Cymbidium* hybrid flowered. Veitch's were more sanguine about the amount of time it had taken to flower and the dull colours that were produced. Eleven years later a second hybrid, *C. grandiflorum* (= *hookerianum*) × *elegans* flowered for Messrs C. Maron et Fils in France. In 1902, two hybrids of *C. lowianum* flowered: *C.* Lowio-grandiflorum for Veitch and

C. Lowio-mastersii for Messrs Charlesworth & Sons of Haywards Heath, Sussex. In the same year Sir Frederick Wigan succeeded in flowering a cross between *C. eburneum* and *C. tracyanum*. The following year, *C.* Lowgrinum (*lowianum* × *tigrinum*) was flowered by R.L. Measures. Two more *C. eburneum* crosses flowered in 1906, one with *C. giganteum* (= *iridioides*) for Charlesworth and the other with *C. grandiflorum* (= *hookerianum*) flowered for Fred Sander. In 1907 Messrs Armstrong & Brown of Tunbridge Wells flowered a plant of *C.* Eburneo-lowianum that had been backcrossed to *C. lowianum*. By the end of the first decade of the 20th century only a dozen or so cymbidium crosses, involving eight species in their parentage, had been flowered, providing little indication of the explosion of breeding that followed shortly afterwards.

The introduction of *C. insigne* revolutionised the breeding in the genus because it produces extremely desirable characteristics in its offspring; tall erect inflorescences with several flowers at a time and borne well above the leaves, and large long-lasting flowers in shades of pearly white to pastel pink (which are still the most important colours for cut flowers today). *Cymbidium insigne* is known in at least four distinctive forms, the two main ones being *C. insigne* 'Sanderae' which offers taller spikes but is less floriferous, while *C. insigne* 'Album' is a more vigorous grower which flowers freely but with slightly shorter stems.

The first artificial hybrids of *C. insigne* started to emerge in 1911, namely *Cymbidium* Gottianum (*insigne* × *eburneum*) produced by Messrs F. Sander & Sons of St. Albans and *C.* Pauwelsii (*insigne* × *lowianum* var. *concolor*) by Pauwels of Ghent in Belgium. In the same year George Holford of Westonbirt produced *C.* Alexanderi (*insigne* × *eburneo-lowianum*) which proved to be the major breakthrough in *Cymbidium* breeding. *Cymbidium* Alexanderi, which set a new standard in flower colour and shape in early hybrid cymbidiums, was named in honour of H.G. Alexander, orchid grower to Sir George Holford at Westonbirt. This grex received three First Class Certificates (FCCs) from the

Fig. 18 (left). *Cymbidium* Rio Rita 'Radiant', an important parent in modern hybridising. **Fig. 19** (right). *Cymbidium* Alexanderi 'Westonbirt'. Probably the most important parent in the history of *Cymbidium* breeding.

43

TABLE 4

EARLY *CYMBIDIUM* CROSSES AND THE FIRST USE OF SPECIES (INDICATED IN BOLD)

Parent 1	Parent 2	Date	Hybridiser	Grex names and Comments
eburneum	**lowianum**	1889	Veitch	C. Eburneo-lowianum—first cross to flower
iridioides (as *giganteum*)	**mastersii**	1900	Maron	C. Maronii
eburneum	**tracyanum**	1902	Wigan	C. Wiganianum
lowianum	mastersii	1902	Charlesworth	C. Lowio-mastersii
lowianum	**hookerianum** (as *grandiflorum*)	1902	Veitch	C. Lowio-grandiflorum
lowianum	**tigrinum**	1903	Measures	C. Lowgrinum
mastersii	tracyanum	1904	Sander	C. Woodlandense
lowianum	tracyanum	?	Colman	C. Gattonense (Fig. 20)
iridioides (as *giganteum*)	eburneum	1906	Charlesworth	C. Eburneo-giganteum
eburneum	hookerianum (as *grandiflorum*)	1906	Sander	C. Holfordianum
Eburneo-lowianum	lowianum	1907	Armstrong & Brown	C. Woodhamsianum —first second-generation cross
elegans	iridioides (as *giganteum*)	1908	Fowler	C. Maggie Fowler (Fig. 21)
Eburneo-lowianum	tracyanum	1908	Colman	C. Lady Colman
devonianum	lowianum	1911	Veitch	C. Langleyense
eburneum	**insigne**	1911	Sander	C. Gottianum
insigne	lowianum	1911	Pauwels	
insigne	**schroederi**	1911	Fowler	C. J. Davis
Eburneo-lowianum	insigne	1911	Holford	The famous C. Alexanderi (Fig. 19)
insigne	tracyanum	1912	McBeans	C. Doris
eburneum	Wiganianum	1912	McBeans	C. Schlegelii
erythrostylum	iridioides (as *giganteum*)	1913	Moss	C. Florinda
insigne	**sanderae** (as *parishii sanderae*)	1914	Holford	C. Dryad
insigne	iridioides (as *giganteum*)	1914	Armstrong & Black	C. Iona
erythrostylum	tracyanum	1914	Hanbury	C. Hanburyanum
hookerianum (as *grandiflorum*)	insigne	1914	Hamilton-Smith	C. Coningsbyanum
Gottianum	insigne	1914	Sander	C. Memoria P.W. Janssen

erythrostylum	*insigne*	1915	McBeans	C. Albanense
erythrostylum	*Wiganianum*	1915	Armstrong & Black	C. Sandhurstiense
Eburneo-lowianum	*sanderae* (as *parishii sanderae*)	1915	Holford	C. Jasper
insigne	*Lady Colman*	1915	Colman	C. Queen of Gatton
Alexanderi	*Lowio-grandiflorum*	1916	Holford	C. Miranda
Eburneo-lowianum	*Pauwelsii*	1916	Hassall	C. Diana
Gottianum	*Pauwelsii*	1916	Sander	C. Egret
Lowgrinum	*lowianum*	1916	Hamilton-Smith & Pitt	C. Vega
insigne	*tigrinum*	1917	Hamilton-Smith	C. Insignigrinum
Alexanderi	*lowianum*	1917	Sander	C. President Wilson
Alexanderi	*hookerianum* (as *grandiflorum*)	1918	McBeans	C. Pearl
Holfordianum	*Pauwelsii*	1918	Schroeder	C. Shillianum
sanderae (as *parishii sanderae*)	*Pauwelsii*	1918	Sander	C. Elfin
eburneum	***roseum***	1921	Hamilton-Smith	C. Juno
roseum	*insigne*	1922	Sander	C. Titania
aloifolium (as *pendulum*)	*tracyanum*	1924	Alexander	C. Mona
eburneum	*erythrostylum*	1926	Hanbury	C. Niveum
devonianum	*insigne*	1928	Lambeau	C. Vogelsang
Alexanderi	*Rosanna*	1934	Rothschild	C. Balkis
insigne	***floribundum*** (as *pumilum*)	1942	Alexander	C. Minuet
aloifolium	***finlaysonianum***	1944	Hirose	C. Hanalei
ensifolium	*Miretta*	1957	Dos Pueblos	C. Peter Pan
canaliculatum	*Eagle*	1961	Stewart	C. Odyssey
goeringii (as *virescens*)	*Alexanderi*	1964	McLennan	C. Nagalex
canaliculatum	*madidum*	1966	Cooper	C. Little Black Sambo
suave	*floribundum* (as *pumilum*)	1969	Andrew	C. Scallywag

Royal Horticultural Society, with the finest 'Westonbirt' (Fig. 19) being awarded a First Class Certificate and the Lindley Medal in 1922. The day it was awarded the robust plant carried a couple of spikes with an average of 12 beautiful ivory-white flowers per stem accentuated by the mauve-spotted lip. The remarkable flower substance, high quality flowers and superior shape were noted in its description. During the next few decades the 'Westonbirt' clone was used to produce hundreds of hybrids and was outstanding as a parent in breeding in the genus. Its characteristics proved to be very dominant and it was later shown to be a natural tetraploid, hence the high quality of the progeny and the number of awards that were to be bestowed upon them.

Flowers of significant pre-Second World War hybrids bred from *C. insigne* ranged in colour from white to vivid pink and dark rose. One of the best is *C.* Ceres, which is *C. lowianum* var. *iansonii* × *insigne* and was bred in 1919 by Hamilton Smith. This hybrid proved to be a significant parent in producing some beautiful offspring—Sander flowered both *C.* Joy Sander (× Pauwelsii) and *C.* Louis Sander (× Alexanderi) in 1924, and *C.* Charm (× *erythrostylum*) in 1930, while McBeans bred *C.* Lyoth (× *insigne*) in 1928, and *C.* Ruskin (× Pearl) in 1938.

In 1927, *C.* Rosanna (Alexanderi × Kittywake) was bred by H.G. Alexander. The cultivar 'Pinkie', flowered by Lionel de Rothschild, proved to be a tetraploid which was surprising because *C.* Kittywake was certainly a diploid. Normally, a diploid crossed with a tetraploid produces a triploid but, in this case, the dominance of the *C.* Alexanderi 'Westonbirt' gametes produced a tetraploid. Later, Rothschild sold his collection to Dos Pueblos Orchids in America and they crossed *C.* Rosanna 'Pinkie' back onto *C.* Alexanderi 'Westonbirt' to remake the beautiful *C.* Balkis. This, in turn, when crossed with *C.* Carisona, makes the delightful and free-flowering *C.* Lilian Stewart, a grex that produces remarkable pink flowers and that is still in production for the cut-flower trade.

In 1961, McBeans bred *C.* Snow Sprite (Pearl-Easter × Alexanderi) which was used by Featherhill Orchids of America in 1982 in a cross with *C.* Rincon (Pearl × Windsor) to produce *C.* Fancy Free. This was remade by several growers and has been awarded on many continents. In 1989 the Eric Young Foundation in Jersey crossed *C.* Rincon with *C.* Angelica's Lady (Angelica × Liliana) to make *C.* Gorey, a highly successful and awarded grex that in turn has gone on to produce *C.* Cotil Point (× Red Beauty). *Cymbidium* Red Beauty 'Rembrandt' FCC/RHS is the clone that has been used in breeding programmes in the UK, but elsewhere different cultivars have been used.

Several generations on in this line of breeding, Alvin Bryant at Valley Orchids in Australia produced *C.* Sleeping Beauty (Sussex Dawn × Durham Castle) which has in turn been the parent of some of the most astonishing pure-coloured cymbidiums bred to date, such as *C.* Sleeping Dream (× Sleeping Glow), of which the clone 'Tetragold' is one of the best known outside Australia. In 1991, Mukoyama Orchids of Japan crossed *C.* Sleeping Beauty with *C.* Yamba (Sleeping Beauty × Coraki) to make the pure bright yellow *C.* Lovely Bunny. The clone 'Othello' of this hybrid won the coveted Orchid Grand Prix trophy in Japan and is continually being recognised by judges for its fine colour and as a floriferous pot plant.

Other significant species used for breeding have been *Cymbidium lowianum, C. erythrostylum, C. sanderae, C. floribundum, C. tracyanum, C. devonianum, C. ensifolium, C. madidum, C. tigrinum* and *C. canaliculatum. Cymbidium lowianum*, described in 1879 from Burma, is variable in colour and form, the two most important being the varieties *album* and *concolor*, the latter having beautiful chartreuse-green flowers with a dark ochre-marked lip. Sander's *Orchid Guide* of 1901 offered *Cymbidium lowianum* var. *concolor* for 105 shillings (or 5 guineas). *Cymbidium lowianum* var. *concolor* crossed with *C.* Alexanderi by Sander in 1917 produced *C.* President Wilson. *Cymbidium lowianum* and its various forms offered hybridists a wide range of colours to choose from and seedlings that grew more easily. It differs from *Cymbidium insigne* not only in flower colour but also in having wider leaves and long arching racemes of flowers. Its hybrids are readily recognised by the distinctive maroon V-shaped bar on the lip of all its progeny and are generally late flowering. Unfortunately, the early hybrids from *C. lowianum* tended to have flowers in shades of dull green to khaki. The best parental clone was *C. lowianum* 'Pitts Variety', very similar to *C. lowianum* 'St. Albans', which was crossed with *C.* Mirabell

Fig. 20. *Cymbidium* Gattonense, an early hybrid (*C. lowianum* × *C. tracyanum*).

Fig. 21. *Cymbidium* Maggie Fowler, bred in 1908 (*C. elegans* × *C. iridioides*).

FCC/RHS by Cooke of Wyld Court Orchids in Berkshire to produce the famous *C.* Blue Smoke in 1946, named after his favourite racehorse.

One of the first pure colour cymbidiums was *C.* Esmeralda (*C. lowianum* var. *concolor* × Venus) made by McBeans in 1937 which had flowers of a clear green. *Cymbidium* Pauwelsii 'Compte de Hemptine' (*C. insigne* × *lowianum* var. *concolor)* was a naturally occurring tetraploid which showed dominance in late-blooming hybrids. When 'Compte de Hemptine' was shown by Lionel de Rothschild to the Royal Horticultural Society it was awarded a First Class Certificate. The spikes carried 20 flowers of a murky shade of pink, but this plant was a natural tetraploid that had produced *C.* Swallow at Sander's in 1916. There were many notable clones from this cross and 'Exbury' FCC/RHS was no exception, providing flowers of a stunning yellow, while the grex as a whole produced a palette of shades: both *C.* Erica Sander (× *hookerianum* (as *grandiflorum*)) and *C.* Joy Sander (× *C.* Ceres) were bred from it in 1924 by Sander's of St. Albans, whereas H.G. Alexander created the beautiful *C.* Babylon (× *C.* Olympus), of which the variety 'Castle Hill' was the most awarded clone. The last also went on to produce a list of notable offspring, including *C.* Burgundian (× Remus), *C.* Vieux Rose (× Rio Rita) and *C.* Runnymede (× Roxanna). All of these can easily be traced in the background of modern clear-coloured orange, pink, red and green hybrids.

Cymbidium erythrostylum from Vietnam is noted for breeding sparkling white, pearly pink and pale hued early-flowering hybrids that flower freely from a plant of intermediate size. It is also noted for its unusual flowers that are taller than broad and have a short and strongly decurved lip. The use of a tetraploid form as a parent has created several masterpieces over the years. An early hybrid, *C.* Albanense, was created from the crossing of *C. erythrostylum* with *C. insigne* in 1915 and subsequently *C.* Albania (× Alexanderi) was made in 1926. The notable hybrid *C.* Early Bird 'Pacific' (× Edward Marshall) (Fig. 23), a significant key to the production of high-quality white cymbidiums as we know them today, was produced in 1946 by Sander of St. Albans. Although there is some doubt as to its true parentage, breeders still continued to use *C.* Early Bird. In 1964 Stewart Orchids in America crossed it with *C.* Balkis to produce *C.* Fred Stewart, a remarkable standard hybrid, a number of cultivars of which are extensively grown for cut-flower production in Europe. Hybrids from *C.* Fred Stewart have been as good as, if not better than, their parents and grandparents.

Cymbidium sanderae, formerly misidentified as *C. parishii* var. *sanderae*, is a beautiful species that produces white flowers with a very dominant dark maroon-spotted lip. It offers to its offspring some warmth tolerance and medium-sized flowers. It has been used extensively in recent times but its use by hybridists in the early part of the 20th Century is difficult to trace. One hybrid that was made as early as 1914 by Holford was *C.* Dryad (*insigne × sanderae*). However, Sander's Lists specifies many recent hybrids where this species has been used as a direct parent, particularly the tetraploid clone, proving that there is renewed interest in this pretty species, especially as it too is known to pass on its heavily marked lip and often some fragrance.

Cymbidium tracyanum, another Burmese species, offers large fragrant flowers with yellowish petals and sepals with interrupted longitudinal lines of crimson, and a creamy yellow lip, spotted and striped with crimson. It was not a successful parent in the early days as it proved to reduce the lasting qualities of the flowers of many of its progeny and reduced the clarity of the flower colours, such as in *C.* Louisiana (*C.* Louis Sander × *tracyanum*). These traits are, unfortunately, passed on through several generations. Its one redeeming feature is that it offers flowers with spots or freckles and this was recognised by an Award of Merit from the Royal Horticultural Society for the unique *Cymbidium* Grand Monarch 'Exquisitum'. Its petals, however, are also narrow and twisted which are undesirable features for breeding. *Cymbidium* Doris (× *insigne*), bred by McBeans in 1912, was a milestone in early breeding and, although it did not have much to commend it, was still used for breeding. From *C.* Grand Monarch a superb range of green cymbidiums emerged and, in 1957, *C.* Vanguard (Atlantes × Sicily) was registered by Sherman. This has subsequently been used extensively as a parent and, in particular, when crossed with *C.* Goldrun (Runnymede × Cariga) by Bart in the Netherlands, produced *C.* Vacaru. The clone 'Big P' is the finest yellow cut-flower of modern breeding. Despite its drawbacks, *C. tracyanum* features in the background of a number of hybrids, such as the beautiful *C.* Solana Beach produced in 1969 by Cobbs' Orchids of Santa Barbara. This has inherited the freckles and spotting which are synonymous with *C. tracyanum* breeding and which it passes onto its progeny. It is also present in the background of the magnificent *C.* Red Beauty (Vanguard × Tapestry), bred by H. Winter in 1979 and renowned for its remarkable size. There are a number of clones of *C.* Red Beauty used by cut-flower growers worldwide as well as some tetraploid clones used by hybridisers to create stunning flowers, as can be seen in those produced by the Eric Young Foundation that has produced *C.* Victoria Village (× Angelicas Lady), *C.* Faldouet (× Palace Angelica), *C.* Maufaunt (× Thurso) to name but a few. A whole range of colours emerged from these hybrids which are proving themselves to be fine parents in modern breeding programmes.

The Chinese *Cymbidium floribundum*, better known in breeding as *C. pumilum*, has been the most significant breakthrough in the production of dwarf and miniature cymbidiums. It is only since 1939 that this species has been extensively used to produce a myriad of miniature cymbidiums. In 1933, Sidney Alexander was the first to cross it with *C. insigne* to produce *Cymbidium* Minuet while Sander bred *C.* Pumander (× Louis Sander), both inheriting the dwarf qualities of *C. floribundum*. A series of new smaller-growing hybrids, which were free-flowering and whose flowers were clear-coloured, were produced from the 1960s onwards.

Other well-known hybrids from this species are *C.* Sweetheart (× Alexanderi*), C.* Oriental Legend (× Babylon), *C.* Mary Pinchess (× Pagaro) by Dos Pueblos Orchids and *C.* Fairy Wand (× Princess Maria). They are still used as miniature *Cymbidium* pot plants in Europe. Dos Pueblos Orchids also produced *Cymbidium* Ivy Fung (× Carasona), *C.* Ann Miller (× Clyde Landers), and *C.* Agnes Norton (× Confection) of which the cultivar 'Show Off' was widely used for breeding. Ray Bilton, while at McBeans, used the petite *C.* Putana (Rutana × *pumilum*) as the backbone of his miniature breeding programme. *Cymbidium* Nip (*pumilum* × Flare) made by Mary Bea Ireland in 1968 was also a great parent when crossed to make the incredible miniature *Cymbidium* Strathdon. The clone 'Cooksbridge Noel' AM/RHS was a perfect pot plant for the Christmas market in the northern hemisphere. The foliage is compact and the flower spikes erect, carrying around 15 beautifully shaped rich pink flowers with excellent separation.

Arno Bowers, an American who in 1955 created *C.* Sweetheart (*pumilum* × Alexanderi), also produced *C.* Geraint (× Jungle), *C.* King Arthur (× Nila) and the incredible *C.* Show Girl (*C.* Sweetheart × Alexanderi), of which there were several clones of note. Sadly, he died before these seedlings flowered. Paul Miller continued in Bowers' footsteps and popularised the *C.* Show Girl progeny.

Mrs Emma Menninger coined the term 'polymin' to describe the progeny of miniature cymbidiums but some of these plants still possessed foliage similar to that of a large-flowered *Cymbidium,* whilst the flowers faded into insignificance amongst the leaves.

Cymbidium floribundum is found in several forms, the most notable being an albino which is a light green rather than white. Some prominent pure colour miniature hybrids have been produced from it. *Cymbidium* Olymilum (× *C.* Olympus), *C.* Dolly (× *C.* Blue Pacific) and most notably *C.* Sarah Jean (× *C.* Sleeping Beauty). *Cymbidium* Sarah Jean 'Ice Cascade', best known for its ease of culture and the number of flowers it produces, is often seen on show benches with a profusion of icy white flowers. *Cymbidium floribundum* hybrids often produce suberect flower spikes and plants which flower in great profusion and have long-lasting qualities.

Cymbidium devonianum, originally introduced from the Assam highlands, is a spring-bloomer with broad leaves and pendant racemes that bear many small dark flowers. The flowers are clustered and the sepals and petals are often deep olive-green spotted with crimson, with a dark purple lip. This delightful miniature species has been used with great success to produce notable cymbidiums: *C.* Bullbarrow (× Western Rose), Devon Carousel (× Carousette), Devon Flute (× Magic Flute) were all produced by Keith Andrew Orchids of Plush in Dorset, England. Ross Tucker from Australia has also created some *C. devonianum* hybrids that are valuable additions to a collection in that they also produce multiple-flowering spikes generally in dark shades, notably *C.* Kiwi Cutie (× Robin*)* and *C.* Pacific Sparkle (×

Fig. 22 (left). *Cymbidium* Gladys Whitesell 'The Charmer', which won the best in Show at the World Orchid Conference in 1990 in Auckland, New Zealand. **Fig. 23 (right).** *Cymbidium* Early Bird 'Pacific', a hybrid of *C. erythrostylum* and an important parent of modern white-flowered hybrids.

Red Beauty) in 1992. Wyld Court Nursery, near Newbury, England, bred the fascinating *C.* Tiny Tiger (× *tigrinum)* in 1979.

Cymbidium tigrinum, originally found in Moulmein, Burma, is a truly dwarf species usually no more than 20 cm tall with drooping spikes, and flowers with olive-green petals and sepals highlighted by a white and striking purple-marked lip. It has some distinctive progeny of which *Cymbidium* Tiger Cub (× Alexanderi 'Westonbirt' FCC/RHS) bred by Stewart Orchids in California is no exception.

Cymbidium ensifolium, found in various varieties throughout east and south Asia and the adjacent islands, offers miniature plants with erect inflorescences and flowers with yellowish green sepals and petals with purplish striations and a lip spotted in purple. Most significant of all is the fresh fragrance that its flowers exude. This variable species has recently become popular in breeding programmes in the Far East, where it has been in cultivation for many centuries. The fragrance is passed onto its progeny as are its compact growth habit, floriferousness and long-lasting flowers. A big boost for hybridisers is the fact that *C. ensifolium* blooms in the late summer, unlike most cymbidiums which flower in the spring. The most famous *C. ensifolium* hybrid ever made is *C.* Peter Pan (× Moretta) produced by Dos Pueblos in the USA in 1957 and still grown as a pot plant today. In this instance the species was crossed with a standard complex hybrid, and whilst many of the seedlings had poor flowers, variety 'Greensleeves' surpassed all its peers. The grex appeared to be infertile even though it was a normal diploid, but it had its chromosomes doubled by Dr. Don Wimber (a pioneer in the use of colchicines, a chemical which can be used to double the chromosomes and create fertile tetraploids) and Dr. Gustav Melquist (who was responsible for counting chromosomes of both species and hybrid cymbidiums). Thus, *C.* Peter Pan 'Greensleeves' had its fertility restored in its tetraploid form, and some quite amazing hybrids have emerged, ranging in flower colour from the pearly white *C.* Sue (× Showgirl) bred by Santa Barbara in 1980, to yellow in *C.* Wild Colonial Boy (× Coraki), and *C.* Summer Pearl (× Trigo Royale), raised by Andy Easton of Geyserland Orchids in New Zealand in 1989, and *C.* Floriflame (× Red Beauty) from Floricultura in Holland in 1988. These hybrids are fragrant, possess upright self-supporting stems, bloom early and are quick growing. Many of them are warm-tolerant and, especially in Europe, have filled the gap for a summer-flowering crop and bridged the time delay between the late flowering plants in March and April and the early flowering cymbidiums in September and October.

One other truly significant *C. ensifolium* hybrid of note is *C.* Golden Elf (× Enid Haupt) made by Rod McLellan in California in 1978. It flowers readily and has the most striking pure yellow flowers and at a time when *Cymbidium* flowers are not usually available. Subsequent hybrids from *C.* Golden Elf are in the pipeline, mainly bred in Taiwan.

Cymbidium kanran and *Cymbidium sinense,* with erect flowering stems and flowers ranging from dark chocolate to almost white, both carry delightful fragrances, and have a lot to offer the hybridists who have now started to use them.

Cymbidium madidum must not be forgotten. Mary Bea Ireland of Santa Barbara in California chose to use this Australian species and produced a group of successful hybrids from it. It was a surprising choice because it is a large-growing plant which produces small green flowers and also unconventional progeny. However, the pendulous racemes are often in excess of 1.5 metres long. The biggest drawback is that the plants are dominated by a fairly large bulb and foliage and the flower spikes are sometimes dwarfed in comparison. Nevertheless, it can be successful as in its crosses with *C. elegans, C. tracyanum* and *C.* Sleeping Dream. One of the most amazing second generation hybrids to date has been *C.* Gladys Whitesell (Fifi × *parishii)* (Fig. 22) raised by Andy Easton in 1983. It was this hybrid that was crowned with the Best-in-Show award at the World Orchid Conference in New Zealand in 1990, a big boost for cymbidiums on the world stage.

Whilst it is impossible to list every hybridiser's achievements and regrettably space precludes the inclusions of all aspects, awards and aspirations, Sander's list provides a catalogue of all the great achievements.

CHAPTER 9

OTHER USES

The ethnobotanical uses of orchids have been comprehensively surveyed by Lawler (1984) and, unless otherwise stated, the original sources of the information in this account are cited there. The uses of cymbidiums can be categorised as follows.

MEDICINAL USES

In China, the mucilaginous root (perhaps also referring to the pseudobulb) of a *Cymbidium* species has been used against rheumatism and neuralgia. An aqueous decoction of the thickened root and rhizome of cultivated specimens of *C. ensifolium* has been used to treat gonorrhoea and syphilis and, in a mixture with fermented glutinous rice, to cure stomach ache (Hu, 1971). This species, along with *C. floribundum*, *C. goeringii* and *C. pendulum* (probably *C. aloifolium*), has been prescribed as herbal medicine by local countryside doctors for the improvement and strengthening of weak lungs and for the relief of coughs (Chen & Tang, 1982).

In northern India, *C. macrorhizon* rhizomes have been used as a diaphoretic and febrifuge, and to treat boils. *Cymbidium ensifolium* has been used in Ayurvedic medicine as a constituent of oils that were applied to tumours, both malignant and benign. In India this species, powdered with ginger and extracted with water to produce a liquid, has been used to incite vomiting and diarrhoea and to cure chronic illnesses, weakness of the eyes, vertigo and paralysis. The Singhalese name for *C. aloifolium*, literally translated, means 'poison dust', but this probably refers to irritations of the eye caused by the fine seed, rather than to any inherent chemical properties.

The flowers of *C. ensifolium* have been used in Indo-China to prepare an eyewash and a diuretic, and the roots form part of a pectoral medicine. *Cymbidium aloifolium* (probably *C. finlaysonianum*) has been used to treat sores and burns, and has also been used to bathe weak infants and to combat menstrual irregularity. The roots of *C. finlaysonianum* are an ingredient of a mixture given as medicine to sick elephants.

In Australia, all of the native *Cymbidium* species have been widely reported to be effective against diarrhoea and dysentery. The pseudobulbs are usually chewed, but they may be dried and powdered and other parts of the plant appear to have been used as well. The seed of *C. madidum* has been used by the Aborigines in Australia as an oral contraceptive, but no antifertility activity has been found.

The active ingredients of these species have not been investigated chemically as far as we can determine.

FOOD USES

In Japan, the flowers of *C. virescens* (= *C. goeringii*) have been made into preserves with plum vinegar, or preserved in salt and made into a drink with hot water. In Bhutan and Nepal, the flower buds of *C. hookerianum*, *C. iridioides* and *C. elegans* are used in curries, but are soaked in water beforehand to remove the bitter flavour. The pseudobulbs of *C. madidum* and *C. canaliculatum* are occasionally eaten by the Australian Aborigines as a rather mucilaginous food. They are eaten raw, or turned into a food resembling tapioca or sago. The grated pseudobulb may also be cooked, and is reputed to be indistinguishable from arrowroot. Starch can be prepared from the pseudobulbs of both of these species. The tender parts of the stem and the base of the leaves may be eaten, and *C. suave* has also been reported as edible.

OTHER USES

The juice of the pseudobulbs of *C. canaliculatum* has been used in Australia to produce a glue which may be used to affix feathers to dancers' bodies or to totemic emblems, or as a fixative for stone and bark painting. The pseudobulb may be chewed to extract the juice, which is then applied directly to the rock or bark surfaces, or over the pigments, or it may be ground together with the colour before it is used. The roasted pseudobulbs of *C. lancifolium* also produce a sticky substance that has been used in western Java to fasten Sundanese knives to their handles.

Fibres from the crushed pseudobulbs of *C. canaliculatum* may be used by the Aborigines of Australia as packing for pipe bowls to retain the smoke, or may be made into brushes. In Malaya brooms made from *C. finlaysonianum* have been used ritually to sprinkle water in the house of a recently deceased person to prevent his spirit from haunting the living.

An unidentified *Cymbidium* species has been used in China for colouring.

CHAPTER 10

DISTRIBUTION AND BIOGEOGRAPHY

*C*ymbidium Sw. is widely distributed in SE Asia, from India, Nepal and Sri Lanka east to Taiwan and from Japan south to northern and eastern Australia (Map 1). The greatest concentrations of species occur in North-east India, South-west China, Indo-China and western Malesia (Table 4). The distributions of individual species are shown in Maps 2–54, and are discussed in the accounts of the species (see Chapter 13).

CENTRES OF DIVERSITY

The greatest diversity of species in the genus is found in the eastern Himalayas across to southern China and northern Indo-China. Half of the species are found in this mountain arc. The number of species tails off rapidly towards the margins of the range. Six species, one of which is endemic, have been recorded in the Philippines. Five species occur in Japan. In Australia only three species are found, all belonging to the same endemic section. Only two species have been recorded in New Guinea and three in southern India/Sri Lanka.

Individual sections also have distinct centres of diversity, and Table 5 lists the number of species that occur in eight geographical regions. Section *Cyperorchis* has its greatest diversity of species in the

TABLE 5

NUMBERS OF SPECIES IN THE SUBGENERA OF *CYMBIDIUM* SW.
THAT OCCUR IN VARIOUS REGIONS OF THE TOTAL DISTRIBUTION

Region / Section	1	2	3	4	5	6	7	8
Floribundum		1	1	1	3			
Austrocymbidium								3
Cymbidium	2	2	3	1	4	3		
Borneense					1	1		
Himantophyllum		1	1	1	1	1		
Annamaea		1						
Cyperorchis		9	7	10	2			
Parishiella		1	1	1				
Bigibbarium		1	1	1				
Jensoa	1	2	2	8		1	1	
Pachyrhizanthe		1	1	1			1	
Total	**3**	**18**	**18**	**24**	**11**	**6**	**2**	**3**

1 = S India/Sri Lanka; 2 = Himalayas/Myanmar; 3 = Indo-China; 4 = China/Korea/Japan; 5 = Malesia; 6 = Philippines; 7 = New Guinea; 8 = Australia

Himalaya of Northeast India and Nepal (nine species) and in South-west China (ten species), with southerly extensions into Indo-China (seven species). Section *Jensoa* is best represented in China (eight species). Taiwan contains a high diversity of species in this section, and some unusual variants have been recorded there (see *C. goeringii*, *C. tortisepalum* and *C. ensifolium*), suggesting that subgenus *Jensoa* may be rapidly evolving in that region at present.

DISTRIBUTION PATTERNS

Cymbidium species usually occur in upland areas, often above about 1000 m (3280 ft) above sea level. Notable exceptions to this are the species in section *Cymbidium* which often occur near sea level. *Cymbidium aloifolium*, for example, is the only species which occurs throughout much of lowland South and South-east Asia, while *C. finlaysonianum* and *C. atropurpureum* have often been reported from coastal areas in Malaysia, Indonesia and the Philippines. Towards the northern extreme of the

Map 1. The distribution of the genus *Cymbidium* in E and SE Asia, Malesia and Australia.

distribution of Japan, and the southern extreme in Australia, *Cymbidium* species often occur at correspondingly lower elevations.

In northern India, Nepal and Bhutan, most *Cymbidium* species occur in the foothills (700–2500 m, 2295–8200 ft) of the Himalaya and in Meghalaya (the Khasia Hills). These two ranges are separated by the Brahmaputra plains, and some species show disjunct distributions in this area (*C. devonianum*, Map 18), although more widespread species are also distributed around the Himalaya of NE Assam and in the hills of Nagaland, linking these two mountainous regions (*C. iridioides*, Map 28). The range of *C. macrorhizon* extends further westwards in the Himalaya than any other *Cymbidium* species (Map 53).

Many of the Himalayan species have distributions that extend eastwards into Myanmar (Burma) and south-western China. In China, Hu (1971) suggested that *Cymbidium* species are generally distributed in a U-shaped pattern in the mountains of the south of China, with the western arm extending northwards into the mountains of Sichuan, and the eastern arm following the coastal hills into Zhejiang. However, the distributions of several Chinese species are now known to be more extensive, notably *C. goeringii*, *C. faberi* and *C. floribundum* (Maps 4, 45 & 46), being recorded from the provinces of central southern China, including Anhui, Hubei, Henan, Hunan and Jiangxi (Wu & Chen, 1980). There is a relatively flat region of lowland plains along the Yangtze River in northern Hunan, central Hubei and northern Jiangxi in which *Cymbidium* is absent. However, these plains are bordered by a chain of hills to the north where several species are found, thus linking the western and eastern arms of the U-shaped distribution pattern suggested by Hu to form a circular pattern instead. The northern limit of *Cymbidium* species in China is about 32°N. The eastward distribution of the genus continues through Taiwan, and the Ryukyu Islands, to Japan and South Korea. The northeastern limit of *Cymbidium* is about 42°N in Japan (Map 1).

The southward distribution of *Cymbidium* extends through Indo-China into western Malaysia and southeast into the Malay Islands. Section *Cyperorchis* (*C. sigmoideum*) has recently been reported in Borneo, but only sections *Jensoa* and *Pachyrhizanthe* are represented in New Guinea. The Philippines can be divided into two regions. The northernmost island, Luzon, has affinities with southern China and Taiwan in that it contains variants of *C. ensifolium* and *C. dayanum* which are typical of the Asiatic mainland. This affinity has also been noted in other orchids by Hu (1971). The southern islands contain *C. atropurpureum* and *C. bicolor* subsp. *pubescens*, species typical of the Malay Islands. *Cymbidium finlaysonianum* is distributed throughout the entire archipelago (Quisumbing, 1940). *Cymbidium aliciae* Quis., endemic to Luzon, is anomalous in that it is closely allied to the Bornean endemic *C. borneense*.

Few species occur south or east of Wallace's Line, and those in Sulawesi and New Guinea have probably spread there from the Malay Islands, using the general west–east wind currents which would carry the dust-like seed. The three *Cymbidium* species which occur in Australia are endemic, and although their distributions extend into northern Australia, they do not occur in New Guinea (Maps 7, 8 & 9). All belong in section *Austrocymbidium*. It is interesting to speculate on the biogeographical history of the Australian species.

THE ORIGINS OF *CYMBIDIUM*

Van den Berg *et al.* (2002) and Yukawa *et al.* (2002) have published phylogenies of *Cymbidium* based upon ITS of nuclear ribosomal DNA and plastid *matK* sequence data (see Chapter 11). They found that phylogenetic relationships attained reasonable support and that both ITS and *matK* datasets were in overall agreement, except for the positioning of section *Floribundum*. Phylogenetic relationships pointed to a South-east Asian rather than a Himalayan or Australian origin for the genus. In van den Berg *et al.*'s strict consensus tree the basal clade was section *Floribundum*, which seems the most likely scenario, whilst for Yukawa *et al.* it was a clade comprising sections *Austrocymbidium*, *Borneense* and

Cymbidium (see Chapter 11). In the latter study, section *Floribundum* was found to be basal to the clade containing sections *Jensoa* and *Pachyrhizanthe*, a rather less likely hypothesis. They also indicated an early branching of the Australian clade, represented by the three species of section *Austrocymbidium*. In both analyses section *Austrocymbidium* was sister to section *Cymbidium*, and in Yukawa *et al.*'s analysis the sister clade also included section *Borneense* (which was not included by van den Berg *et al.*).

CYMBIDIUM IN AUSTRALIA

Hooker (1860) was the first to recognise an Indo-Malesian element in the Australian tropical and subtropical rainforest flora. Although many genera and species are distinct from those found in Malesia, the families represented are similar. Barlow (1981) used plate tectonic theory to show that the Australian plate collided with the Sunda Island arc system to the north during the Miocene epoch, about 15 million years BP, forming New Guinea and a chain of islands linking Australia with Malesia. During the following millennia, intrusive elements of the Malesian flora used this chain of islands to colonise the lowland tropical regions of northern Australia, which at that time were extensively covered in tropical forest (Burbridge, 1960). The temperature was probably 5°C (9°F) higher than at present (Specht, 1981). The Orchidaceae would have been able to spread rapidly by their wind-dispersed seed, and this route through New Guinea and Cape York has been suggested as the main route for colonisation by the epiphytic orchids (Lavarack, 1981). This route was taken by such genera as *Dendrobium* (Lavarack, 1981), and subsequent evolution has produced new taxa that are very distinct, certainly at the sectional level. The presence of the three *Cymbidium* species in the north and on the east coast appears to support this pattern, and the distribution of the complete genus does not refute it (Map 1). However, an examination of the distribution of *Cymbidium* shows that no species related to section *Austrocymbidium* occurs in New Guinea, and the closest locality of the related *C. chloranthum* is in Java (Map 2). It seems more likely, therefore, that *Cymbidium* entered north-western Australia from Java, perhaps using the Sunda Islands and Timor in the process, rather than New Guinea. This probably occurred during a wetter period when tropical forest extended throughout northern Australia. At the present day, *C. canaliculatum* ranges across northern Australia as far west as the Kimberleys. The efficiency of *Cymbidium* seed dispersal suggests that Java could have been the source despite its present distance from north-western Australia. Those two land masses were probably closer than at present, but continued northerly movement of the Australian plate has subsequently created a broader ocean barrier between them (Barlow, 1981). It is probable that *Cymbidium* colonised Australia at about the same time as most of the *Dendrobium* species, allowing the evolution of distinct endemic species that differ substantially from related species in adjacent land masses.

Increasingly frequent periods of aridity are thought to have started about 15 million years ago, causing great periodic reductions of the rainforest area, and consequently natural selection has produced the evolution of xerophytic characteristics in the plant communities of northern Australia (Specht, 1981). *Cymbidium madidum* and *C. suave* are confined to the damper forest refugias of eastern Australia (Maps 8 & 9). *Cymbidium canaliculatum* is widespread over northern Australia (Map 7), undoubtedly because it has evolved xerophytic characteristics such as very thick, leathery leaves. These can resist desiccation through the channelled leaf shape directing any available water to the base of the plant, a rough epidermal surface that helps to maintain a boundary layer around the leaf, and CAM photosynthetic pathways. Its somewhat spiny seeds have a large surface/volume ratio making them more buoyant in the air and increasing the probability of a seed encountering a suitable tree on which to grow. Its roots are usually well protected from desiccation as they grow into the moist rotten heartwood of *Eucalyptus* and *Melaleuca* trees. This species is perhaps the most successful of the Australian epiphytic Orchidaceae in its adaptation to arid environments.

CHAPTER 11

PHYLOGENY

Considerable insight has been gained in recent years into the phylogenetic relationships of orchids and this has somewhat clarified both the generic and infrageneric relationships of *Cymbidium*. Cameron *et al.* (1999), in a study based on a chloroplast gene *rbcL*, suggested a basal placement of *Cymbidium* in relation to the Maxillarieae / Cymbidieae, being sister to *Grammatophyllum*. The morphological cladistic analysis of Freudenstein and Rasmussen (1999) placed *Cymbidium* together with *Thecostele* and *Acriopsis*, and this group sister to *Catasetum*.

Van den Berg *et al.* (2002) and Yukawa *et al.* (2002) have published phylogenetic analyses of *Cymbidium* based upon sequence data from ITS (Internal Transcribed Spacer gene) of nuclear ribosomal DNA and *matK*, a chloroplast gene. Both based their interpretation of the results on the classification published by Du Puy & Cribb (1988). Du Puy (2005) has published a critical review of these analyses. Van den Berg and his co-workers sampled 35 accessions of 27 species of *Cymbidium* and six outgroup species while Yukawa and his team sampled 34 species, one with two subspecies, and five outgroup species, to evaluate the infrageneric classification of the genus and phylogenetic relationships of the species. Van den Berg *et al.* noted that the levels of variation found were higher than in other Epidendroideae studied previously, and phylogenetic relationships attained reasonable support, pointing to a South-east Asian rather than a Himalayan or Australian origin for the genus.

Van den Berg *et al.*'s study (2002) included species from each of the subgenera and most sections except for subg. *Cymbidium* section *Borneense* (*C. borneense*), subgenus *Cyperorchis* section *Parishiella* (*C. tigrinum*) and subgenus *Jensoa* section *Pachyrhizanthe* (*C. macrorhizon*). Outgroups were drawn from tribes Cymbidieae and Maxillarieae, and they showed the three subgenera of *Cymbidium* to be a monophyletic group. Yukawa *et al.*'s (2002) study included more species, including *C. aberrans* (section *Pachyrhizanthe*), *C. aliciae*, *C. bicolor*, *C. borneense* (section *Borneense*), *C. canaliculatum*, *C. cochleare*, *C. faberi*, *C. hartinahianum*, *C. macrorhizon* (section *Pachyrhizanthe*), *C. roseum*, *C. sanderae*, *C. tigrinum* (section *Parishiella*), *C. tortisepalum*, and *C. tracyanum*. Those with the section indicated represent sections not included in the van den Berg study.

In both studies, analyses of the data from both genes suggest that *Cymbidium* is monophyletic in relation to the outgroups. Both ITS and *matK* datasets were in overall agreement, except for the anomalous results concerning the position of section *Floribundum,* which was shown to be basal to *Cymbidium* as a whole in the ITS analyses, but in a clade alongside subgenus *Jensoa* in the *matK* analyses. Although subgenera *Cyperorchis* and *Jensoa* were reasonably well supported within the genus, subgenus *Cymbidium* included two distinct clades. There was little molecular information to assess the current sectional delimitation.

In the van den Berg *et al.* analysis, four main groups appeared on the combined tree, besides *C. devonianum*, placement of which was unresolved. From the base upwards, the first clade contained subgenus *Cymbidium* section *Floribundum* plus *C. chloranthum*. The second clade contained subgenus *Cymbidium* section *Cymbidium,* sister to two accessions of *C. madidum* (section *Austrocymbidium*). A third clade included all species of subg. *Jensoa*, and a fourth included all species of subgenus *Cyperorchis* plus *C. dayanum* (subgenus *Cymbidium* section *Himantophyllum*). Subgenus *Jensoa* was therefore monophyletic, and if it were not for the position of *C. dayanum,* subgenus *Cyperorchis* would also be monophyletic. There was more overall support in the clades of the combined analysis than in individual datasets. However, a higher number of collapsed branches was found within subgenus *Cyperorchis*; little support for the sections could be discerned.

A similar result was obtained in the strict consensus tree based on the *matK* and ITS sequences published by Yukawa *et al.* (2002). In their tree a clade that included species of subgenus *Cymbidium*

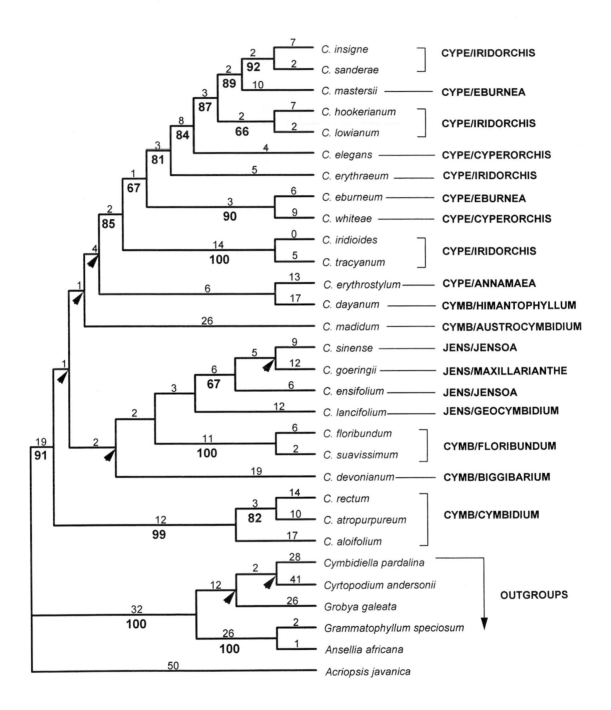

Fig. 24 One of the 32 most parsimonious trees from the analysis of *Cymbidium* including only *matK* sequence data. L=1276, CI=0.7257, RI=0.6403. Numbers above branches are Fitch branch-lengths, whereas bootstrap percentages are given below. Arrows indicate branches that collapse in the strict consensus (van den Berg *et al.*, 2002).

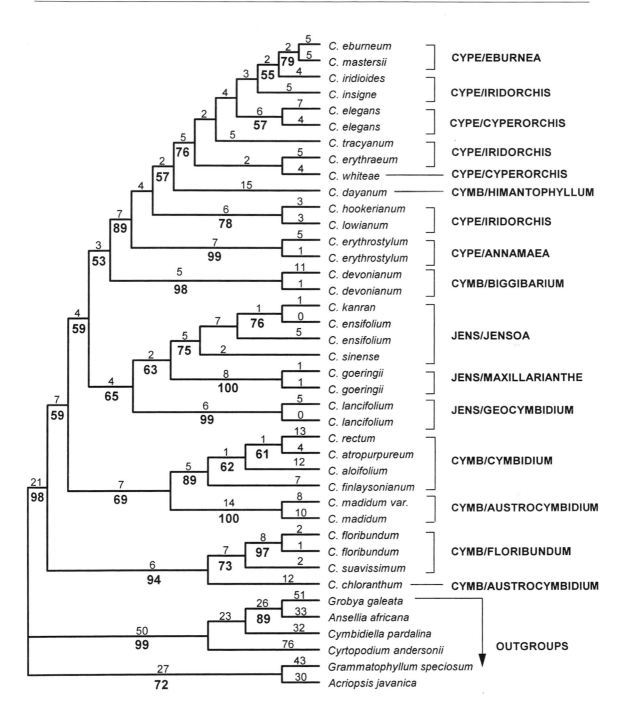

Fig. 25. One of the 702 most parsimonious trees from the analysis of *Cymbidium*, including only ITS sequence data. L=1380, CI=0.6920, RI=0.6252. Numbers above branches are Fitch branch-lengths, whereas bootstrap percentages are given below. (van den Berg *et al.*, 2002).

Combined ITS and *matK* cladogram

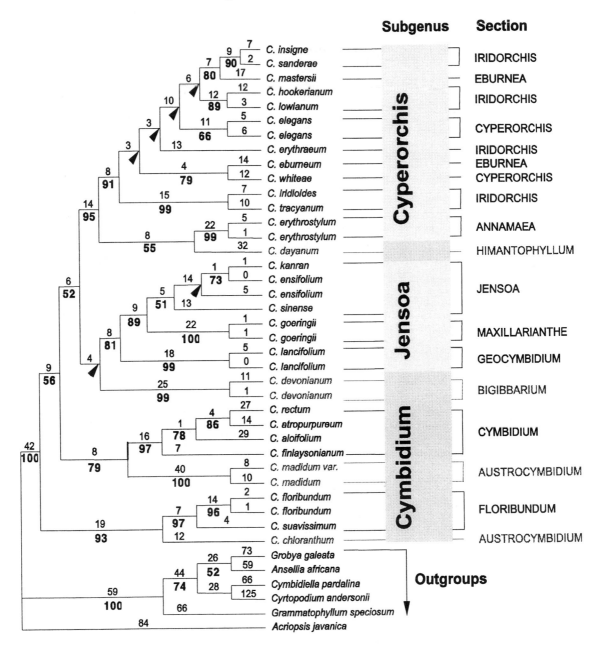

Fig. 26. One of the 42 most parsimonious trees from the analysis of *Cymbidium,* combining ITS and *matK* sequence data and not excluding missing taxa from either dataset. L=1332, CI=0.718, RI=0.674. Numbers above branches are Fitch branch-lengths, whereas bootstrap percentages are given below. Arrows indicate branches that collapse in the strict consensus (van den Berg *et al.*, 2002).

Strict consensus of ITS and *matK* cladograms

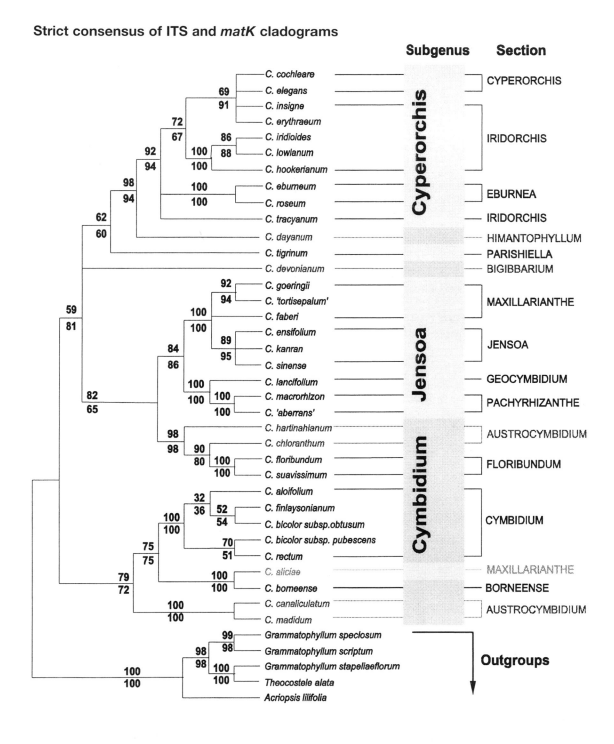

Fig. 27. Strict consensus of 168 most-parsimonious Fitch trees based upon *matK* and ITS sequences: numbers above the internodes indicate bootstrap values (from 1000 replicates of Fitch parsimony analysis); numbers below the internodes also indicate bootstrap values (from 1000 replicates of neighbour-joining distance analysis). (Yukawa *et al.*, 2002).

sections *Cymbidium, Borneense* and the Australian members of section *Austrocymbidium*, along with *C. aliciae* (previously considered to belong to subgenus *Jensoa*), was sister to the rest of *Cymbidium* with 79% bootstrap support. *Cymbidium aliciae* is sister to *C. borneense* with 100% bootstrap support. Subgenus *Cyperorchis* is monophyletic with 98% bootstrap support if *C. dayanum* (previously placed in subgenus *Cymbidium*) was included and with 60% support if *C. tigrinum* was also included. As in van den Berg *et al.*'s analyses the sectional delimitation within subgenus *Cyperorchis* remained unclear: in Yukawa *et al.*'s analyses only sections *Eburnea* and *Parishiella* received significant support. Subgenus *Jensoa* (excluding *C. aliciae*) was also monophyletic with 82% support. The clade containing section *Floribundum* again included *C. chloranthum*, and also *C. hartinahianum* (the non-Australian members of section *Austrocymbidium*). The position of section *Floribundum* was, however, anomalous, being in the same clade as subgenus *Jensoa*. Finally, the placement of *C. devonianum* was again unresolved.

Cymbidium dayanum* has some peculiar morphological characters, and for this reason it has always been placed in its own section (*Himantophyllum*) in all classifications following that of Schlechter (1924). It is clear that *C. dayanum* should be placed somewhere at the base of subgenus *Cyperorchis*, closest to *C. erythrostylum* (section *Annamaea*), but more data will be necessary to find out its exact position. It cannot remain in subgenus *Cymbidium*. However, when it is included with subgenus *Cyperorchis*, then this subgenus can no longer be characterised by a fusion of the base of the lip and column which is absent in *C. dayanum*.

The position of *C. devonianum* also poses problems. In both combined analyses its position is unresolved along the spine of the cladogram. ITS data suggest that it is sister to subgenus *Cyperorchis* but with only 40 to 53% bootstrap support, whereas in the *matK* tree it is either unresolved (van den Berg *et al.*, 2002) or sister to a clade comprising subgenus *Jensoa* and section *Floribundum* and the Indonesian species of section *Austrocymbidium* (Yukawa *et al.*, 2002). Some morphological characters, nevertheless, indicate a close relationship to subgenus *Cyperorchis*, with which it shares narrow, acuminate leaf margins in transverse section and seeds that are shorter than those in subgenus *Jensoa*.

Both analyses indicated a well-supported cluster with subgenus *Cymbidium* section *Cymbidium* (*C. aloifolium / atropurpureum / bicolor / rectum / finlaysonianum*) and section *Borneense* sister to *C. madidum* and *C. canaliculatum* (section *Austrocymbidium*). This would suggest that section *Austrocymbidium* (sensu Cribb & Seth, 1984) is polyphyletic, since *C. chloranthum* and *C. hartinahianum* were placed with strong support in a separate clade as sister to subgenus *Cymbidium* section *Floribundum* (*C. floribundum / C. suavissimum*) by Yukawa *et al.* (2002). The cluster of section *Cymbidium* presents several typical anatomical characters, such as stomata with an elliptical cover and slit-shape pores, and a complete layer of sub-epidermal schlerenchyma cells. The first two characters are also found in *C. borneense,* while neither is present in *C. madidum*, supporting the recognition of sections *Cymbidium, Borneense* and *Austrocymbidium*.

Subgenus *Jensoa* appeared as a monophyletic group in all analyses if *C. aliciae* was excluded. The species of this subgenus are also characterised by some defining seed morphology and leaf anatomical characters, such as filiform seeds with transverse striations on the testa cells, which germinate to form microrhizomes rather than protocorms, stomatal covers raised above the surrounding epidermis and with circular-shaped pores, and a papillose epidermal cell surface. Within subgenus *Jensoa*, relationships still need some clarification, with sections *Maxillarianthe* and *Jensoa* sister to *Geocymbidium* and *Pachyrhizanthe* in Yukawa *et al.* (2002), while *Maxillarianthe* and *Geocymbidium* are successive sisters to section *Jensoa* in the van den Berg *et al.* (2002) analysis. Given the number of new taxa recognised in this subgenus, none of which are represented in these analyses, it is difficult at this stage to provide grounds for maintaining sections *Jensoa* and *Maxillarianthe* apart.

Phylogenetic relationships point to a South-east Asian rather than a Himalayan or Australian origin for the genus. The geographical origin of *Cymbidium* in Asia is supported by the relationships presented in both combined analyses. In van den Berg *et al.* (2002), the first clade comprises three species (section

Floribundum), two of which are Chinese and Indo-Chinese, the other being found in Sumatra and Java (*C. chloranthum*). *Cymbidium madidum*, the only Australian species of section *Austrocymbidium* examined, falls in the next clade and sister to species of section *Cymbidium* from South-east Asia and the eastern Malay Archipelago. In Yukawa *et al.* (2002), the basal clade contains both South-east Asian (sections *Cymbidium* and *Borneense*) and the Australian species (section *Austrocymbidium*) in a sister relationship. Thus, it appears likely that *Cymbidium* was an early coloniser of Australia from South-east Asia via the western Malay Archipelago (see also the discussion in Chapter 10). An analysis of Cyrtopodiinae (van den Berg, in preparation) placed *Acriopsis* as the sister group of *Cymbidium*, and the distribution of this genus, which is also predominantly South-east Asian but with one species extending into Australasia, supports this pattern.

CONCLUSIONS

Although less than three-quarters of the species of *Cymbidium* were represented in the molecular systematic analyses discussed above, it is apparent that the classification used by Du Puy and Cribb (1988) requires modification. Only two of the three subgenera, *Cyperorchis* and *Jensoa*, are monophyletic and then only with modifications, whereas subgenus *Cymbidium* is a grade composed of two or more clades.

In subgenus *Cymbidium*, the Asiatic species ascribed to section *Austrocymbidium* (*C. chloranthum*, *C. elongatum* and *C. hartinahianum*) are removed and placed in section *Floribundum*: section *Austrocymbidium* is restricted to those species found in Australia. Section *Floribundum* may be basal in the *Cymbidium* clade. The monotypic section *Himantophyllum* (*C. dayanum*) is obviously misplaced in subgenus *Cymbidium*: there is strong evidence that it is sister to *Cyperorchis*, probably closest to *C. erythrostylum*. The position of the other monotypic section *Bigibbarium* (*C. devonianum*) is entirely unresolved.

In this treatment, therefore, we propose a simplified classification of eleven sections (see pages 74–75). At the same time, we recognise that further studies will certainly resolve some of the uncertainties in this classification. The subgeneric divisions have been removed, although subgeneric groups are still broadly recognisable and can provide useful insights into the relationships between the sections. Subgenus *Jensoa* is particularly strongly defined when morphological characters are taken into account. The alternative of having subgenera which cannot be distinguished on morphological grounds seems untenable, and, moreover, the position of section *Bigibbarium* (*C. devonianum*) remains unresolved. The divisions (formerly sections) within sections *Cyperorchis* and *Jensoa* are not supported in the molecular systematic analyses.

A. LAMB

CHAPTER 12

CONSERVATION

Cymbidiums are seldom mentioned in talks on endangered species, yet their plight is probably as great as that of many slipper orchids. They are useful plants: the species are popular in horticulture both in the Eastern and Western Hemispheres and, in parts of the Himalayas, the plants are collected for use of their flowers as a spice in local cuisines. The scale of *Cymbidium* collection from the wild is unknown, but anecdotal evidence indicates that some species are the subject of massive collection pressure (Liu & Nakayama, 2005; Luo, 2005).

Chinese species are particularly badly exploited by over-collection. In southern and western China, local villages are visited by dealers willing to pay for wild-collected terrestrial species by the 100 kilograms. In Taiwan, wild populations of terrestrial species have been brought to the edge of extinction by collectors seeking unusual forms. In China, nearly every town and city has an orchid collection or orchid park. Native cymbidiums feature prominently in these collections, often by the thousands. All appear to be wild-collected, either in the past or more recently as the hobby has grown. In Guizhou in 1997, the local villagers reported that a Korean dealer had recently visited them and purchased a ton and a half of cymbidiums from their immediate region. The dealers are not concerned with typical forms but are searching for the rare mutations, such as variegated and peloric forms, that can fetch high prices in the Far East. The typical plants are discarded or sold in the marketplace for low prices. The species targeted by dealers are *C. goeringii*, *C. sinense*, *C. ensifolium*, *C. kanran*, *C. faberi* and to a lesser extent *C. floribundum*. Until recently, the epiphytic and lithophytic species with larger flowers were not prized, but tastes are changing rapidly with a growing propensity for Western-style orchids and hybrids. Consequently, the large-flowered species, such as *C. eburneum*, *C. lowianum*, *C. hookerianum*, *C. wilsonii*, *C. erythraeum*, *C. elegans* and *C. devonianum* are also collected for market in cities such as Kunming, Baoshan and Wenshan in Yunnan.

Newly described species are also greatly sought after. The discovery of *C. wenshanense* and *C. nanulum* in Yunnan in the 1990s led to their immediate large-scale collection. A visit to the type locality of the former, the year after it was described, revealed that all of the plants had been collected. At the same time, a local nursery in Wenshan had a collection of several hundred plants potted up for sale.

The thirst for cymbidiums in China has also stimulated their collection in neighbouring countries. The markets in Yunnan exhibit Vietnamese, Laotian and Burmese species for sale in quantity. Several of the recently described "Chinese" orchid species originated in this manner, including possibly some cymbidiums. Wild-origin plants of the showy Vietnamese *C. insigne*, *C. sanderae* and *C. schroederi* are again appearing in trade, through both Chinese and Vietnamese nurseries. Vietnamese collectors have also found populations of *C. suavissimum*, previously thought to be Burmese, and *C. sinense*, *C. mastersii* and *C. devonianum*, all known from China but new to Vietnam.

In eastern Bhutan, the local people collect plants of *C. iridioides* and *C. elegans* and plant them by their houses. They collect the flowers and use them as a bitter-tasting spice in curries. In the Nu Shan of western Yunnan, the local people plant clumps of *C. lowianum* on the thatched roofs of their huts as decoration.

Although collecting from the wild is a threat to some species, by far the greatest threat to cymbidiums and most other orchids is the rapid rate of habitat destruction in range countries. Lowland

---

Fig. 28 (opposite). *Cymbidium finlaysonianum*, Papar beach, Sabah.

forests in Borneo and Sumatra, for example, have virtually disappeared over the past 30 years through unrestricted logging activities followed by transmigration and agricultural practices that are often unsustainable. Once orchid habitats have disappeared, so have niches for orchid seed germination.

Fortunately, many cymbidiums are widespread in nature, and most are probably not endangered, although a number are vulnerable. Others with narrow distributions are, however, at risk. *Cymbidium wenshanense* is certainly endangered in the wild and may be near extinction. The same may be true of other narrow endemics such as *C. nanulum*, *C. parishii* and *C. sanderae*. Others, such as the Bornean *C. elongatum* and *C. borneense*, are found within national parks and forest reserves and are protected. However, their small populations make them vulnerable, particularly to forest fires and illegal logging activities. A conservation rating, following the IUCN Red List criteria version 3.1 (IUCN 2003), is given in each species account.

CHAPTER 13

THE CLASSIFICATION
OF CYMBIDIUM

THE TAXONOMIC POSITION OF *CYMBIDIUM* SW.
WITHIN THE ORCHIDACEAE

John Lindley, who was the first to produce a working classification of the Orchidaceae (1830–40), placed *Cymbidium* in the tribe Vandeae. The work of Rudolf Schlechter (1924) incorporated many of Lindley's and later botanists' ideas and has provided the basis for most modern classifications. He divided the Orchidaceae into two subfamilies: the Monandreae, with one anther, comprising the majority of orchid species, and the Diandreae with two, or rarely three, anthers. He further split the Monandreae into three tribes, placing *Cymbidium* in tribe Kerosphaerae. He used the names given by Pfitzer (1887) rather than those published by earlier authors such as Lindley. Consequently, because of priority under the *International Code of Botanical Nomenclature*, many of these names were synonyms. Dressler & Dodson (1960) and Dressler (1974) published classifications based on Schlechter's work, but correcting the nomenclatural mistakes. These were updated again by Dressler (1981, 1993). Dressler (1981) placed *Cymbidium* in a more broadly defined subtribe Cyrtopodiinae, which he divided into five tentative alliances: Alliance 1 – *Bromheadia, Claderia*; Alliance 2 – *Eulophia, Eulophiella*; Alliance 3 – *Cymbidiella, Grammangis*; Alliance 4 – *Ansellia, Grammatophyllum, Cymbidium* (and perhaps also *Porphyroglottis* and *Poicilanthe*); Alliance 5 – *Dipodium*.

All of these related genera have two cleft pollinia, but can usually be distinguished from *Cymbidium* by the structure of the pollinarium, especially by the presence of a stipe. *Ansellia* and *Grammatophyllum* are considered to be closely related to *Cymbidium*, and have similar chromosome numbers (*Ansellia*, $2n = 42$; *Grammatophyllum*, $2n = 40$; *Cymbidium*, $2n = 40$ – Tanaka & Kamemoto, 1974). The pollinarium structure of *Ansellia* is close to that of *Cymbidium*, with two cleft pollinia attached by elastic caudicles almost directly on to a narrow viscidium (Fig. 29G). It is an African genus, and can be distinguished by its paniculate rather than racemose scape, which is produced terminally rather than basally or laterally on the pseudobulb. *Grammatophyllum*, however, has two long strap-shaped stipes connecting the two cleft pollinia to the narrow viscidium (Fig. 29H). It is otherwise similar to *Cymbidium*, with a basal, racemose scape, and the short fusion of the base of the lip and the base of the column resembles especially section *Cyperorchis*. A third genus with cane-like pseudobulbs is *Porphyroglottis*. It has a motile, hairy lip on a long column-foot. Its pollinarium is highly distinct, with the two cleft pollinia placed directly in the centre of a disc-shaped viscidium (Fig. 29N).

Dipodium also has a distinctive pollinarium (Fig. 29I), with a broad, bilobulate stipe from the centre of a disc-shaped viscidium, resembling that of *Grammatophyllum*, from which it differs in its lateral scape. The structure of the lip is highly specialised and distinct from *Cymbidium*: the lip mid-lobe is hairy, and the side-lobes are reduced. The base of the lip is again fused to the column.

Grammangis, Cymbidiella, Eulophia and *Eulophiella* all have a similar type of pollinarium which is distinct from that of *Cymbidium*. A broad, weakly bilobed stipe arises from the centre of a disc-shaped viscidium (Fig. 29J–M). *Cymbidiella* is otherwise similar to *Cymbidium*, although it does have a column-foot, an unusual character in *Cymbidium*, and a distinctive chromosome number of $2n = 54$. *Eulophia* is easily distinguished by its spurred lip. *Eulophiella* has a short mentum formed by a fusion of the lateral sepals to the column-foot.

More recently, a new classification of the orchids is in the process of being published by Pridgeon *et al.* (1999, 2001, 2003, 2005). The phylogeny and classification upon which it is based have been published recently in outline by Chase *et al.* (2003). This latter is the classification used in this study. Five subfamilies are recognised in the Orchidaceae, with *Cymbidium* included in Epidendroideae. Within this subfamily, 13 tribes are recognised, including Cymbidieae Pfitzer, which contains all of the sympodial vandoid orchids with two pollinia. By this criterion, sections *Jensoa*, *Pachyrhizanthe* and *Borneense*, with four pollinia, should not be included here. Dressler (1981) noted, however, that four pollinia is probably the primitive condition, with the fusion along the inner margin of each pair producing the more advanced condition of two deeply cleft pollinia. He stated that this reduction has occurred at least twice in the Vandeae, and perhaps more often. The change from four to two pollinia appears to have happened twice within the genus *Cymbidium* itself, and such criteria as compatibility in hybridisation, morphological and cytological similarity, and molecular data support the case for maintaining *Cymbidium*. There is no justification, therefore, for removing sections *Jensoa*, *Pachyrhizanthe* or *Borneense* to the tribe Maxillarieae simply on the basis of their pollinium number.

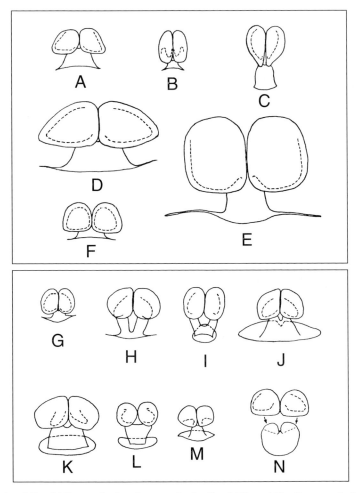

Fig. 29. The pollinaria of *Cymbidium* and related genera in the Cymbidieae (all × 5 approx.). **A.** *Cymbidium bicolor* (Cult. Kew no. 434-63.43407). **B.** *C. canaliculatum* (Cult. Kew no. 040-83.00318). **C.** *C. elegans* (Cult. Kew no. 120-81.01685). **D.** *C. tracyanum* (Cult. Kew no. 269-80.02507). **E.** *C. eburneum* (Cult. Kew no. 120-81.01673). **F.** *C. ensifolium* (Cult. Kew no. 481-81.06516). **G.** *Ansellia gigantea* (*Philcox and Leppard* 8865, Kew spirit colln. no. 29047). **H.** *Grammatophyllum wallisii* (Cult. Burnham Nurseries, Kew spirit colln. no. 45825). **I.** *Dipodium scandens* (*Lamb* SAN 91508, Kew spirit colln. no. 45489). **J.** *Grammangis ellisii* (*Mason* 24, Kew spirit colln. no. 46442). **K.** *Eulophiella roempleriana* (*Pottinger* s.n., Kew no. 548-82.05641). **L.** *Eulophia keithii* (Cult. Kew no. 400-65, Kew spirit colln. no. 37630). **M.** *Cymbidiella flabellata* (*Andrew* s.n., Cult. Kew no. 340-82.03516). **N.** *Porphyroglottis maxwelliae* (*Schage* s.n., Kew spirit colln. no. 33762).

Cymbidium is placed, with *Grammatophyllum*, *Graphorkis* and *Porphyroglottis*, in the subtribe Cymbidiinae, characterised by having pseudobulbs of several internodes, articulated leaves, usually a lateral inflorescence, the column usually with a prominent column-foot and a pollinarium with two or four pollinia and a viscidium. It is evident that pollinium number has limited usefulness in differentiating tribes and subtribes within the Epidendroideae.

Summary of the classification of *Cymbidium* (Chase *et al.*, 2003)
Family **Orchidaceae**
Subfamily **Epidendroideae**
Tribe **Cymbidieae** Pfitzer
Subtribe **Cymbidiinae**
Genera *Cymbidium*, *Grammatophyllum*, *Graphorkis*, *Porphyroglottis*

THE INFRAGENERIC CLASSIFICATION OF *CYMBIDIUM* SW.

The genus *Cymbidium* was established in 1799 by Olof Swartz, who included 44 species in his broadly defined genus. Many of these are not now considered to belong to the genus as currently circumscribed. Only three of them are still recognised within *Cymbidium*: *C. aloifolium*, *C. ensifolium* and *C. pendulum (= C. aloifolium)*, the first selected as the type of the genus. Lindley (1833) also included many species that are now placed in other genera, and noted that the genus would probably be split when sufficient information became available, stating 'I presume that each section will be hereafter recognised as distinct, for which reason I have given names which may be retained either as generic or sectional'. He placed the south and east Asian plants in subgenus *Eucymbidium*, giving the genus its present circumscription.

Blume (1848, 1849, 1858) removed *C. elegans* to form the genus *Cyperorchis* on the basis of its connivent perianth, sessile lip that is parallel to the column and is shortly trilobed at the apex, and elongated column with a beaked rostellum and a pair of pyriform pollinia on the middle of a flat, transversely ovoid viscidium. He emphasised the importance of the structure of the pollinarium, and the shape of the pollinium. Also in 1858, Blume removed *C. giganteum (= C. iridioides)* to the genus *Iridorchis*, this latter being distinguished from *Cymbidium* and *Cyperorchis* by its lip and column which are fused at the base for several millimetres, and the erroneous observation that the anther contained four pollinia. Blume saw the lip and column fusion as a link with *Grammatophyllum*, which has a similar structure, and distinguished these two genera on the mistaken basis of their differing pollinium number.

Reichenbach (1852) initially accepted the separation of *Cyperorchis* from *Cymbidium*. However, he returned *C. elegans* to *Cymbidium* in 1864, noting that although the pollinarium and pollinium shape are different, they were not sufficient to merit the separation of a distinct genus.

Hooker (1891) recognised *Cyperorchis*, and included in it *C. elegans*, *C. cochleare* and *C. mastersii* from northern India, but the discovery of more species blurred the distinction between *Cymbidium* and *Cyperorchis*. He stated that 'except by the narrow lip, long hypochile and small usually orbicular epichile (or mid-lobe), it is not easy to separate this genus from *Cymbidium*, for the pollinia vary in both genera, and *Cyperorchis mastersii* resembles very much *Cymbidium*'.

Schlechter (1924) recognised *Cyperorchis* as a distinct genus, but emphasised the fusion of the base of the lip and the base of the column as the distinguishing feature, rather than the pollinium shape, thereby extending the limits of *Cyperorchis* to include all of the large-flowered species, including those of Blume's *Iridorchis*. He further distinguished *Cyperorchis* by its large pseudobulbs, its thinner and more pointed leaves, larger flowers and the presence of a squarish stalk joining the pollinia to the viscidium. This latter character is not, in fact, evident in any of the species in *Cyperorchis*. He

TABLE 6

A COMPARISON OF THE CLASSIFICATIONS OF THE GENUS *CYMBIDIUM* SW. (ROMAN TYPE INDICATES SECTIONS)

Blume (1848, 1849, 1858)	Schlechter (1924)	Hunt (1970)	Seth & Cribb (1984) and Du Puy & Cribb (1988)	Sections proposed here
Cymbidium Sw.	*Cymbidium*	*Cymbidium*	*Cymbidium* subgen. *Cymbidium*	*Cymbidium*
	Eucymbidium	Cymbidium	Cymbidium	Cymbidium
			Borneense	Borneense
	Himantophyllum	Himantophyllum	Himantophyllum	Himantophyllum
	Austrocymbidium	Austrocymbidium	Austrocymbidium	Austrocymbidium
			Floribundum	Floribundum
			Suavissimum	
	Bigibbarium	Bigibbarium	Bigibbarium	Bigibbarium
Cyperorchis Bl.	*Cyperorchis* Bl.	Cyperorchis	*Cymbidium* subgen. *Cyperorchis*	Cyperorchis
Iridorchis Bl.	Eucyperorchis	Iridorchis	Cyperorchis	
	Iridorchis		Iridorchis	
			Eburnea	
	Annamaea	Annamaea	Annamaea	Annamaea
	Parishiella	Parishiella	Parishiella	Parishiella
	Cymbidium Sw.		*Cymbidium* subgen. *Jensoa*	
	Jensoa	Jensoa	Jensoa	Jensoa
	Maxillarianthe	Maxillarianthe	Maxillarianthe	
	Geocymbidium	Geocymbidium	Geocymbidium	
	Macrorhizon	Macrorhizon	Pachyrhizanthe	Pachyrhizanthe

recognised four sections within the genus (Table 6), including section *Eucyperorchis* in which he placed *C. elegans, C. cochleare, C. mastersii, C. roseum* and *C. whiteae*, and section *Iridorchis,* including *C. eburneum, C. iridioides*, and most of the other related species. *Cymbidium tigrinum* and *C. erythrostylum* were placed in separate, monotypic sections.

In his revision of *Cymbidium* and *Cyperorchis,* Schlechter (1924) proposed a total of 12 sections. His framework is the basis of the modern infrageneric classification of *Cymbidium*, most of his sections still being recognised more or less in their original form (Table 6). He included 79 species in the two genera, but he apparently did not critically examine all of them.

Hunt (1970) included *Cyperorchis* within *Cymbidium* but maintained Schlechter's sectional divisions (Table 6). He listed the characters that have been used to split *Cyperorchis sensu* Blume *non* Schlechter from *Cymbidium* as 'the narrower labellum, relatively longer hypochile, smaller epichile, and broader stipes (viscidium) of the pollinia ... the very short column-foot adnate to the column, the fusion of the lip and column, the flowers not opening very wide ... and its much denser, pendulous racemes'. He also noted that some species are morphologically intermediate, and that the characters mentioned are 'not of generic status'. This treatment is now generally accepted.

Seth & Cribb (1984) divided the genus *Cymbidium* into three subgenera. They distinguished subgenus *Cyperorchis*, which corresponds to Schlechter's *Cyperorchis*, from subgenus *Cymbidium* by the fusion of the lip and column-base. Blume's more restricted delimitation of *Cyperorchis*, and his *Iridorchis*, were maintained as sections within this subgenus. They emphasised that the four pollinia found in the species in Schlechter's sections *Jensoa, Maxillarianthe, Geocymbidium* and *Macrorhizon* (= *Pachyrhizanthe*) could equally be regarded as evidence for the splitting of the genus, and they placed these in a third subgenus, *Jensoa*. They also created three new sections (Table 6). Du Puy & Cribb (1988) followed this treatment but added section *Borneense* for the recently described *C. borneense* J.J. Wood. They presented evidence from seed and leaf undersurface micromorphology to further support subgenus *Jensoa*, and characters from leaf anatomy and micromorphology to distinguish section *Cymbidium*.

The current treatment reflects the information gained from the cladistic analyses based upon DNA studies by van den Berg *et al.* (2002) and Yukawa *et al.* (2002). Both concluded that the subgenera of Du Puy and Cribb (1988) were not monophyletic, although with minor alterations (notably in the removal of *C. dayanum* from subgenus *Cymbidium* to subgenus *Cyperorchis*) there remains strong evidence that subgenera *Cyperorchis* and *Jensoa* are monophyletic. Unfortunately, monophyly could not be demonstrated in subgenus *Cymbidium*, and there is evidence of two distinct clades within this subgenus, while the position of *C. devonianum* (section *Bigibbarium*) is anomalous and remains unresolved. In the light of this data it seems better to retain the sections but dispense with the subgenera.

Nevertheless, the simplified classification adopted here must be regarded as provisional and interim, pending more comprehensive and critical molecular systematic studies (Du Puy, 2005). The loss of information concerning the inter-relationships of the sections within subgenera is cause for regret, and the lack of consensus on the relationships and position of *C. devonianum* is particularly unsatisfactory.

Several floras have dealt with the species of *Cymbidium* on a regional basis. These include the works by Hooker (1891), King & Pantling (1898) and Pradhan (1979) for the Himalayan and Indian species; Hu (1973) and Wu & Chen (1980) and Chen (1999) for China; Lin (1977) for Taiwan; Maekawa (1971) for Japan; Seidenfaden (1983) for Thailand; Holttum (1953, 1957, 1964) for western Malaysia; Smith (1905), Backer & Bakhuizen (1968) and Comber (1980, 2001) for Java and Sumatra respectively; Du Puy & Lamb (1984) for Sabah; and Dockrill (1969) for Australia. However, the first comprehensive revision of *Cymbidium* was published by Du Puy and Cribb (1988).

A. LAMB

TAXONOMY

Cymbidium Sw. in *Nov. Acta Soc. Sci. Upps.* 6: 70 (1799); Lindley, *Gen. Sp. Orchid. Pl.*: 161 (1833); Schltr. in *Fedde, Repert. Sp. Nov. Regni Veg.* 20: 96 (1924); Hunt in *Kew Bull.* 24: 93 (1970); Seth & Cribb in J. Arditti (ed.), *Orchid Biology: Reviews and Perspectives,* 3: 283 (1984); Du Puy & Cribb, *The Genus Cymbidium* (1988). Type: *C. aloifolium* (L.) Sw.

Jensoa Raf., *Fl. Tellur.*, 4: 38 (1836). Type: *Jensoa ensata* (Thunb.) Raf. (= *C. ensifolium*) (L.) Sw.).
Cyperorchis Blume, *Rumphia* 4: 47 (1848), *Mus. Bot. Lugd.*1: 48 (1849) & *Orchid. Arch. Ind.* 1: 92 (1858). Type: *Cymbidium elegans* Lindl..
Iridorchis Blume, *Orchid. Arch. Ind.* 1: 91, t.26 (1858). Type: *Cymbidium giganteum* Wall. ex Lindl. (= *C. iridioides* D.Don).
Arethusantha Finet in *Bull. Soc. Bot. France* 44: 178–180, t. 15 (1897). Type: *Arethusantha bletioides* Finet (= *C. elegans* Lindl.).
× *Cyperocymbidium* A. Hawkes in *Orchid Rev.* 72: 420 (1964).

Epiphytic, lithophytic or terrestrial *herbs*, with vegetative growth from the base or lower nodes of the persistent pseudobulb, which is usually produced annually, but may grow indeterminately for two or occasionally many more years; autotrophic or rarely heteromycotrophic. *Pseudobulbs* ovoid to spindle-shaped, often laterally compressed, occasionally absent and replaced by a slender stem, often inconspicuous and concealed within the leaf bases. *Roots* thick, white, velamen-covered, branching, usually arising from the base of the new growth. *Cataphylls* several, surrounding the young growth, often becoming scarious and fibrous with age. *Leaves* up to 13, distichous, linear-elliptic or narrowly ligulate to elliptic, acuminate to strongly bilobed at the apex, articulated close to the pseudobulb to their persistent, broadly sheathing bases. *Inflorescences* racemose, densely to laxly flowered, usually arising from within the cataphylls, but occasionally from within the axils of the leaves; peduncle erect, arching or pendulous, usually covered with inflated, cymbiform sheaths. *Flowers* 1–many, up to 12 cm in diameter, often large, showy and sometimes fragrant, subtended by a small bract, pedicel and ovary. *Sepals* and *petals* free, subsimilar, often spreading or with the petals porrect and covering the column. *Lip* 3-lobed, free or fused at the base to the base of the column for 3–6 mm; side-lobes erect and weakly clasping the column; mid-lobe often recurved; callus usually two parallel ridges from near the base of the lip to the base of the mid-lobe, usually swollen towards the apex, sometimes broken in the middle, convergent towards the apices or reduced to a pair of small swellings at the base of the mid-lobe, rarely absent and replaced by a glistening shallow depression or fused into a single cuneate ridge. *Column* arcuate, weakly or strongly winged, semi-terete in cross-section, concave on the ventral surface. *Pollinia* usually 2, deeply cleft but sometimes 4 in two unequal pairs, triangular, quadrangular, ovoid or club-shaped, subsessile and attached by short, elastic caudicles to a usually triangular viscidium with the lower corners usually drawn out into short, thread-like appendages. *Capsule* fusiform to ellipsoidal or oblong-ellipsoidal, narrowing at the base to a short pedicel and at the apex to a short beak formed by the persistent column.

DISTRIBUTION. Some 52 species distributed from the north-west Himalaya to Japan, and south through Indo-China and Malesia to the Philippines, New Guinea and Australia (Map 1).

(Opposite) **Fig. 30.** *Cymbidium atropurpureum* Sepilok, Sabah.

Sectional treatment of *Cymbidium*

Section *Floribundum* Seth & P.J. Cribb
1. *C. chloranthum* Lindl. (p. 83)
2. *C. elongatum* J.J. Wood, Du Puy & P.S. Shim (p. 88)
3. *C. floribundum* Lindl. (p. 93)
4. *C. hartinahianum* J.B. Comber & Nasution (p. 97)
5. *C. suavissimum* Sander ex C. Curtis (p. 99)

Section *Austrocymbidium* Schltr.
6. *C. canaliculatum* R. Brown (p. 102)
7. *C. madidum* Lindl. (p. 110)
8. *C. suave* R. Br. (p. 114)

Section *Cymbidium*
9. *C. aloifolium* (L.) Sw. (p. 118)
10. *C. atropurpureum* (Lindl.) Rolfe (p. 125)
11. *C. bicolor* Lindl. (p. 130)
      subsp. *bicolor* (p. 131)
      subsp. *obtusum* Du Puy & P.J. Cribb (p. 132)
      subsp. *pubescens* (Lindl.) Du Puy & P.J. Cribb (p. 136)
12. *C. finlaysonianum* Lindl. (p. 138)
13. *C. rectum* Ridley (p. 142)

Section *Borneense* Du Puy & P.J. Cribb
14. *C. aliciae* Quisumb. (p. 147)
15. *C. borneense* J.J. Wood (p. 150)

Section *Bigibbarium* Schltr.
16. *C. devonianum* Paxton (p. 153)

Section *Himantophyllum* Schltr.
17. *C. dayanum* Rchb.f. (p. 159)

Section *Annamaea* (Schltr.) P.F. Hunt
18. *C. erythrostylum* Rolfe (p. 167)

Section *Cyperorchis* (Blume) P.F. Hunt
19. *C. banaense* Gagnep. (p. 173)
20. *C. cochleare* Lindl. (p. 175)
21. *C. eburneum* Lindl. (p. 179)
22. *C. elegans* Lindl. (p. 184)
23. *C. erythraeum* Lindl. (p. 189)
24. *C. hookerianum* Rchb.f. (p. 192)
25. *C. insigne* Rolfe (p. 197)
26. *C. iridioides* D. Don (p. 203)
27. *C. lowianum* (Rchb.f.) Rchb.f. (p. 207)
      var. *lowianum* (p. 209)
      var. *iansonii* (Rolfe) Du Puy & P.J. Cribb (p. 211)

28. *C. mastersii* Griffith ex Lindl. (p. 212)
29. *C. parishii* Rchb.f. (p. 217)
30. *C. roseum* J.J. Sm. (p. 220)
31. *C. sanderae* (Rolfe) Du Puy & P.J. Cribb (p. 224)
32. *C. schroederi* Rolfe (p. 229)
33. *C. sigmoideum* J.J. Smith (p. 234)
34. *C. tracyanum* L. Castle (p. 236)
35. *C. wenshanense* Y.S. Wu & F.Y. Liu (p. 240)
36. *C. whiteae* King & Pantl. (p. 243)
37. *C. wilsonii* (Rolfe ex Cook) Rolfe (p. 247)

Section *Parishiella* (Schltr.) P.F. Hunt
38. *C. tigrinum* Parish ex Hook. (p. 250)

Section *Jensoa* (Raf.) Schltr.
39. *C. cyperifolium* Wall. ex Lindl. (p. 254)
   subsp. *cyperifolium* (p. 256)
   subsp. *indochinense* Du Puy & P.J. Cribb (p. 259)
40. *C. defoliatum* Y.S. Wu & S.C. Chen (p. 261)
41. *C. ensifolium* (L.) Sw. (p. 263)
   subsp. *ensifolium* (p. 268)
   subsp. *haematodes* Du Puy & P.J. Cribb (p. 270)
   subsp. *acuminatum* (M.A. Clem. & D.L. Jones) P.J. Cribb & Du Puy (p. 276)
42. *C. faberi* Rolfe (p. 277)
   var. *faberi* (p. 281)
   var. *szechuanicum* (Y.S. Wu & S.C. Chen) Y.S. Wu & S.C. Chen (p. 281)
43. *C. goeringii* (Rchb.f.) Rchb.f. (p. 283)
44. *C. kanran* Makino (p. 290)
45. *C. munronianum* King & Pantl. (p. 294)
46. *C. nanulum* Y.S. Wu & S.C. Chen (p. 296)
47. *C. omeiense* Y.S. Wu & S.C. Chen (p. 299)
48. *C. qiubeiense* Y.S.Wu & S.C. Chen (p. 301)
49. *C. sinense* (Jackson in Andr.) Willd. (p.302)
50. *C. tortisepalum* Fukuyama (p. 308)
   var. *tortisepalum* (p. 310)
   var. *longibracteatum* (Y.S. Wu & S.C. Chen) S.C. Chen & Z.J. Liu (p. 311)

Section *Pachyrhizanthe* Schltr.
51. *C. lancifolium* Hook. (p. 313)
   var. *lancifolium* (p. 319)
   var. *aspidistrifolium* (Fukuy.) S.S. Ying (p. 319)
   var. *papuanum* (Schltr.) P.J. Cribb & Du Puy (p. 321)
52. *C. macrorhizon* Lindl. (p. 323)
   var. *macrorhizon* (p. 330)
   var. *aberrans* (Finet) P.J. Cribb & Du Puy (p. 330)

Key to the Sections of *Cymbidium*

1. Lip fused at the base to the base of the column for about 3–6 mm **2**
    Lip free, attached to the base of the column, or rarely to a short column-foot **4**

2. Lateral sepals curved downwards, giving the flower a narrow, triangular appearance; callus ridges united apically into a single, 3-lobed, cuneate (wedge-shaped) structure; column strongly hairy beneath, bright deep pink .... *section* **Annamaea** (p. 167)
    Lateral sepals porrect or spreading; callus ridges 2, not fused at the apex (except in *C. eburneum*); column not strongly hairy beneath, not as above **3**

3. Leaves 2–4, short, to 17(22) cm, elliptic, at the apex of an exposed, lens-shaped pseudobulb; side-lobes of the lip almost uniformly purple-brown; mid-lobe with transverse dashes of purple-brown .... *section* **Parishiella** (p. 250)
    Leaves 5 or more, longer than above, slender, linear-elliptic, their bases covering an ovoid, usually lightly compressed pseudobulb; lip markings not as above .... *section* **Cyperorchis** (p. 171)

4. Lip with callus ridges not forming a tube at the apex (except in *C. aliciae*); pollinia 2, each deeply cleft, or rarely 4 (sect. *Borneense* only) **5**
    Lip with callus ridges converging towards the apex and forming a short tube at the base of the mid-lobe; pollinia 4, in two unequal pairs **12**

5. Pollinia 4, in two pairs .... *section* **Borneense** (p. 145)
    Pollinia 2, deeply cleft behind **6**

6. Leaves elliptic, petiolate; callus ridges reduced to two small swellings at the base of the mid-lobe of the lip .... *section* **Bigibbarium** (p. 153)
    Leaves ligulate to linear-elliptic, without a distinct petiole; callus not as above **7**

7. Scape erect; callus ridges crenulate .... *section* **Floribundum** (in part) (p. 83)
    Scape suberect to pendulous; callus ridges not crenulate **8**

8. Leaves thick, coriaceous, somewhat rigid, narrowly ligulate with an obtuse to emarginate and unequally bilobed apex .... *section* **Cymbidium** (p. 118)
    Leaves usually thinner textured, linear-elliptic, acute to obtuse, usually oblique at the apex (if thick and rigid then acute at the apex) **9**

9. Lip callus a glistening depression .... *section* **Austrocymbidium** (in part) (p. 102)
    Lip callus not as above **10**

10. Leaves relatively thin-textured, long, slender, linear-elliptic, acute to acuminate; callus ridges pubescent, with white glandular hairs .... *section* **Himantophyllum** (p. 159)
    Leaves not as above; callus ridges usually papillose, and lacking glandular hairs **11**

11. Leaves coriaceous 3–4, elliptic, acute; flowers heavily spotted and blotched with red-brown on the sepals and petals; lip with narrow, reduced side-lobes ....
    .... *section* **Austrocymbidium** (in part) (p. 102)

Leaves not coriaceous, 5–10, linear-ligulate, oblique or bilobed at the apex; flowers unspotted on the sepals and petals; lip with broad side-lobes which weakly clasp the column .... *section* **Floribundum** (in part) (p. 83)

12. Leaves linear; pseudobulbs ovoid, rarely absent .... *section* **Jensoa** (p. 254)
    Leaves elliptic or absent; pseudobulbs fusiform or absent .... *section* **Pachyrhizanthe** (p. 313)

Artificial Key to the Species of *Cymbidium*

1. Margin of the lip fused at the base to the base of the column for *c*. 3–6 mm, forming a short sac **2**
    Lip free, attached to the base of the column or occasionally a short column-foot **22**

2. Callus composed of a single, wedge-shaped ridge **3**
    Callus in two parallel ridges **4**

3. Flowers usually 4–8; lip strongly veined with deep red; side-lobes of the lip with long hairs in lines over the veins; callus tapering into the mid-lobe of the lip, cream with pink mottling; column bright pink, with a dense indumentum of long hairs beneath; pollinia triangular .... **18. C. erythrostylum**
    Flowers usually solitary; lip white, sometimes with a few pink spots; side-lobes of the lip papillose or minutely hairy; callus ending abruptly behind the mid-lobe base, bright yellow; column white, almost glabrous beneath; pollinia quadrangular .... **21. C. eburneum**

4. Leaves 2–4 per pseudobulb, short, to 17(22) cm long, elliptic, at the apex of an exposed, lens-shaped pseudobulb; side-lobes of the lip almost entirely purple-brown .... **38. C. tigrinum**
    Leaves 5 or more, usually much longer than above, slender, linear-elliptic, their broad, sheathing bases covering an ovoid, somewhat compressed pseudobulb; side-lobes not coloured as above **5**

5. Sepals, petals and lip green, spotted with dark or purple-brown, glossy; mid-lobe of the lip 2–2.5 mm broad, strap-shaped, glabrous, waxy .... **33. C. sigmoideum**
    Sepals, petals and lip not as above; mid-lobe of the lip 4–30 mm broad, not as above **6**

6. Lip white, with numerous fine maroon or red-brown spots; callus a broad, glabrous, slightly raised ridge behind, slightly swollen at the margins and forming two raised ridges which become strongly inflated and confluent at the apices, white .... **36. C. whiteae**
    Lip and callus not as above **7**

7. Flowers pendulous, bell-shaped, the sepals and petals not or slightly spreading; anther-cap with a distinct backwards-pointing beak; pollinia clavate **8**
    Flowers held erect, the lip and column almost horizontal, not bell-shaped; anther-cap without a distinct beak; pollinia triangular or quadrangular **9**

8. Leaves up to 1.4(2) cm broad; flower cream or yellowish with a cream to pale green lip, unspotted or occasionally sparsely spotted with pale pink; mid-lobe strap-shaped at the base, expanded into two lobes apically and with an emarginate apex; callus ridges extending to the base of the lip, usually with a pair of auricles at the base which form a trough-shaped depression .... **22. C. elegans**

Leaves narrower, up to 0.8 cm broad; flower greenish-brown with a yellowish lip covered with numerous, fine, red spots; mid-lobe cordate to elliptic, mucronate; callus ridges short, quickly tapering from an inflated apex and terminating well behind the base of the lip, and lacking auricles at the base .... **20. C. cochleare**

9.   Scape produced from the axils of the leaves; leaf apex usually finely forked, rarely obtuse or rounded (in *C. banaense*), often with a short mucro in the sinus; sepals and petals white or pink; pollinia quadrangular **10**
Scape produced from near the base of the pseudobulb, leaf apex acute; sepals and petals white, green or brown; pollinia triangular **13**

10.   Stem not pseudobulbous, growing indeterminately for many years, eventually forming an elongated, strap-shaped base covered by scarious, persistent leaf bases with fresh leaves present only towards the apex; scape often with 5–10 flowers; petals narrow, 0.5–0.7 cm broad; mid-lobe of the lip up to 1.3 cm broad .... **28. C. mastersii**
Stem weakly pseudobulbous, growing for 2–4 years before a new growth is produced, not as above; scape usually with 2–5 flowers; petals wide, 1.1–1.3 cm broad; mid-lobe larger, 1.3–1.7 cm broad **11**

11.   Flowers not opening widely; dorsal sepal 4.4–4.8 cm long; side-lobes of the lip with rounded to subacute apices, shortly pubescent .... **30. C. roseum**
Flower opening widely; dorsal sepal about 5.9 cm long **12**

12.   Side-lobes of the lip with acute, porrect apices, with minute, papillose hairs; lip mid-lobe as long as wide .... **29. C. parishii**
Side-lobes of lip rounded, lacking papillose hairs; lip mid-lobe twice as long as wide .... **19. C. banaense**

13.   Sepals and petals white or pinkish; side-lobes of the lip with broadly rounded apices; pseudobulbs weakly compressed; scape erect to suberect; bracts up to 15 mm long; capsule almost spherical **14**
Sepals and petals yellowish to green, sometimes heavily lined and stained red-brown; side-lobes of the lip with triangular apices; pseudobulbs strongly bilaterally compressed; scape suberect to subpendulous, arching; bracts less than 5 mm long; capsule cylindrical **16**

14.   Scape 100–150 cm long, with a very long, erect peduncle up to 125 cm long, and the flowers closely spaced in the apical portion; mid-lobe of the lip acute, usually with maroon veining and spotting; terrestrial .... **25. C. insigne**
Scape 30–50 cm long, with a shorter, suberect peduncle about 20 cm long; plant epiphytic **15**

15.   Flowers white with the mid-lobe of the lip rounded and with heavy maroon blotches; column glabrous .... **31. C. sanderae**
Flowers white with faint pink specks on the obovate mid-lobe of the lip; column hairy on the upper surface, especially in the basal half .... **35. C. wenshanense**

16.   Flowers less than 8 cm across; dorsal sepal less than 11 mm broad; petals very narrow, up to 7 mm broad, ligulate and strongly curved; leaves long, narrow, usually about 1 cm broad, but up to 1.6 cm broad; scape very slender; mid-lobe of the lip white or cream with sparse, irregular, red-brown spots .... **23. C. erythraeum**

Flowers lacking the above combination of characters, more than 8 cm across; dorsal sepal more than 11 mm broad; petals more than 7 mm broad, usually slightly dilated towards the apex and less strongly curved; leaf wider, up to 4 cm broad; scape more robust; mid-lobe of the lip strongly marked with red or red-brown spots or a single V-shaped blotch **17**

17. Mid-lobe of the lip with a large, V-shaped, red to pale chestnut, submarginal blotch, porrect, almost flat, margin slightly undulating, indumentum in two zones with short, silky hairs at the base and in the centre, with the V-shaped mark densely covered with minute hairs; side-lobes with right-angled apices; flower lacking an obvious scent **18**
    Mid-lobe of the lip spotted or blotched with red or red-brown, recurved, margin strongly undulating; indumentum not in two distinct zones as above, sometimes with 2–3 lines of long hairs in the centre; side-lobes with acute apices; flowers sweetly scented **19**

18. Side-lobes of the lip strongly veined red-brown; sepals and petals yellow-green, striped red-brown on the veins; flowers about 8 cm across; dorsal sepal 4.5–4.9 cm long, 1.4–1.6 cm broad; callus long, two-thirds of the length of the disc .... **32. C. schroederi**
    Side-lobes of the lip unmarked; sepals and petals usually clear green, or lightly veined or shaded red-brown; flowers 8–10 cm across; dorsal sepal 4.8–5.7 cm long, 1.6–1.8 cm broad; callus short, about half of the length of the disc .... **27. C. lowianum**

19. Callus on the lip with long cilia which continue in 2 or 3 lines well into the mid-lobe; sepals and petals heavily marked red-brown; leaves up to 4 cm broad, plants very large; flowering August to December **20**
    Callus on the lip papillose to pubescent, but the hairs not continuing beyond the apices of the callus ridges; sepals and petals clear green, or very lightly flushed with red-brown; leaves up to 2.5 cm broad, plants less robust; flowering January to April **21**

20. Petals sickle-shaped, often reflexed; lip large, prominent, cream, irregularly spotted and dashed red-brown; side-lobes fringed with cilia more than 1 mm long, and with hairs confined to the veins; callus of two ciliate ridges with a third line of cilia between; flower very large, 12–15 cm across; dorsal sepal 5.7–7.8 cm long, 1.4–2.9 cm broad, column 3.4–4.4 cm long, evenly winged to the base .... **34. C. tracyanum**
    Petals lightly curved, spreading; lip not prominent, yellow, marked with a submarginal ring of red-brown spots; side-lobes fringed with short hairs less than 1 mm long, and with an even indumentum of short hairs; callus of two hairy ridges only; flower smaller, about 9–10 cm across; dorsal sepal 4.4–4.7 cm long, 1.2–1.8 cm broad; column 2.6–2.9 cm long with wings which taper to the base .... **26. C. iridioides**

21. Side-lobes of the lip spotted dark red-brown along the veins, fringed on the margins with cilia more than 1 mm long; mid-lobe large, striking; flower up to 15 cm across; dorsal sepal 5.6–6.0 cm long, 1.7–1.9 cm broad .... **24. C. hookerianum**
    Side-lobes of the lip with red-brown veins, with some spots towards the margins, fringed with short hairs; mid-lobe smaller, not prominent; flowers smaller, 9–10 cm across; dorsal sepal 4.4–5.7 cm long, 1.2–1.9 cm broad .... **37. C. wilsonii**

22. Callus ridges separated, not forming a tube in apical half, or absent; pollinia 2 (except in *C. borneense*) **23**
    Callus ridges strongly convergent in the apical half forming a short tube at the base of the mid-lobe of the lip; pollinia 4 .... **39**

23. Callus ridges absent, replaced by a glistening viscid stripe **24**
    Callus ridges present or reduced to small cushions **25**

24. Plant with large, ovoid pseudobulbs; flowers distantly spaced on a 30–80 cm or more long, pendulous scape, produced from the pseudobulb base; mid-lobe of the lip narrowly obovate, with a truncated apex .... **7. C. madidum**
    Plant with an elongated stem; flowers densely spaced on a *c.* 15–24 cm long, arching scape, produced towards the stem apex, usually in the axils of fallen leaves; mid-lobe of the lip ovate, with an obtuse to subacute apex .... **8. C. suave**

25. Callus ridges reduced to 2 small swellings at the base of the mid-lobe of the lip **26**
    Callus ridges well-defined, not as above **27**

26. Leaves elliptic, narrowing to a distinct petiole, 3.5–6.2 cm broad; scape sharply pendulous, usually with 15–35 flowers; lip rhomboid when flattened, the side-lobes not sharply demarcated from the mid-lobe; mid-lobe of the lip with 2 large, rich purple blotches near the base; pollinia 2, cleft behind .... **16. C. devonianum**
    Leaves linear-ligulate 0.5–2.1 cm broad; scape suberect, with 3–5 flowers; lip distinctly 3-lobed, the side-lobes angled at their apices; mid-lobe of the lip lightly spotted with maroon; pollinia 4, in two pairs .... **15. C. borneense**

27. Callus ridges strongly S-shaped and often broken in the middle, swollen towards the base and at the apex **28**
    Callus ridges entire, straight or slightly curved **29**

28. Side-lobes of the lip not exceeding the column and anther-cap, obtuse to subacute, finely mottled with maroon or red-brown .... **11. C. bicolor**
    Side-lobes of the lip longer than the column and anther-cap, acute, veined with maroon.................
    .... **9. C. aloifolium**

29. Callus ridges and mid-lobe of the lip pubescent **30**
    Callus ridges and mid-lobe of the lip glabrous or minutely papillose **31**

30. Leaves 3–4 on each pseudobulb, broad, rigid and very coriaceous; scape usually with 20–60 flowers; dorsal sepal 11–25 mm long, obtuse to subacute; mid-lobe of the lip white, pale pink or pale green, lightly spotted with red or purple .... **6. C. canaliculatum**
    Leaves usually 5–8 on each pseudobulb, slender, flexible and only slightly coriaceous; scape usually with 5–15 flowers; dorsal sepal usually 25–34 mm long, acute to shortly acuminate, often mucronate; mid-lobe of the lip deep maroon, with a basal, pale yellow, triangular patch ....
    .... **17. C. dayanum**

31. Dorsal sepal 25–36 mm long; mid-lobe of the lip 10–14 mm broad; column 15–18 mm long **32**
    Dorsal sepal 11–25 mm long; mid-lobe of the lip 5–10 mm broad; column 6–15 mm long **34**

32. Leaves well-spaced along an elongated stem, with slightly oblique apices; scape with 1–5 flowers, produced from the axils of the upper leaves; side-lobes of the lip not angled at the apex, confluent with the base of the mid-lobe .... **2. C. elongatum**
    Leaves covering a short, swollen or pseudobulbous stem, usually with strongly oblique, bilobed apices; scape with (7)10–33 flowers, produced from within the cataphylls at the base of the stem; side-lobes of the lip angled at the apex, making the lip strongly 3-lobed **33**

33. Side-lobes of the lip with porrect, triangular, acute apices, which are longer than the column and anther-cap; callus ridges strongly raised and terminating abruptly at the base of the mid-lobe of the lip .... **12. C. finlaysonianum**
  Side-lobes of the lip with truncated, obtuse apices which are shorter than the column; callus ridges weakly raised, tapering gradually into the base of the mid-lobe of the lip ....
  .... **10. C. atropurpureum**

34. Scape pendulous or arching downwards; callus ridges often with a short, oblique groove near the apex .... **11. C. bicolor**
  Scape erect, suberect, or arching upwards; callus ridges not as above **35**

35. Callus ridges crenulate; sepals and petals olive-green to bright yellow **36**
  Callus ridges smooth; sepals and petals usually red-brown, with a narrow to broad cream margin
  .... **37**

36. Plant epiphytic; leaves 23–38 mm broad, oblique to unequally bilobed at the apex; peduncle 15–22 cm long, covered by overlapping sheaths; petals narrower than the sepals; side-lobes of the lip subacute at the tip .... **1. C. chloranthum**
  Plant terrestrial; leaves 9–15 mm broad, acute, slightly hooded at the apex; peduncle 35–60 cm long, incompletely covered by distant sheaths; petals almost as broad as the sepals; side-lobes of the lip rounded at the tip .... **4. C. hartinahianum**

37. Mid-lobe of the lip 5–5.5 mm broad, ligulate, pale yellow with a single maroon spot near the apex; column 8 mm long; pollinia connate at the top, forming a single pollinium .... **13. C. rectum**
  Mid-lobe of the lip 7–8 mm broad, ovate, white or cream with several maroon or pink blotches; becoming deep red on pollination or with age; column 12–15 mm long; pollinia 2, not connate
  .... **38**

38. Leaves narrow, up to 1.5(2) cm broad; cataphylls green; dorsal sepal 1.8–2.1 cm long; column lacking auricles at the base .... **3. C. floribundum**
  Leaves broader, 3–3.8 cm broad; cataphylls purple; flower slightly larger, the dorsal sepal 2–2.5 cm long; column with small auricles at the base .... **5. C. suavissimum**

39. Plant saprophytic, entirely lacking leaves; roots more or less absent; rhizome off-white, warty ....
  .... **52. C. macrorhizon**
  Plant autotrophic, with 2 to several green leaves; roots well-developed, not warty **40**

40. Leaves elliptic, petiolate; pseudobulb cigar-shaped; scape from near the middle of the pseudobulb
  .... **51. C. lancifolium**
  Leaves linear-elliptic, without a distinct petiole; pseudobulbs ovoid, often inconspicuous, rarely absent; scape from near the base of the pseudobulb **41**

41. Flowers solitary, rarely paired; sepals often spatulate .... **43. C. goeringii**
  Flowers usually 3–26; sepals not spatulate **42**

42. Leaves annually deciduous, absent in winter; pseudobulbs small, forming a rhizome-like string ......
  .... **40. C. defoliatum**
  Leaves evergreen; pseudobulbs absent or, if present, not rhizome-like **43**

43. Rhizome present, subterranean, slightly flattened-cylindric, thick, multi-noded; plants usually solitary, 2- to 3-leaved; flowers small; sepals 13–16 mm long .... **46. C. nanulum**
Rhizome absent; plants often in colonies, flowers larger; sepals 20 mm or more long **44**

44. Leaves 2–4 on each pseudobulb **45**
Leaves 5–10 or more on each pseudobulb **51**

45. Leaf bases slenderly petiolate, wiry, purple-spotted .... **48. C. qiubeiense**
Leaf bases not as above **46**

46. Leaves dark green, glossy above; flowers large, dorsal sepal usually longer than 30 mm; petals porrect and covering the column; mid-lobe large, (10)12–16 mm long **47**
Leaves mid-green, not glossy; flowers smaller, dorsal sepal usually less than 26(30) mm long; petals somewhat spreading, not covering the column; mid-lobe shorter, 6–10(12) mm long **48**

47. Leaves narrow, usually less than 1.5 cm broad; bracts long, equalling the pedicel and ovary in the lower flowers; sepals very slender, about 7 times as long as broad, finely acuminate ....
.... **44. C. kanran**
Leaves broader, usually greater than 2 cm broad; bracts shorter, usually shorter than the pedicel and ovary in the lower flowers; sepals broader, about 4–5 times as long as broad, acute ....
.... **49. C. sinense**

48. Scape usually with 8–13 flowers; sheaths distant, amplexicaul, closely clasping the peduncle; mid-lobe of the lip small, much narrower than the side-lobes when the lip is flattened, 4.2–7.1 mm broad; column 7–11 mm long .... **45. C. munronianum**
Scape usually with 3–8 flowers; sheaths overlapping, cymbiform, somewhat spreading from the peduncle; mid-lobe of the lip almost as broad as the side-lobes when the lip is flattened, usually 6–10 mm broad; column 10–15 (18) mm long **49**

49. Floral bracts as long as the ovary; leaves arcuate, 15–30 cm long .... **47. C. omeiense**
Floral bracts usually half the length of the pedicel and ovary or less; leaves erect to suberect, 30–60 cm long **50**

50. Lip side-lobes truncate; lip mid-lobe papillose, tapering to acute apex; bracts a quarter the length of the pedicel and ovary .... **14. C. aliciae**
Lip side-lobes narrowly elliptic; lip mid-lobe not papillose; bracts half the length of the pedicel and ovary .... **41. C. ensifolium**

51. Mid-lobe of the lip covered with glistening, inflated papillae; mid-lobe margin minutely fimbriate and strongly undulating; leaves with translucent veins .... **42. C. faberi**
Mid-lobe of the lip minutely papillose or with a few inflated papillae; mid-lobe margin entire, lightly kinked or weakly undulating; leaves lacking translucent veins **52**

52. Leaf lamina 0.9–1.5 cm broad; sepals and petals green or yellowish green, often stained with red-brown on the veins, especially towards the base of the mid-vein .... **39. C. cyperifolium**
Leaf lamina usually less than 0.9 cm broad; sepals and petals usually cream, with pale greenish veins .... **50. C. tortisepalum**

SECTION FLORIBUNDUM

Section **Floribundum** Seth & P.J. Cribb in Arditti (ed.), *Orchid Biol., Rev. Persp.* 3: 298 (1984). Type: *C. floribundum* Lindl.

Cymbidium section *Suavissimum* Seth & P.J. Cribb, *loc. cit.*: 297 (1984). Type: *C. suavissimum* Sander ex Curtis.

C. section *Austrocymbidium* sensu Du Puy & P.J. Cribb, *The Genus Cymbidium* 88 (1988), *pro parte*.

Species of section *Floribundum* are characterised by having numerous (15–50), closely spaced flowers on short, suberect scapes. The sepals and petals are obtuse at the apex except in *C. elongatum*, placement of which here is debatable. The lip has broad side-lobes that are rounded at the apex and a broadly ovate mid-lobe. The section has affinities with section *Austrocymbidium*, both having coriaceous leaves, many closely spaced flowers on a relatively short scape, obtuse sepals and petals that have a greenish background colour and often ellipsoidal rather than triangular pollinia. It is distributed in southern China, Vietnam, Burma, peninsular Malaysia, Borneo, Sumatra and Java.

The section was established for *C. floribundum* by Seth and Cribb (1984) and was expanded by Du Puy and Cribb (1988) to include *C. suavissimum*. Based on DNA sequence data, a further three species, namely *C. chloranthum*, *C. hartinahianum* and *C. elongatum,* all formerly placed in sect. *Austrocymbidium*, are now added. The inclusion of *C. elongatum*, a rare endemic to northern Borneo, is tentative pending further investigation. It has an elongated monopodial stem and short lateral inflorescences, somewhat reminiscent of the Australian *C. suave*.

1. CYMBIDIUM CHLORANTHUM

This attractive species from the Malay Peninsula and Archipelago has distinctive, erect, many-flowered inflorescences, whereas most of the other epiphytes have arching or pendulous scapes. Another noticeable feature is the strong colour change in the flower induced by pollination, removal of the anther, or even by mechanical disturbance of the stigmatic surface, the flower becoming suffused with deep carmine pink. This 'blushing' is well known in other species in this genus, such as in *C. floribundum, C. suavissimum*, section *Cyperorchis* and in the modern, commercial hybrids, but it is particularly strong in *C. chloranthum*.

The pseudobulbs of *C. chloranthum* are large and slightly flattened, with the abscission zone of the leaf close to the pseudobulb, strongly resembling the pseudobulbs of the Australian *C. madidum* and *C. canaliculatum*. The strap-shaped leaves are unequally bilobed or strongly oblique at the apex, resembling those of the species in section *Cymbidium*, but are thinner, less rigid and lack a lignified layer of cells below the epidermis. In fact, *C. chloranthum* is somewhat intermediate in these respects between sections *Floribundum* and *Cymbidium*, but the flowers and scape are closer to the former. The flowers most strongly resemble those of *C. hartinahianum*, a closely related species that is rare and known only from a small area in Sumatra. The latter is a terrestrial, with narrower, acute

DAVID DU PUY

Fig. 31. *Cymbidium chloranthum*, Sabah, cult. Kew.

Plate 4. *Cymbidium chloranthum*. Curtis's Botanical Magazine t. 4907, del. W.H. Fitch.

leaves, petals and sepals of almost equal breadth, a narrower and shorter mid-lobe of the lip and a darker olive-green flower colour. *Cymbidium chloranthum* and *C. hartinahianum* both have distinctively gnarled and lumpy callus ridges and winged capsules.

Historically, some confusion has surrounded the distribution of *C. chloranthum*. Lindley (1843) described it from a plant cultivated at Loddiges' Nursery in Hackney which was said to have been originally collected in Nepal. William Hooker (1856) later figured this plant in *Curtis's Botanical Magazine*. Since then, no specimens have been collected in or near the Himalaya of northern India, and this provenance must be regarded as an error. Indeed the closest authentic locality appears to be in West Malaysia (Pahang, *Carr 290*).

Reichenbach (1856) described *C. variciferum* from cultivated material, but Joseph Hooker (1891) noted that it was conspecific with *C. chloranthum*, and suggested that it was Australian, a provenance that must also be regarded as erroneous. Rolfe (1919) also sank *C. variciferum* into *C. chloranthum* and gave the first confirmed locality for this species as Java, from a specimen imported and cultivated by Sander.

In the same publication, *C. sanguineolentum*, described by the Dutch botanists Johannnes Teijsmann & Simon Binnendijk in 1862 from material collected on Gunong Salak in Java, was also reduced to synonymy in *C. chloranthum*. The habit of *C. sanguineolentum* was compared with that of *Iridorchis gigantea* Bl. (= *C. iridioides* D. Don), and the large, ovoid, bilaterally flattened pseudobulbs of *C. chloranthum* do indeed resemble those of *C. iridioides*, but the leaves, scape and flowers differ markedly. Smith (1905) accepted *C. sanguineolentum* in his *Orchid Flora of Java*, but did not discuss how it differed from *C. chloranthum*, which his detailed description and later figure (Smith, 1911) closely match. Backer & Bakhuizen (1968) included *C. sanguineolentum* as a synonym of *C. chloranthum*, and their treatment is followed here. The name *C. sanguineum* is a misprint of *C. sanguineolentum*, a mistake made by Teijsmann & Binnendijk (1866).

Schlechter (1910) described *C. pulchellum* from Sarawak as a relative of *C. pubescens* (= *C. bicolor* Lindl.) because of their similarity in flower size and structure, but did not compare it with *C. chloranthum* or any of its later synonyms. The description closely resembles that of *C. chloranthum*. Although the type has been destroyed, his illustration of it, published in 1934, is unmistakenly of *C. chloranthum*. *Cymbidium pulchellum* is, therefore, included here in the synonymy of *C. chloranthum*.

1. Cymbidium chloranthum Lindl. in *Bot. Reg.* 29: 68 (1843). Type: cult. *Loddiges* (holotype K!).

C. variciferum Rchb.f. in *Bonplandia* 4: 324 (1856). Type: cult. *Booth* (holotype W).

C. sanguineolentum Teijsm. & Binnend. in *Tijdschr. Neder. Ind.* 24: 14–15 (1862). Type: Java, Gunong Salak, *Teijsmann & Binnendijk 902* (holotype BO!).

C. sanguineum Teijsm. & Binnend., *Cat. Hort. Bog.*: 51 (1866) [sphalm. pro *C. sanguineolentum* Teijsm. & Binnend.].

C. pulchellum Schltr. in *Fedde, Repert. Sp. Nov. Regni Veg.* 8: 570–571 (1910). Type: Borneo, Sarawak, Kuching, *Schlechter 15846* (holotype B).

A medium-sized, perennial, epiphytic *herb*. *Pseudobulbs* large, up to 11 × 4 cm; ovoid, strongly bilaterally flattened, covered by the persistent, sheathing leaf bases and surrounded by 3–5 cataphylls up to 10 cm long, the longest like the true leaves, with an abscission zone near the apex and a short lamina, becoming scarious and eventually fibrous with age. *Leaves* 5–6(7) per pseudobulb, distichous, up to 40–60 × 2.3–3.8 cm, ligulate to ligulate-elliptic, obtuse, unequally bilobed at the apex, with a small mucro in the sinus, coriaceous, but thinner and more flexible than those of species in section *Cymbidium*, narrowing and becoming conduplicate towards the base of the lamina, articulated close to the pseudobulb to a broadly sheathing base 2–7 cm long, with a 2–4 mm broad, membranous margin. *Scape* (30)36–47 cm long, erect, usually produced in the axil of the lowest leaf at the base of the pseudobulb; peduncle 15–22 cm long, covered in the basal half by about 6 overlapping, keeled, cymbiform, acute sheaths, up to 5.5 cm long, spreading apically and with cylindrical bases; rachis about 25–29 cm long, with (15)20–25 closely spread flowers; bracts short, about 1–2 mm long, triangular, acute. *Flowers* about 3 cm in diameter; not scented; rachis, pedicel and ovary bright green; sepals and petals pale yellow to yellow-green, with some red speckling at the base of the petals; lip pale yellow-green, the side-lobes mottled and barred with red, especially towards the margin, the mid-lobe sparsely spotted with red and with a broad white margin, callus ridges yellow; column yellow-green, lightly speckled with red; anther-cap yellow. The whole flower, but especially the lip, becomes strongly suffused with crimson after pollination. *Pedicel and ovary* long, slender, (19)25–38 mm long. *Dorsal sepal* 17–20 × 5.5–7 mm, oblong-elliptic to obovate, obtuse, erect; lateral sepals similar, slightly broader, up to 8 mm broad, spreading. *Petals* narrower than the sepals, 17–19 × 4–5 mm, narrowly oblong-elliptic, subacute, slightly oblique, weakly porrect but not closely covering the column. *Lip* 12–14.5 mm long, 3-lobed, minutely pubescent; side-lobes small, erect, with

Fig. 32. *Cymbidium chloranthum*, Sabah, cult. Kew.

porrect, subacute apices; mid-lobe (5.5)6.5–7.5 × (4)6–7.5 mm, broadly ovate with a broadly rounded apex, decurved, margin undulate; callus of 2 glabrous, crenulate ridges extending from the base of the lip into the base of the mid-lobe. *Column* 9–10 mm long, weakly winged towards the apex; pollinia 2, 1.2 mm across, ellipsoidal, deeply cleft, on a crescent-shaped viscidium extending into two, short, thread-like appendages at the tips. *Capsule* about 2.5 × 1.3 cm, ellipsoidal, strongly winged and acute-angled in transverse section. Pl. 4; Figs. 31–33 & 48: 2a–f.

DISTRIBUTION. W Malaysia, Sumatra, Java, Borneo (Sarawak and Sabah) and Palawan (Map 2); 250–1000 m (820–3280 ft). In Sabah it is found between 500–1000 m (1640–3280 ft) elevation, and flowers less well when cultivated at sea level (Du Puy & Lamb, 1984). In Java, Comber (1980) reported it growing at about 800 m (2625 ft) and Smith (1933) recorded it from 250 m (820 ft) in Sumatra (Map 2). Although no reference has been found to specimens collected in the rest of Borneo, it is likely that this species also occurs there. It appears to be uncommon throughout its range, although it may be locally more abundant.

HABITAT. It grows epiphytically in shade on trees in evergreen moist tropical forest, flowering sporadically throughout the year. It probably flowers throughout the year in the tropics at relatively low elevations where climatic seasonality is not pronounced.

CONSERVATION STATUS. VU A1c; B1b.

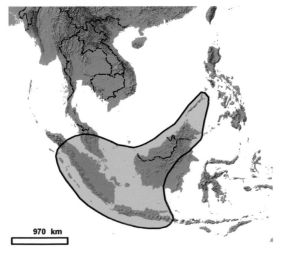

970 km

Map 2. Distribution of *C. chloranthum*.

Fig. 33. *Cymbidium chloranthum*, Sabah.

A. LAMB

2. CYMBIDIUM ELONGATUM

Cymbidium elongatum is perhaps the most unusual species in the genus, in being usually terrestrial and in having a monopodial habit, with indeterminately growing stems which tend to lean on to and scramble over the surrounding vegetation as they elongate. It was described in 1988 by Jeffrey Wood, David Du Puy and P.S. Shim from a plant collected on Mt. Kinabalu, Sabah by Sheila Collenette.

In its distinctive vegetative habit and its few-flowered scape with greenish flowers with purple backs to the sepals, *C. elongatum* is distinguished from all other species of *Cymbidium* in Malesia. The habit is unusual in *Cymbidium*, and is reminiscent of certain climbing species of *Dipodium*, such as *D. pictum* (Lindl.) Rchb.f. However, the flowers of *Dipodium* species differ markedly in having distinct dark spots on the reverse of the tepals, a hairy lip and bilobulate stipes on the viscidium. This latter character similarly prevents the inclusion of *C. elongatum* in *Grammatophyllum*.

The two pollinia and free lip are similar to those of other species in sect. *Floribundum*, but in its indeterminate growth habit and leaf shape it closely resembles the Australian *C. suave* in section *Austrocymbidium*. It is, however, easily distinguished from the latter by its shorter, coriaceous, more distant leaves, much fewer-flowered scape, larger flowers and the presence of two distinct callus ridges on the lip. We have tentatively placed it in section *Floribundum*, pending its inclusion in molecular systematic analyses, on account of its more or less erect inflorescence, relatively broad, oblong-elliptic sepals, the obscurely 3-lobed lip with a two-ridged, convergent callus, the broad wings at the apex of the column and the ellipsoidal pollinia.

2. Cymbidium elongatum J.J. Wood, Du Puy & P.S. Shim in Du Puy & P.J. Cribb, *The Genus Cymbidium*: 103 (1988). Type: Sabah, *Collenette A47* (holotype BM!, isotype K!).

A medium-sized, perennial, terrestrial or epiphytic *herb*. *Pseudobulbs* absent; stem elongated, 30–130 cm or more long, growing and flowering indeterminately, erect when young, becoming reclinate as it lengthens, the base of the shoot covered by the scarious or fibrous remains of the leaf bases, through which new roots emerge in the basal portion. *Leaves* 4–9, distichous, 10–19(37) × 1.2–2.3(3.5) cm, broadly linear, linear-elliptic or ligulate, conduplicate, curved, obtuse and mucronate to acuminate, oblique, cucullate at the apex, coriaceous, eventually deciduous, articulated to distichous, overlapping, sheathing, persistent bases (3.5)6–7 cm long, with scarious margins 1 mm broad. *Scape* 9–28 cm long, suberect, produced in the upper leaf axils; peduncle 8–23 cm

Fig. 34 (above). *Cymbidium elongatum*, Kinabalu Park. **Fig. 35** (below). Sabah, *Bailes & Cribb* 847. **Fig. 36** (opposite). Sabah, *Bailes & Cribb* 847.

P.J. CRIBB

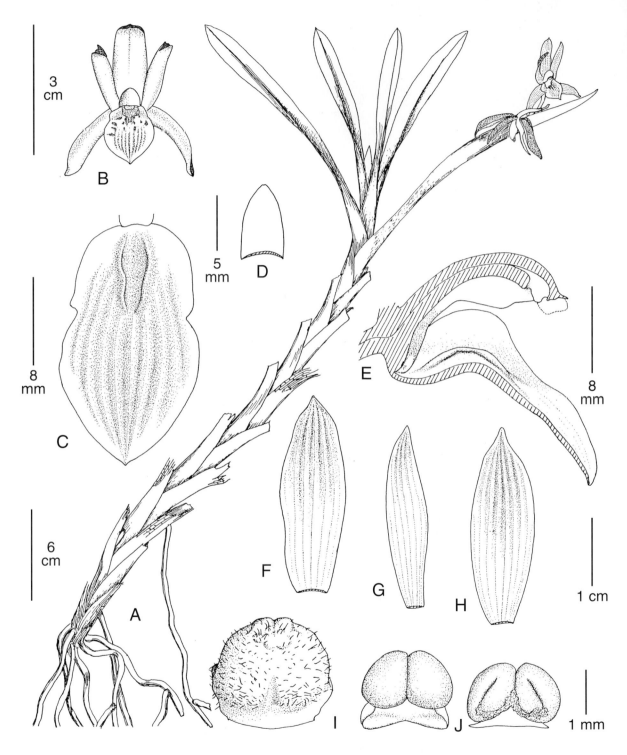

Fig. 37. *Cymbidium elongatum.* **A.** Habit. **B.** Flower. **C.** Lip. **D.** Bract and petal. **E.** Lip and column, longitudinal section. **F.** Dorsal sepal. **G.** Petal. **H.** Lateral sepal. **I.** Anther-cap. **J.** Pollinarium. All drawn from *Shim* s.n. by P.S. Shim.

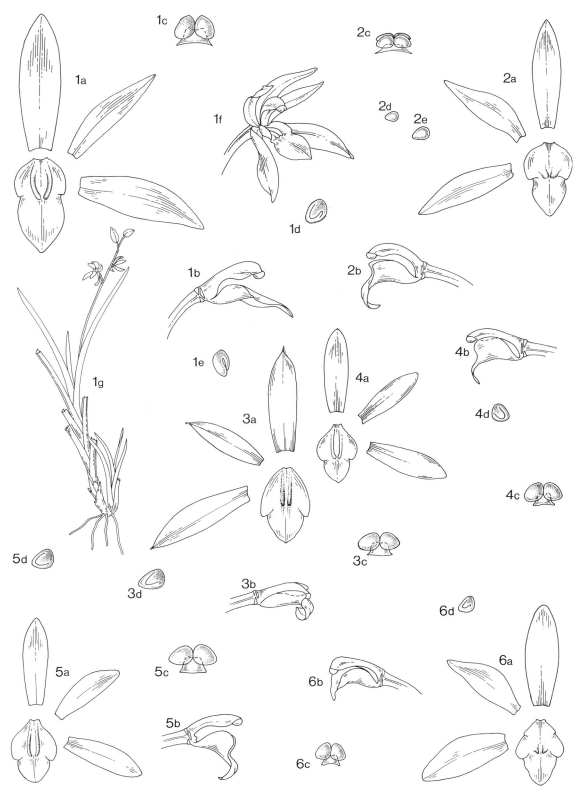

Fig. 38. *Cymbidium elongatum* (Kew spirit no. 47629/29462). **1a.** Perianth, x1. **1b.** Lip and column, ×1. **1c.** Pollinarium, ×3. **1d.** Pollinium (reverse), ×3. **1e.** Pollinium (side), ×3. **1f.** Flower, × 0.8. **1g.** Flowering plant, ×0.1. *C. borneense* (Kew spirit no. 48015). **2a.** Perianth, ×1. **2b.** Lip and column, ×1. **2c.** Pollinarium, ×3. **2d & e.** Pollinium (1 pair), ×3. *C. dayanum* (Kew spirit no. 47962). **3a.** Perianth, ×1. **3b.** Lip and column, ×1. **3c.** Pollinarium, ×3. **3d.** Pollinium (reverse), ×3. *C. floribundum* (Kew spirit no. 44942). **4a.** Perianth, ×1. **4b.** Lip and column, ×1. **4c.** Pollinarium, ×3. **4d.** Pollinium (reverse), ×3. *C. suavissimum* (Kew spirit no. 47448). **5a.** Perianth, × 1. **5b.** Lip and column, ×1. **5c.** Pollinarium, ×3. **5d.** Pollinium (reverse), ×3. *C. devonianum* (Kew spirit no. 47905). **6a.** Perianth, ×1. **6b.** Lip and column, ×1. **6c.** Pollinarium, ×3. **6d.** Pollinium (reverse), ×3. All drawn by Claire Smith.

Fig. 39. *Cymbidium elongatum*, cult. Mountain Garden, Kinabalu Park, Sabah.

long, suberect, covered in the basal portion by about 6 cymbiform, acute, overlapping sheaths up to 8 cm long; rachis 1–8 cm long, with 1–5 (perhaps more) distant flowers; bracts 4–7 mm long, ovate to triangular, acute, scarious. *Flowers* about 4 cm in diameter; scented; rhachis, pedicel and ovary apple-green; sepals and petals purplish-red outside, olive-green to cream and sometimes stained with red-brown inside; lip pale yellow-green or green to cream, sometimes suffused with pink, usually spotted and blotched with red; callus ridges yellow; column greenish-yellow to red-brown, flushed with purple at the base. *Pedicel and ovary* 2–4.3 cm long. *Dorsal sepal* 26–36 × 10–12 mm, oblong-elliptic, acute, erect; lateral sepals similar, oblique, spreading. *Petals* narrower than the sepals, 22–31 × 6–8 mm, narrowly elliptic, acute, porrect. *Lip* 20–24 × 11–13 mm, obscurely 3-lobed, minutely papillose; side-lobes erect, weakly differentiated; mid-lobe 12–14 × 10–12 mm, ovate, acute, porrect, entire; callus ridges 2, inflated towards the apex, convergent, from near the base of the lip to the base of the mid-lobe. *Column* 15–16 mm long, with a short foot, curved, strongly winged towards the apex; pollinia 2, 1.8 mm long, triangular-ellipsoidal, deeply cleft, on a broadly triangular viscidium extending into two short, thread-like appendages at the lower corners. Figs. 34–37, 38: 1a–g, 39.

DISTRIBUTION. Sabah, Sarawak (Map 3); 1200–2300 m (3940–7545 ft).

HABITAT. Terrestrial in seepages in open woodland; flowering during September–October, but also in March and June so this is perhaps not strictly seasonal. On Mt. Kinabalu in Sabah, it grows in open,

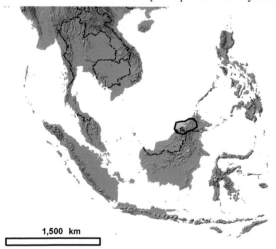

scrubby *Leptospermum* woodland of stunted trees at between 1200 and 1750 m in slurry of ultramafic shale through which water percolates all the time at the base of *Leptospermum* or amongst rattans, sedges, Ericaceae and *Begonia* on sandstone or ultrabasic serpentine rock. It is commonly associated with *Nepenthes* species, including *N. burbidgei*, *N. edwardsiana* and *N. rajah*. It has also been found growing epiphytically at higher elevations in eastern Sarawak. On Mts Murud and Mulu in north-eastern Sarawak it grows in upper montane, mossy forest from 2000 to 2300 m elevation (Beaman *et al.*, 2001).

CONSERVATION STATUS. VU B1a,b.

1,500 km

Map 3. Distribution of *C. elongatum*.

3. CYMBIDIUM FLORIBUNDUM

Cymbidium floribundum was described by Lindley (1833) in a footnote appended to his description of *C. sinense*. The short Latin diagnosis was made from a painting by a Chinese artist, now in the possession of the Royal Horticultural Society. Lindley distinguished *C. floribundum* from *C. sinense* by its smaller, more numerous flowers, its obtuse sepals and the red lip with a yellow centre. Although its leaf is not unlike *C. sinense*, the two cleft pollinia, rather than four found in section *Jensoa*, indicate that it is not closely related. It has recently gained importance as one of the parents of the modern miniature hybrids.

Cymbidium floribundum is allied to *C. suavissimum*, but the latter can be easily distinguished by its broader leaves and the small auricles at the base of the column. A more detailed discussion can be found under *C. suavissimum*.

The name *C. floribundum* was unfortunately ignored by later authors and the species became well known in cultivation under the later synonym *C. pumilum*, which was described by Rolfe in 1907, based on a specimen imported from Japan and cultivated by Barr, and a second collected by the French missionary Monbeig in Yunnan, China. This latter specimen was selected by Du Puy and Cribb (1988) as the lectotype. The description of *C. pumilum* is far more detailed than that of *C. floribundum*, but

Plate 5. *Cymbidium floribundum*. China, *Andrew* s.n., cult. Kew, del. Claire Smith.

93

Fig. 40. *Cymbidium floribundum*, China, Guizhou, on limestone.

there can be little doubt that they are the same species and that *C. pumilum* is therefore a later synonym. The strongly red-coloured lip in the painting of *C. floribundum* is unusual in that the species usually has a spotted lip, but the lip turns deep red, obscuring the spots, when the rostellum is disturbed following pollination, or when the flower ages.

Cymbidium illiberale was described in 1914, by Hayata, from a plant cultivated in Taiwan. He differentiated it from *C. pumilum* (= *C. floribundum*) by the 'light reddish green petals and sepals, and by the lips which are light red with red maculatum blotches on the front lobe, and numerous minute red spots on the side-lobes'. These colour patterns are typical of *C. floribundum*, and it appears that the two taxa are not distinct. Specimens from Taiwan cannot be distinguished from those collected on mainland China (Lin, 1977).

A variant lacking the red-brown pigment in the flowers, which are consequently green with a white lip, is commonly cultivated as var. *album*, and has been named f. *virescens* by Makino (1912). The lip of this colour variant also turns red on pollination, or as the flower ages. Specimens with pale pink markings on the lip are also known (Fig. 41).

The Japanese name 'Kinryohen' means 'Golden-margined' (Nagano, 1955). This probably refers to the narrow yellowish margin on the sepals and petals, but may also refer to the highly prized specimens with variegated leaves cultivated in Japan, although the Japanese name for the latter is 'Jitsugetsu'. There are many named variants maintained in cultivation in Japan and China, where *C. floribundum* has been grown for several centuries. Long *et al.* (2003) redescribed this alba variant as *C. chawalongense* based on a collection from south-east Xizang (Tibet). Their diagnosis contrasts it with *C. floribundum*, distinguishing it on its 22-cm long scape and albino flowers. Both fall within the range of variation of *C. floribundum*. Wu & Chen (1980) recognised two varieties differentiated by the shape of the side-lobes of the lip: var. *floribundum* with broad, rounded side-lobes, and var. *pumilum* with narrower, more acute side-lobes. This difference is not apparent in the specimens we have seen, but further study may uphold this distinction.

3. Cymbidium floribundum Lindl., *Gen. Sp. Orchid. Pl.*: 162 (1833). Lectotype: Icon. in R. Hort. Soc. London.

C. pumilum Rolfe in *Kew Bull.*: 130 (1907). Type: Yunnan, Tsekou, *Monbeig* (lectotype K!, isolectotypes P, K!).

C. illiberale Hayata in *Icon. Pl. Formosa* 4: 78 (1914). Type: Taiwan (Formosa), cult. Taihoku, *B. Hayata* (holotype TI).

C. pumilum Rolfe f. *virescens* Makino in *Iinuma, Somoku-Dzusetsu* 18: 1185 (1912).

C. floribundum Lindl. var. *pumilum* (Rolfe) Y.S. Wu & S.C. Chen in *Acta Phytotax. Sin.* 18: 301 (1980).

C. chawalongense C.L. Long, H. Li & Z.L. Dao in *Novon* 13: 203 (2003). Type: China: Xizang, Chayu Xian, Chawalong, *Gaoligongshan Exped. 13727* (holotype KUN).

A perennial, lithophytic or epiphytic, often clump-forming *herb*. *Pseudobulbs* small, up to 3.3 cm long, 2.2 cm in diameter, ovoid, slightly bilaterally compressed, covered by persistent sheathing leaf bases with a 1 mm wide scarious margin, and surrounded by about five cataphylls that become scarious, and eventually fibrous with age. *Leaves* 5–6, up to 20–55 × 0.8–1.5(2) cm, the shortest merging with the cataphylls, linear-elliptic, arching, acute, the apex usually oblique, articulated 1.5–6 cm from the pseudobulb. *Scape* usually 15–25(40) cm long, robust, suberect, arching upwards from the base of the pseudobulb, with (6)15–30(45) closely spaced flowers; peduncle about 10(–15) cm long, covered in 6–8 sheaths; sheaths up to 6 cm long, becoming scarious, cylindrical in the basal half, expanded and cymbiform in the upper half, acute; bracts short, 2–6(17) mm long, triangular, acute. *Flowers* 3–4 cm across, red-brown or occasionally green, not scented; rhachis apple-green; pedicel and ovary often stained with red-brown; sepals and petals strongly flushed red-brown, with a narrow, yellow or green margin; lip white, mottled purple-red on the side-lobes and blotched with purple-red on the mid-lobe (occasionally with pink markings instead), yellow at the base, becoming bright red on pollination or

Fig. 41. *Cymbidium floribundum*, pale-flowered form. Cult. Kew.

95

Fig. 42. *Cymbidium floribundum*, China, cult. Kew.

with age; callus ridges bright yellow; column yellow-green, flushed red-brown above, pale green below, dark red-purple at the base; anther-cap yellow. *Pedicel and ovary* 1.5–3.3 cm long. *Dorsal sepal* suberect, 1.8–2.1 × 0.5–0.8 cm, oblong-elliptic, obtuse, margins recurved; lateral sepals similar, spreading or porrect. *Petals* 1.6–2 × 0.5–0.8 cm, elliptic, obtuse to subacute, curved, weakly spreading. *Lip* about 1.8 cm long, 3-lobed; side-lobes up to 7 mm broad, erect, rounded or obtuse, minutely papillose; mid-lobe 0.7 × 0.8 cm, broadly ovate, subacute, weakly recurved, minutely papillose, the margin sometimes undulating; callus of two parallel ridges that tend to converge at their apices, with a shallow channel between them. *Column* 1.2–1.5 cm long, curved, broadening into two narrow wings near the apex, minutely papillose below; pollinia 2, triangular, cleft, on a small rectangular viscidium. *Capsule* about 3 cm long, cylindrical, tapering at each end, apex with a short beak. Pl. 5; Figs. 38: 4a–d, 40–42.

DISTRIBUTION. Southern China, Taiwan, northern Vietnam (Map 4); 800–2800 m (2625–9185 ft). It is not native to Japan, although it may now be found naturalised in the warmer regions (Nagano, 1955). In Taiwan and China it grows in the southern provinces, Fujian, Guangxi, Guangdong, Guizhou, Hubei, Jiangxi, Sichuan, Yunnan, Xizang and Zhejiang (Wu & Chen, 1980). Plants from Taiwan and eastern China usually have broader leaves (1.5–2 cm, 0.6–0.8 in) and flower earlier, in March and April. It is a montane plant, usually found between 1000 and 2800 m (3280 and 9185 ft) elevation, although it has been recorded from as low as 800 m (2625 ft) in Taiwan (Lin, 1977).

HABITAT. On rocks and trees in shaded gorges and ridge-tops, often in coniferous or mixed forest, in light shade or in full sun. It is normally lithophytic in mountain gorges, in shaded situations on hard

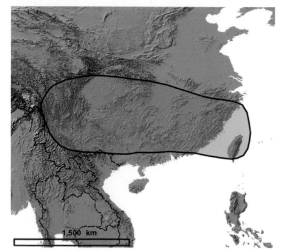

limestone, often in conifer forest, although it may grow epiphytically, and has been observed in *Liquidambar formosana* (Hamamelidaceae) in eastern China. One of us (PC) saw flowering specimens in April 2001, growing on lightly shaded limestone ridgetops under *Podocarpus* in karst country in south-east Guizhou. He also saw it growing epiphytically in monsoon forest on karst limestone at *c.* 1200 m in northern Vietnam close to the Chinese border. It has a high tolerance of dry conditions, growing in open situations, such as on cliffs, where it can form extensive clumps. It flowers between March and June.

CONSERVATION STATUS. VU A1d; A2.

Map 4. Distribution of *C. floribundum*.

4. CYMBIDIUM HARTINAHIANUM

Cymbidium hartinahianum was described by Comber & Nasution in 1977 from a collection from the western side of Lake Toba in northern Sumatra, on a plateau at about 2000 m (6560 ft), 1100 m (3610 ft) higher than the lake itself.

It is closely related to *C. chloranthum*, both having pseudobulbs of similar size and shape, erect scapes, and numerous small, slender-stalked, greenish flowers with similar broad, rounded floral segments and crenulate callus ridges. This last, and their winged capsules, are characters found only in this species and in *C. chloranthum*. However, *C. hartinahianum* has narrower, usually shorter, more coriaceous leaves that are V-shaped in section and lack the bilobed apex usually found in *C. chloranthum*. It also has a longer scape with a much longer, erect peduncle only partially covered by the distant sheaths. Its flowers are slightly larger than those of *C. chloranthum*, and are darker in colour, with a much darker column. The petals and sepals are almost equal in size, whereas in *C. chloranthum* the petals are narrower than the sepals. In *C. hartinahianum* the lip has broader, rounded side-lobes that are not angled at their apices, and a smaller, truncated mid-lobe, which has an undulating margin at the base only. Comber & Nasution (1978) and Comber (2001) illustrated flowers of both these species.

The leaves of *C. hartinahianum* resemble those of *C. canaliculatum* but are much narrower. The similarity in vegetative morphology between these two species would indicate a close relationship between them, but that is not supported by molecular data (Yukawa *et al.*, 2002).

4. Cymbidium hartinahianum J.B. Comber & Nasution in *Bull. Kebun Raya* 3: 1–3, + fig. (1977) & in *Orchid Dig.* 42: 55–57, + fig. (1978). Type: Sumatra, Sidikalang, *Nasution & Bukit 6* (holotype BO!).

A medium-sized, perennial, terrestrial *herb. Pseudobulbs* conspicuous, up to 7 × 3.5 cm, ovoid, bilaterally compressed, covered by persistent, sheathing leaf bases and 3–4 cataphylls up to 9 cm long which become scarious and fibrous with age. *Leaves* 7–10 per pseudobulb, distichous, 13–30(60) × 0.9–1.5 cm, ligulate, acute, with a hooded, oblique apex, coriaceous, V-shaped in section, articulated 2–6 cm from the pseudobulb to a broadly sheathing base. *Scape* 50–80(100) cm long, erect, produced in the axil of the cataphyll which is ruptured by the developing scape; peduncle 35–60 cm long, with 8–9 keeled, distant, cymbiform, loose sheaths up to 6 cm long, which are inflated and lack a cylindrical basal region; rhachis about 15–30 cm long, with (9)14–21 flowers; bracts 1–8 mm long, triangular, acute, scarious. *Flowers* about 3.5 cm in diameter; not scented; rhachis, pedicel and ovary green; petals and sepals olive-green to purple-brown, with some brownish staining towards the base; lip white, the side-lobes barred with red, and the mid-lobe sparsely blotched with red; callus ridges pale yellow; column dark purple-brown; anther-cap pale yellow. *Pedicel and ovary* slender, 2–3.5 cm long. *Dorsal sepal* 15–20 × (4)6–8 mm, oblong-elliptic to obovate, obtuse, erect; lateral sepals similar, spreading. *Petals* almost as broad as the sepals,

Fig. 43. *Cymbidium hartinahianum*, Sumatra.

J. COMBER

97

14–18 × 6–8 mm, oblong-elliptic, oblique, spreading or weakly porrect. *Lip* about 21 mm long, 3-lobed, minutely pubescent; side-lobes erect, rounded, with rounded apices; mid-lobe 7 × 8–10 mm, oblong, with a broadly rounded or weakly mucronate apex, porrect or slightly decurved, margin undulating at the base; callus in 2 glabrous, somewhat crenulate ridges, extending from the base of the lip into the base of the mid-lobe, convergent towards their apices. *Column* 11–12 mm long, narrowly winged; pollinia 2, ellipsoidal, deeply cleft, on a crescent-shaped viscidium. *Capsule* about 3.2–4 × 1.8–2.5 cm, ellipsoidal, strongly winged and acute-angled in transverse section, stalked, pendulous, with an 11 mm long apical beak formed by the persistent column. Figs. 43–44, 48: 3a–e.

DISTRIBUTION. Northern Sumatra (Map 5); 1700–2700 m (5580–8860 ft).

HABITAT. It grows on a plateau that has a vegetation of broken forest and rough, damp grassland, known locally as 'blang', on poor soil dominated by the ubiquitous tropical grass *Imperata cylindrica*, with tufts of *Themeda villosa* and patches of a coarse, bracken-like fern, *Gleichenia*. These are species that will tolerate poor soils. *Cymbidium hartinahianum* grows in this grassland as a terrestrial. Although this habitat is unusual for a *Cymbidium* species, the environment resembles that in which many epiphytic species survive. The soil is so poor that it cannot support vigorous vegetation, or a wide diversity of species, allowing the orchid to grow in damp conditions among small ferns and mosses, in good light, and without competition from other strongly growing plants. Comber (2001) reported further collections from the grasslands of the Gunong Leuser range at elevations up to 2600 m and provided a photograph of a plant growing on Gunong Bandahara at 2700 m. It flowers between February and June.

CONSERVATION STATUS. VU B1a,b.

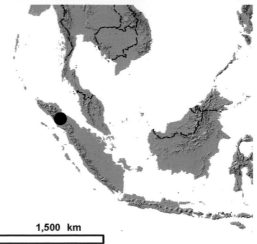

Fig. 44 (top). *Cymbidium hartinahianum* growing in its natural habitat on Gunong Bandahara, Aceh, North Sumatra, 2700 m.

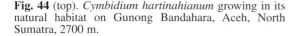

1,500 km

Map 5. Distribution of *C. hartinahianum*.

5. CYMBIDIUM SUAVISSIMUM

Cymbidium suavissimum was first described in the *Gardeners' Chronicle* by its editor C.H. Curtis (1928), after it had been exhibited at the Royal Horticultural Society by Messrs Sander in August of that year. The name refers to the sweet-smelling flowers (Latin: *suaveolens*). No type specimen was preserved, so the figure given with the original description is taken as the type. A specimen at Kew (spirit collection no. 12858), taken from a plant bought from Sander in 1933, may be from the type plant.

This species has seldom been collected. Until recently, it was thought that all plants in cultivation derived from the original introduction from Myanmar (Burma). It is not known for certain where these plants originated, but Ghose (1960) stated that Kohn collected the plants near Bhamo in northern Burma close to the border with Yunnan, China, and shipped them to him in Darjeeling. Sander probably obtained plants from this source. Ghose was expecting a consignment of *C. tracyanum* and grew the plants under conditions similar to those of *C. tracyanum* and *C. lowianum* that he had acquired from the same region. Unfortunately, they were less hardy, and many plants were lost during the winter. Therefore, it seems probable that *C. suavissimum* was collected in the valleys at lower elevations than *C. tracyanum* or *C. lowianum*. More recently, plants have been collected in north-western Vietnam by Averyanov and Hiep (pers. comm.).

Vegetatively, the plant appears close to the larger-flowered species in section *Cyperorchis*, but the flowers are different, lacking the fusion at the base of the lip and column, and are much smaller. The

Fig. 45 (left). *Cymbidium suavissimum*, Cult. Kew, *Seth 95*. **Fig. 46** (right). *Cymbidium suavissimum*, Vietnam, Vinh Phu.

flower is almost identical to *C. floribundum*, albeit slightly larger, and the scape has a similar structure, but is longer, more robust, and has more flowers. Table 7 gives the characters by which these two species can be separated. These characters are almost all concerned with the larger size and stronger growth of *C. suavissimum*, but the species are similar in the manner of growth.

Cymbidium suavissimum appears to be distinct, but the possibility of a hybrid origin involving *C. floribundum* must be considered, especially as it appears to have such a restricted distribution. The leaves and pseudobulbs strongly resemble those found in the sympatric species in section *Cyperorchis*: *C. lowianum*, *C. tracyanum* and *C. iridioides*. Either of the last two is a possible candidate as one of the putative parents, as both of these have scented flowers and the red-brown colour. The flowering time of *C. suavissimum* is intermediate between those of the possible parents. However, its flowers and inflorescences strongly resemble those of *C. floribundum* and it is difficult to imagine a hybrid flower that is so similar to one parent and almost unaffected by the other. Nevertheless, the small auricles at the base of the column in *C. suavissimum* are unique in the genus, and may be the rudimentary expression of the lip and column fusion found in *C. tracyanum* and *C. iridioides*. Furthermore, *C.* Pumilow, the primary hybrid between *C. floribundum* and *C. lowianum* (closely related to *C. iridioides* and *C. tracyanum*), is similar in many respects to *C. suavissimum*.

TABLE 7

A COMPARISON OF THE CHARACTERS USED TO DISTINGUISH
BETWEEN *C. FLORIBUNDUM* AND *C. SUAVISSIMUM*

Character	*C. suavissimum*	*C. floribundum*
Pseudobulb size	up to 6.0 × 3.0 cm	up to 3.3 × 2.0 cm
Cataphyll colour	purple	green
Leaf length	up to 70 cm	up to 50 cm
Leaf width	3.0–3.8 cm	1.5(2.0) cm
Scape length	about 50 cm	15–25(40) cm
Number of flowers	about 50	15–30(40)
Ovary length	3.0–4.3 cm	1.5–3.3 cm
Scent	fruitily scented	unscented
Dorsal sepal length	2.0–2.5 cm	1.8–2.1 cm
Petal length	1.8–2.2 cm	1.6–2.0 cm
Auricle at the column base	present	absent
Flowering period	July–August	(March)April–June

5. Cymbidium suavissimum Sander ex C. Curtis in *Gard. Chron.* 84: 137, 157, fig. 67 (1928). Lectotype: Icon. in *Gard. Chron.*, *loc. cit.*: fig. 67 (1928), selected by Du Puy & Cribb (1988).

A large perennial *herb*. *Pseudobulbs* about 6 cm long, 3 cm in diameter, ovoid, lightly bilaterally compressed, covered by persistent, sheathing leaf bases with a membranous margin up to 3 mm wide, and surrounded by about 5 purple cataphylls which become scarious and eventually fibrous with age. *Leaves* 5–7, up to 70 × 3–3.8 cm, the shortest merging with the cataphylls, linear-elliptic, arching, acute, the apex usually oblique; articulated 2–6 cm from the pseudobulb. *Scape* usually about 50 cm long, robust, suberect, arching upwards from the base of the pseudobulb, with about 50 closely spaced flowers; peduncle about 15 cm long, covered by about 7 sheaths; sheaths up to 7 cm long, inflated, cymbiform, acute. *Flower* about 3.5 cm across, red-brown; sweetly fruit-scented; sepals and petals green or yellow at the margins, strongly flushed red-brown in the centre; lip white, mottled purple-red on the side-lobes and blotched with purple-red on the mid-lobe, yellow at the base, turning bright red on pollination, callus ridges bright yellow; column yellow-green, speckled red-brown above, pale green below, with a dark purple-red blotch at the base; anther-cap yellow. *Pedicel and ovary* 3–4.3 cm long. *Dorsal sepal* suberect, 2–2.5 × 0.6–0.7 cm, oblong-elliptic, obtuse, margins recurved; lateral sepals similar, spreading. *Petals* 1.8–2 × 0.6–0.7 cm, elliptic, obtuse to subacute, somewhat spreading or weakly porrect and almost covering the column. *Lip* about 1.6 cm long, 3-lobed; side-lobes 6 mm broad, erect, rounded, apex obtusely angled, minutely hairy; mid-lobe 0.6 × 0.7 cm, broadly elliptic, rounded, minutely acuminate, recurved, minutely papillose, sometimes with an undulating margin; callus of 2 parallel ridges, which tend to converge at their apices and are not strongly raised but have a shallow channel between them. *Column* 1.3–1.4 cm long, curved, broadening into 2 narrow wings near the apex, and with 2 small (1–2 mm), incurved auricles on the margins at the base, minutely papillose; pollinia 2, triangular, cleft, on a small rectangular viscidium. Figs. 38: 5a–d, 45–47.

DISTRIBUTION. Northern Myanmar (Burma) and north-west Vietnam (Map 6); about 800–1000 m (2625–3280 ft).

HABITAT. In Vietnam it has reportedly been found in lower montane evergreen forest, growing lithophytically on moss-covered hard limestone rocks where leaf litter accumulates and the soil is rich in humus. Its roots find a precarious purchase and eventually form a basket that catches leaf litter from the shading forest to fertilise the orchid as it grows. Plants can develop to a considerable size. The pH of the friable soil around the roots ranges from 7.50–7.86 (Averyanov *et al.*, 2003). It flowers in July and August.

CONSERVATION STATUS. EN A1d; A3, B1a,b; E.

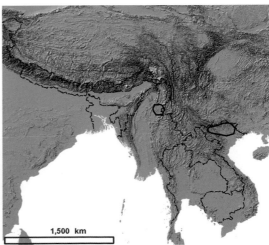

Fig. 47. *Cymbidium suavissimum*, Cult. Kew, *Seth* 95.

Map 6. Distribution of *C. suavissimum*.

SECTION AUSTROCYMBIDIUM

Section **Austrocymbidium** Schltr. in *Fedde, Repert. Sp. Nov. Regni Veg.* 20: 104 (1924). Type: *C. canaliculatum* R. Br., lectotype chosen by P.F. Hunt (1970).

Schlechter (1924) established this section for the Australian species of *Cymbidium*, distinguished by their flowers that have green or yellowish-green sepals and petals and a lip often marked with red and usually bearing a shiny depressed callus. He noted that some species had strongly inflated pseudobulbs, whereas *C. suave* had a shortly elongated stem. Seth & Cribb (1984) and Du Puy and Cribb (1988) extended the limits of this section to include the three Malaysian species *C. chloranthum*, *C. elongatum* and *C. hartinahianum*, and further characterised the section by its broad, often obovate sepals and petals. In this treatment, following the phylogeny established using molecular data (see Chapter 11), we return to Schlechter's circumscription of the section.

This section is difficult to characterise. The species usually have leaves articulated close to the pseudobulbs and many densely crowded flowers on a pendent scape. The flower colour is yellow or greenish, except for some variants of *C. canaliculatum* that are greenish but are heavily marked with dark red-brown. The sepals are characteristically broad, obovate to oblong-elliptic and rounded or obtuse at the apex and are usually much broader than the petals. The lip has small, narrow side-lobes. The column is short, and the two pollinia are ellipsoidal rather than triangular, and parallel to each other (Fig.48: 1a–d).

The species in this section can be divided into two groups on the basis of their vegetative habit. *Cymbidium canaliculatum* and *C. madidum* have large, ovoid, bilaterally flattened pseudobulbs that are produced annually. Conversely, *C. suave* lacks pseudobulbs and has instead a slender stem that grows indeterminately, giving it an almost monopodial habit. This division appears to be artificial, however, when the flowers are examined. The flowers of *C. suave* and *C. madidum* are similar in colour and shape, and they both have a highly unusual, single, glistening, viscid, ligulate depression on the lip in place of the usual paired callus ridges. This latter character is so distinctive that it might also be used to subdivide the section. The unusual flower shape, the hairy callus ridges, the thick, rigid leaves and other extreme xerophytic adaptations of *C. canaliculatum* could also be used to separate it in a section of its own. The most satisfactory treatment, however, seems to be to maintain this group as a single section. Its origins are discussed in Chapter 10.

6. CYMBIDIUM CANALICULATUM

Cymbidium canaliculatum was described by Robert Brown in 1810 based on a specimen that he collected at Broad Sound, Queensland, during Matthew Flinders' circumnavigation of Australia on which he was the ship's botanist. It is vegetatively distinctive and can form large clumps, each new pseudobulb often producing more than one scape with 20–50 or more densely crowded flowers each. A single large plant may carry several hundred flowers.

It can be distinguished from the other Australian species by its highly characteristic leathery leaves, its densely crowded scapes, and by the presence of two, distinct, parallel callus ridges on the lip. It can be further distinguished from *C. suave* by its inflated pseudobulbs.

The size of the flowers and the length of the bracts are variable. The sepal, petal and lip apices vary from obtuse to subacute to acute, and the lip apex may even be shortly acuminate. The breadth of the side-lobes of the lip and the length of the column are also extremely variable. This variation appears to be continuous, and is significant even between different flowers on the same scape. Flower colour is also variable from greenish, variously marked with red-brown, to almost black. The variation in flower colour is so great that it has led some authors to separate the more distinctive extremes at specific or varietal level. A discussion of Rupp's (1934) treatment serves to illustrate the variation involved. Rupp divided this species into three varieties, based mainly on the colour of the sepals and petals: the typical

Plate 6. *Cymbidium canaliculatum*. Curtis's Botanical Magazine t. 5851, del. W.H. Fitch.

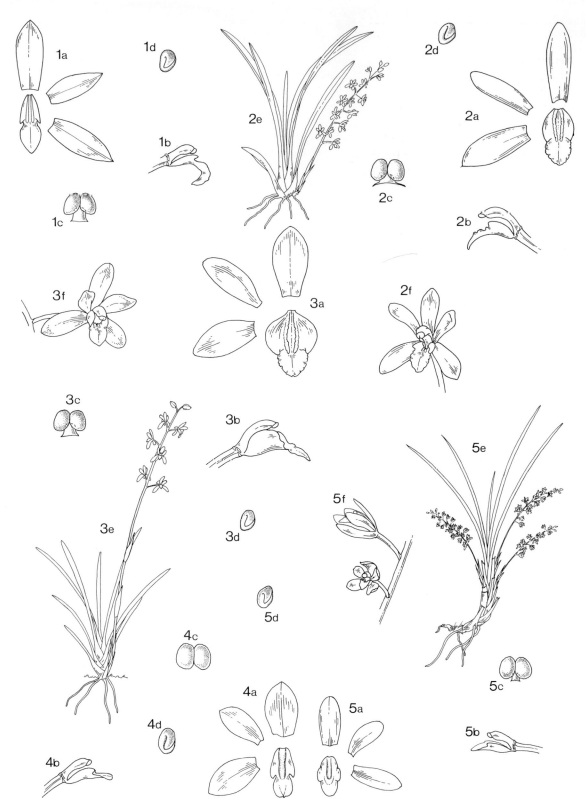

Fig. 48. *Cymbidium canaliculatum* (Kew spirit no. 47247). **1a.** Perianth, × 1. **1b.** Lip and column, × 1. **1c.** Pollinarium, × 3. **1d.** Pollinium (reverse), × 3. *C. chloranthum* (Kew spirit no. 47268). **2a.** Perianth, × 1. **2b.** Lip and column, × 1. **2c.** Pollinarium, × 3. **2d.** Pollinium (reverse), × 3. **2e.** Flowering plant, × 0.1. **2f.** Flower, × 0.8. *C. hartinahianum* (Kew spirit no. 39722). **3a.** Perianth, × 1. **3b.** Lip and column, × 1. **3c.** Pollinarium, × 3. **3d.** Pollinium (reverse), × 3. **3e.** Flowering plant, × 0.1. *C. madidum* (Kew spirit no. 45606). **4a.** Perianth, × 1. **4b.** Lip and column, × 1. **4c.** Pollinarium, × 3. **4d.** Pollinium (reverse), × 3. *C. suave* (Kew spirit no. 8818). **5a.** Perianth, × 1. **5b.** Lip and column, × 1. **5c.** Pollinarium, × 3. **5d.** Pollinium (reverse), × 3. **5e.** Flowering plant, × 0.1. **5f.** Flowers, × 0.8. All drawn by Claire Smith.

variety (var. *canaliculatum*), var. *marginatum* and var. *sparkesii*. He also recognised two forms of each of the first two of these. His var. *canaliculatum* has flowers with red-brown spots and blotches on the sepals and petals, over a dull green to bright yellow-green background colour. He recognised f. *inconstans*, which included most of the different shades of colour within this variety, and f. *aureolum,* which included the brightest yellow-green variants. Variety *canaliculatum* is widespread in New South Wales and southern Queensland, from the coast well inland over the Great Dividing Range, and perhaps into the Northern Territory and the north of Western Australia, with f. *aureolum* more or less confined to the western plains of New South Wales.

Variety *marginatum* included all specimens with more-or-less uniformly brown or reddish flowers, the sepals and petals usually with a pale green margin. The two forms of this variety that Rupp recognised were f. *fuscum*, brown in colour, and f. *purpurascens,* which was described by Rupp as 'bright or deep magenta'. He gave the distribution of this variety as northern Queensland and Cape York, and included the variant illustrated by Hooker (1870) in *Curtis's Botanical Magazine* (Plate 6). Rupp depicted these varieties and forms, and also included an intermediate with blotched sepals and uniformly coloured sepals. We have seen a specimen with the opposite colour combination of blotched petals and uniformly coloured petals. In reality, there seems to be a complete intergradation between these taxa, and Rupp himself states that the taxa are 'more or less connected by intermediates'.

A striking extreme of the colour variation described in var. *marginatum* was recognised by Rupp as var. *sparkesii*, which had previously been described as *C. sparkesii* by Rendle (1898, 1901), and later reduced to varietal status within *C. canaliculatum* by Bailey (1913). Rendle distinguished this variant by 'the deep, dark crimson colour of the flower, which in reflected light appears almost black', and by its slightly longer perianth segments (sepals 20 × 5 mm (0.8 × 0.2 in), petals 16 × 4.5 mm (0.6 × 0.18 in)). Bailey noted that he could not distinguish it from *C. canaliculatum*, except by the dark red flower colour, and certainly the dimensions given by Rendle are no different from those commonly found in *C. canaliculatum*. The lip of var. *sparkesii* is usually suffused with pink on the margins and green on the disc, and Rupp suggested that the lip is usually smaller than normal, although this variation is continuous. Variety *sparkesii* occurs in tropical Queensland, north of Townsville. The colour remains its only distinguishing character, but the darker specimens included by Rupp in his var. *marginatum* f. *purpurascens* approach var. *sparkesii*, and are also found in northern Queensland. Dockrill (1966, 1969), recognising the continuous nature of this variation, sank all of these taxa into *C. canaliculatum*, and his treatment is followed here.

Rupp (1934) also included a note concerning several reports of an alba variant of *C. canaliculatum*, and indicated that the species also has a tendency to produce abnormal floral variants such as peloric and compound flowers where, for example, the backs of two lateral sepals from otherwise separate flowers are fused.

In 1942, Nicholls described var. *barrettii* from the Northern Territory, distinguished by its pale greenish-yellow colour but lacking any brown markings on the sepals and petals, and occasionally with some basal red spotting at the back of the sepals. This, then, represents the opposite extreme of colour variation to the dark red variants recognised by Rupp as var. *sparkesii*.

Mark Clements (pers. comm.) has reported specimens with exceptionally large flowers from the Northern Territory (Arnhem Land) and also perhaps from the Atherton Tableland. These, and other large-flowered variants, should be investigated cytologically, and might prove to be natural polyploids. Herraman (2005) shows a range of colour variants, including a striking albino form in which the sepals and petals are pale yellow-green and the lip pure white.

Ferdinand von Mueller described *C. hillii* in 1879 from material cultivated in Brisbane, but originally collected by Hill from *Eucalyptus* trees in the coastal forests of the Mulgrave Range in northern Queensland, probably near Cairns. Mueller differentiated this from *C. canaliculatum* in several characters, including its thinner, less rigid, flat (not canaliculate) leaves with 3 prominent nerves below, fewer-flowered inflorescences, longer bracts, shorter pedicel and ovary, longer and more acute, oblong-lanceolate sepals and petals, glabrous lip with weakly defined side-lobes, and long, semi-

Fig. 49. *Cymbidium canaliculatum.* Queensland, Coen, growing on *Eucalyptus* sp.

Fig. 50 (left). *Cymbidium canaliculatum*, Cult. Kew. **Fig. 51** (right). *Cymbidium canaliculatum* 'Sparkesii', Queensland.

lanceolate, acuminate mid-lobe that is nearly three times as long as broad and much larger than the rest of the lip. Bailey (1902) repeated some of these characters, and maintained it as a separate species. Rupp (1937) claimed that it had been re-collected by Barrett on the Daly River in the Northern Territory. He described this new collection as being similar to *C. hillii* in its leaf with three prominent nerves, its longer perianth segments (sepals 21 × 4.5 mm (0.8 × 0.18 in), petals 19 × 4 mm (0.7 × 0.16 in) and its long lip, which agreed well with the type description (lip 16.5 × 5 mm (0.7 × 0.2 in), mid-lobe 10 × 4 mm (0.4 × 0.16 in), acuminate). He added a further character, that of the capsule, which he described as 5.5 cm (2.2 in) long and 1.5 cm (0.6 in) broad at its broadest point, 2 cm (0.8 in) from the apex. Although narrow, this does not seem to differ greatly from typical *C. canaliculatum*. However, the leaf characters certainly appear to be distinct, and some of the lip characters such as the length/breadth ratio of the mid-lobe of the lip appear to be outside of the wide range of variation known in *C. canaliculatum*. However, in most of its characters it agrees well with *C. canaliculatum*, and Dockrill (1969) considered it to be a synonym of that species. Further research may show this to be a separate taxon, or the result of hybridisation with *C. madidum* or *C. suave*.

6. Cymbidium canaliculatum R. Br., *Prodr. Fl. Nov. Holl.*: 331 (1810). Type: Australia, *R. Brown 5303* (holotype BM!, isotype K!).

C. hillii F. Muell. in Regel, *Gartenflora* 28: 138–9 (1879). Type: Australia, N Queensland, Mulgrave Mts., *Hill*, cult. Brisbane Botanic Gardens (holotype MEL).

C. sparkesii Rendle in *J. Bot.* 36: 221 (1898) & 39: 197 (1901). Type: Australia, NE Queensland, *Jones*, cult. *Sparkes* (holotype BM!, isotype K!).

C. canaliculatum R. Br. var. *sparkesii* (Rendle) F.M. Bailey, *Compr. Cat. Queensland Pl.*: 845 (1913).

C. canaliculatum R. Br. var. *canaliculatum* f. *inconstans* Rupp in *J. Linn. Soc. New South Wales* 59: 98–9 (1934).

C. canaliculatum R. Br. var. *canaliculatum* f. *aureolum* Rupp, *loc. cit.* 59: 99 (1934).

C. canaliculatum R. Br. var. *marginatum* Rupp, *loc. cit.* 59: 99 (1934).

C. canaliculatum R. Br. var. *marginatum* Rupp f. *fuscum* Rupp, *loc. cit.* 59: 99 (1934).

C. canaliculatum R. Br. var. *marginatum* Rupp f. *purpurascens* Rupp, *loc. cit.* 59: 99 (1934).

C. canaliculatum R. Br. var. *barrettii* Nicholls in *Australian Orchid Rev.* 7: 40 (1942). Type: Australia, Northern Territory, Groote Eylandt, *Barrett* (holotype MEL).

A medium-sized, perennial, epiphytic *herb*. *Pseudobulbs* large, up to 12 × 3.5 cm, elongate-ellipsoidal, bilaterally compressed, covered by persistent, sheathing leaf bases and 3–4 cataphylls up to about 8 cm long, the longest of which have an abscission zone near the apex and a short lamina, soon becoming scarious and eventually fibrous. *Leaves* 3–4 per pseudobulb, distichous 15–46(65) × 1.5–3(4) cm, narrowly elliptic, tapering to an acute, cucullate apex, coriaceous and stiff, canaliculate, glaucous, narrowing and becoming conduplicate towards the base, articulated close to the pseudobulb to a broadly sheathing base 2–10 cm long, with a 2–4 mm broad, membranous margin. *Scape* (15)25–55 cm long, suberect to horizontal, arching, usually from the axils of the upper cataphylls, often more than one per pseudobulb; peduncle 12–24 cm long, with 5–8 distant, closely sheathing, scarious, often purple-tinted sheaths up to 3.5 cm long, rhachis (3)13–35 cm long, with (13)20–60 closely spaced flowers; bracts 2–4(8) mm long, triangular, acute to acuminate. *Flowers* 1.8–4 cm across, variable in colour; not scented; rhachis, pedicel, ovary and bracts usually stained purple-brown; sepals and petals greenish or brown to almost black on the outer surface, dull to golden-yellow or green within, usually spotted and blotched or uniformly coloured with red-brown to deep magenta, usually with a narrow greenish margin, or occasionally uniformly deep reddish-black; lip white or cream, sometimes tinged with green or pink, lightly spotted with red or purple; callus ridges cream to pale green; column yellowish-green or green, often blotched with red-brown, to uniformly dark maroon, paler at the base, white streaked maroon below, anther-cap cream or green to dark red-brown or maroon. *Pedicel and ovary* slender, 2.3–3.5 cm long. *Dorsal sepal* 11–25 × 4.5–6(10) mm, narrowly oblong-elliptic, obtuse to subacute, often mucronate, erect; lateral sepals similar, slightly broader, oblique, spreading. *Petals* shorter and narrower than the sepals, 10–20 × 4.5–5.5(8) mm, narrowly elliptic, obtuse to acute, often mucronate, oblique, porrect or spreading. *Lip* 9–15 × 4–10 mm when flattened, 3-lobed, minutely pubescent; side-lobes well-defined, with acute to

M. CLEMENTS

Fig. 52. *Cymbidium canaliculatum*, Queensland.

obtuse, porrect apices; mid-lobe 5–7(9) × 5–7.5(10) mm, broadly elliptic to ovate, with a rounded, obtuse to shortly acuminate apex, decurved, margin entire or weakly undulating; callus of two parallel, well-defined, slightly pubescent ridges, from the base of the lip to the base of the mid-lobe. *Column* 6–13 mm long, winged towards the apex; pollinia 2, about 1 mm long, ellipsoidal, deeply cleft; viscidium quadrangular, about 0.7 mm across, extending into two short, thread-like appendages at the lower corners. *Capsule* 4–6 × 1.5–2.5 cm, fusiform-ellipsoidal, tapering at the base to a short pedicel, and at the apex into a short beak formed by the persistent column. Pl. 6; Figs. 48: 1a–d, 49–52.

DISTRIBUTION. Australia, from northern Western Australia across to Cape York in Queensland, and New South Wales (Map 7); from sea level to about 1000 m (3280 ft). The most southerly localities in the distribution of *C. canaliculatum* are in New South Wales: the Hunter River near the coast, Forbes, Ardleton and Hay. It is most commonly found on the drier western slopes of the Great Dividing Range, and the western plains, and is uncommon on the coast, which has higher rainfall. Its distribution extends northwards through Queensland to the tip of Cape York, and although it is found in the coastal forest it does not occur in the rainforest. The capacity of this species to withstand drought is reflected in its wide range over the north of Australia into Arnhem Land and the Kimberleys, to Roebuck Bay and Derby in the north of Western Australia, where it occurs in tropical savannah and scrubland, but again is uncommon on the coast. Indeed, Mueller (1879) noted that this is often the only orchid found over much of this arid area, such as near Sturt's Creek and the Victoria River, where the annual rainfall is less than 50 cm (20 inches).

Its drought tolerance is partly due to its thick, leathery leaves, which are resistant to desiccation, and the shape of these leaves, which are erect, hooded at the apex and strongly V-shaped in section, and channel any available water directly to the base of the plant. Furthermore, Winter *et al.* (1983) have shown that the leaves are also xerophytically adapted in their physiology, using crassulacean acid metabolism (CAM), which allows photosynthesis without the usual debilitating loss of water.

HABITAT. Epiphytic on *Eucalyptus* and *Melaleuca* trees, growing in rotting wood in hollow trees and in hollows formed by fallen branches, often in dry areas, usually in partial shade.

Cymbidium canaliculatum is almost always found growing in rotting wood in hollows in tree trunks and branches of *Eucalyptus* and *Melaleuca* species that often have a decaying central core in the trunk (Fig. 49). In this situation, the surrounding wood of the tree helps to maintain the moisture in the substrate into which the orchid forms an extensive root system, up to 12 m (39 ft) or more in length. Clements (pers. comm.) has successfully used a wood-rotting fungus in the symbiotic germination of seeds of this orchid. It is often found around creek beds, but this is probably due to the environmental preferences of the trees on which it grows, rather than as a consequence of any direct benefit to the orchid. It prefers some shade, but usually occurs in open woodland and will withstand exposure to strong, direct sunlight. When conditions are dry, such as in the plains of New South Wales where the annual rainfall is less than 50 cm (20 in), it can tolerate summer temperatures in excess of 35°C (95°F) and freezing winter night temperatures.

It flowers from September until December.

CONSERVATION STATUS. NT.

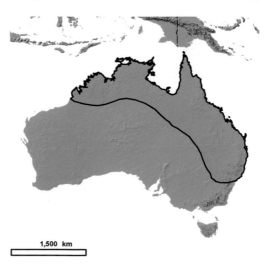

Map 7. Distribution of *C. canaliculatum*.

1,500 km

7. CYMBIDIUM MADIDUM

Alan Cunningham described this fine orchid, the last of the three endemic Australian species of *Cymbidium* to be discovered, in Edwards' *Botanical Register* as *C. iridifolium* in 1839 based on his own collection from the Brisbane River in Queensland. Unfortunately, the name had already been used for a distinct Himalayan species by William Roxburgh (1814, 1832). Thus, John Lindley's *Cymbidium madidum*, described the following year from a specimen cultivated by Messrs Rollinsons that was mistakenly said to have originated in the East Indies, has priority.

Cymbidium madidum may attain greater dimensions in the wild than those given here in the description, which have been taken from cultivated specimens. The pseudobulbs are as large as any in the genus. The leaves are articulated close to the pseudobulb, so that when the leaf eventually falls, the pseudobulb does not have the prominent tips of the leaf bases projecting beyond it. This character appears to be typical of section *Austrocymbidium*. The tip of the leaf base is, however, sharply spiny when the lamina falls. It has a pendulous scape which is long and has been reported to reach over 120 cm (47 in) (Leaney, 1966), with numerous, well-spaced, yellowish or greenish flowers.

Fig. 53. *Cymbidium madidum.* Cult. Kew.

The colour variant most commonly found in the wild has olive-green flowers, with brownish staining on the backs of the sepals and a paler yellow lip, but selected clear-coloured variants are commonly found in cultivation. Those are apple-green to bright yellow in colour. A yellow-green variant is also found occasionally which lacks the typical dark red pigmentation on the disc of the lip (M. Clements, pers. comm.). Rupp (1937) stated that there is an unusual variant that is clear, pale green in colour with a yellowish lip, found in the region of Cairns. This has straighter, more rigid racemes, more densely spaced flowers and narrower, more widely spreading floral segments. He also states that 'the two are so nearly identical that it would be unwise to separate them', and that opinion is followed here.

The callus ridges normally found in most *Cymbidium* species are absent in *C. madidum* and in *C. suave*, and are replaced by a tongue-shaped, shiny, secretory depression that extends from the base of the lip into the base of the mid-lobe. This most unusual and easily observed character immediately separates them from all other species of *Cymbidium*. *Cymbidium suave*, however, is easily differentiated as it has smaller leaves with an obtuse, strongly oblique apex, and no pseudobulb, the stem

Fig. 54. *Cymbidium madidum.* Australia, ex *Easton* s.n., Cult. Kew.

growing in an indeterminate fashion, and flowering for several seasons on the same growth. The stem eventually becomes long and has fresh leaves only towards the apex, the lower part being covered by the fibrous remains of the leaf bases. The scape is much shorter, usually only about 15–20 cm (6–8 in) long, and is arching rather than pendulous, and the flowers are more densely spaced. The flowers are similar in these species, but the mid-lobe of the lip of *C. suave* is ovate and obtuse, not obovate and truncated, and the flowers are smaller than those of *C. madidum*.

Cymbidium madidum has had an unusually convoluted taxonomic history besides the discounting of Cunningham's name *C. iridifolium*, and it was not until 1961 that the name was correctly applied. Previously, the name commonly used for this species was *C. albuciflorum*, a name given by Mueller (1859) and used by both Bentham (1873) and Bailey (1902). Rupp (1930) also used this name, but in 1934 he resurrected the name *C. iridifolium*. Rolfe had previously noted, as early as 1889 on the type sheet of *C. madidum*, that it was the same species as *C. albuciflorum*, which was therefore a later synonym. Rupp (1934) concluded that because *C. iridifolium* was published before *C. albuciflorum* and *C. madidum*, it should be given priority and accepted as the correct name for this species (Macpherson & Rupp, 1936; Rupp, 1937, 1943). Hawkes (1961) rectified his mistake in 1961 and correctly applied the name *C. madidum*, more than 120 years after its original publication.

Three further synonyms have been included in *C. madidum* by Dockrill (1966). In 1898 Rendle described *C. leai*, which he thought was close to *C. canaliculatum*. He distinguished it by its broader, blunter sepals, its 10–11 mm (0.4 in) long lip, with a 5-mm-long mid-lobe which was almost as broad at the tip, and had a broad, tongue-like depression on the disc that passed into the base of the mid-lobe between shallow, lateral crests, and had a button-like tubercle at the base. This unusual callus structure is similar to that of *C. madidum* and *C. suave*, except for the lateral crests and the tubercle, which are not present in either of these species. The rest of the description and the type specimen agree with *C. madidum*, except that the leaf apex was described as obtuse, suggesting *C. suave*. Rupp (1937) was sent scapes of a plant that appeared to fit the description of *C. leai*. He described the colour as 'brownish, with darker blotches on the perianth', and the lip 'appeared to agree precisely with *C. leai*'. During the following two years, further scapes were sent from the same plant, but the flowers were identical to those of *C. madidum*, and none had the peculiar lip characters of *C. leai*. The flower colour, with blotching on the sepals and petals, and the presence of some rudimentary callus ridges on the lip indicate that *C. leai* might have been a hybrid of *C. madidum* or *C. suave* with *C. canaliculatum*. However, Dockrill (1966) included *C. leai* in the synonymy of *C. madidum*, and his treatment is followed here. Tierney (1957) stated that '*C. leai* has not been recorded since its discovery in the Mulgrave Ranges when mining was active at Goldsborough. Sugar-growing has since destroyed much of [the natural vegetation of] this range, and may have destroyed this species.'

Similarly, little is known about *C. queeneanum*, which Klinge described in 1899 without any locality or other collection data. There does not appear to be any outstanding difference between this and some of the darker-flowered variants of *C. madidum* with olive-green and brownish flowers. We have therefore followed Dockrill (1966) and included it as a synonym of *C. madidum*.

Dockrill (1966, 1969) recognised *C. leroyi* St Cloud (1955) as distinct at varietal rank. Menninger (1961) reduced it to varietal rank but did not explain this move. It was described by Dockrill as having a lip with a pointed, cymbiform mid-lobe and a flowering time during November and December, slightly later than would be expected in northern Queensland where it was collected, but not outside of the normal flowering period in more southerly localities. The unusual lip can be understood as a failure of the lip to expand fully when the flower opened, possibly due to environmental factors. Indeed, when specimens have been removed from the wild and are cultivated, they appear to be indistinguishable from the typical variety. This variety is not, therefore, accepted as distinct. Dockrill gave its distribution as 'apparently confined to the area between the Barron River and the Endeavour River', in Queensland, but further field work into this variant is required to establish its geographical and morphological range.

Cymbidium madidum prefers a more humid habitat than the other Australian species. It is found in eastern Australia from near the Clarence River in northern New South Wales, northwards probably to the tip of Cape York, growing in tropical rainforest or in adjacent more open forest with high rainfall. It occurs from sea level up to about 1300 m (4265 ft), preferring the coastal plains and the eastern slopes of the coastal range that have high rainfall. It is an epiphyte, often reported growing in the bases of staghorn or elkhorn ferns (*Platycerium* spp.). It also grows commonly on rotting trees or in hollows in branches or trunks of trees such as *Eucalyptus* or *Melaleuca*, usually in conjunction with rotting wood, and can eventually form huge clumps. It occasionally grows on the rough, fibrous covering on the trunks of palm trees, and has been observed growing as a terrestrial on roadside cuttings on the Atherton Plateau, and near Cape Tribulation in northern Queensland (M. Clements, pers. comm.). It usually occurs in positions where it has full sun for at least part of the day, but will tolerate even quite heavy shade.

It usually flowers from August to October in the more northerly, tropical regions, and until December in its more southerly localities. The unusually late (October to December) flowering of a population found between the Barron and Bloomfield/Endeavour Rivers in northern Queensland has already been discussed (see var. *leroyi*).

7. Cymbidium madidum Lindl. in *Bot. Reg.* 26: misc. 9–10 (1840). Type: cult. *Rollinsons* (holotype K!).

C. iridifolium Cunn. in *Bot. Reg.* 25: misc. 34 (1839); *non* Roxb., *Hort. Bengal.*: 63 (1814) & *Fl. Indica* 3: 458 (1832) = *Oberonia iridifolia*; *non* Sw. ex Steud. *Nom.*, ed. 2, 1: 460 (1840) = *Oncidium iridifolium*. Type: Australia, Brisbane R., *Cunningham* (holotype BM!).
C. albuciflorum F. Muell., *Fragm. Phyt. Austr.* 1: 188 (1859). Type: Australia, Queensland, Moreton Bay, *Hill s.n.* (holotype MEL!).
C. leai Rendle in *J. Bot.* 36: 221–222 (1898). Type: Australia, Queensland, Sonata, *Lea* (holotype BM!).
C. queeneanum Klinge in *Acta Hort. Petrop.* 17: 137–138, t. 2 (1899). Type: Australia, *Persich* (holotype LE).

C. leroyi St. Cloud in *North Queensland Nat.* 24 (112): 3–5, + fig. (1955). Type: Australia, North
  Queensland, Emmagen Creek, *Le Roy*, cult. Cairns (holotype QRS).
C. madidum var. *leroyi* (St. Cloud) Menninger in *Amer. Orchid Soc. Bull.* 30: 870–871 (1961).

A medium-sized to large, perennial, epiphytic *herb. Pseudobulbs* large and conspicuous, up to 10–16 ×
4–6 cm, ellipsoidal, bilaterally flattened, covered by persistent, sheathing leaf bases, and by 4–5
cataphylls up to 10 cm long, the longest like true leaves, with an abscission zone near the apex, and a
short lamina, becoming scarious and eventually fibrous with age. *Leaves* 6–8(9), distichous, up to
50–80(90) × 2.3–3.4(4) cm, linear-elliptic, acute to cuspidate, sometimes oblique at the apex, erect,
flexible, mid-green, channelled at the base, articulated close to the pseudobulb to a broadly sheathing,
spiny base 1.5–9.0 cm long, with a membranous margin up to 4 mm broad, the base persistent and with
sharp apical spines. *Scape* (30)40–80 cm long, pendulous, produced at the base of the pseudobulb in the
axil of the upper cataphylls, which are often ruptured to allow the scape to develop; peduncle usually
11–25 cm long, horizontal to pendulous, covered in the basal half by about 6–7 overlapping, keeled,
cymbiform, acute sheaths up to 6–7 cm long, which are spreading apically with a short cylindrical base,
the peduncle exposed in the apical half, but with some sterile bracts up to 1 cm long; rhachis (20)30–56
cm or more long, with (17)22–60 flowers; bracts 1–4 mm long, triangular, acute. *Flowers* 2.6–2.8 cm
across, although sometimes not opening fully; sweetly scented; rhachis, pedicel and ovary purple-
brown, occasionally apple-green; sepals and petals straw-yellow stained with pale brown, especially on
the backs of the sepals, producing an olive-green effect inside, to clear yellow or yellow-green; lip
primrose-yellow with a broad deep yellow to red-brown stripe from the base of the lip into the base of
the mid-lobe, bordered with brownish-red and with two deep red-brown blotches at the base of the mid-
lobe; column clear yellow or greenish, paler at the base; anther-cap pale yellow. *Pedicel and ovary*
(10)16–33 mm long. *Dorsal sepal* 11–17 × (4)6–9 mm, elliptic to obovate, rounded to obtuse, weakly
mucronate, erect; lateral sepals similar, spreading. *Petals* narrower than the sepals, 10.5–15 × (4)5–7
mm, elliptic, obtuse, weakly mucronate, oblique, usually porrect, usually almost parallel with the
column, spreading at the tips. *Lip* 10.5–13.5 × 5–6 mm when flattened, 3-lobed, minutely pubescent;
side-lobes small, erect, almost as long as the column, apex acute, porrect; mid-lobe 5.5–7 × (3.6)4–5.5
mm, narrowly obovate, truncate, mucronate, porrect, margin entire; callus a shallow, viscid, secretory
depression extending from the base of the lip well on to the base of the mid-lobe. *Column* 7.5–8 mm
long, winged towards the apex; pollinia 2, ellipsoidal, cleft, on a crescent-shaped viscidium about 0.7
mm across, extending into two hair-like processes at the lower corners. *Capsule* about 4 cm long,
sphaerical-ellipsoidal, with an apical beak. Figs. 48: 4a–d, 53–54.

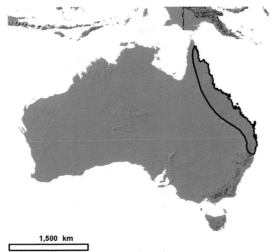

1,500 km

Map 8. Distribution of *C. madidum.*

DISTRIBUTION. Eastern Australia, from
northern Cape York to northern New South Wales
(Map 8); from sea level to 1300 m (4265 ft).

HABITAT. In the damper, tropical forests on the
eastern side of the coastal ranges in areas of high
rainfall, often in clumps of the epiphytic staghorn
or elkhorn fern (*Platycerium* spp.), or in rotting
wood in hollows of trees and branches; flowering
August to October in tropical regions, and until
December in the south. Flowering about March in
cultivation in north-temperate regions of the world.

CONSERVATION STATUS. NT.

8. CYMBIDIUM SUAVE

Robert Brown described *Cymbidium suave* in 1810 based on his own collection from near Sydney (Port Jackson) in New South Wales. It has an unusual growth habit that is distinct from any other species in section *Austrocymbidium*. Instead of producing a new pseudobulb annually, the same stem continues to grow and flower indeterminately for many years, eventually reaching 50 cm (20 in) or more in length.

The scape is usually about 15–25 cm (6–10 in) long, with many closely spaced flowers. Rupp (1937) suggested that plants from northern Queensland have shorter scapes, averaging only about 13 cm (5 in) in length, whereas those from southern New South Wales may reach 30 cm (12 in) or more. Usually the flower colour varies between bright green and yellow-green, often tinged with brown on the backs of the sepals. The colour of the flowers of plants from northern Queensland is paler, but Rupp (1937) noted that he had seen a plant with dull green flowers with indistinct red blotches at Illawarra, New South Wales.

The shape of the lip is variable. Brown (1810) described the lip as undivided. Bentham (1873) and Bailey (1902) described the variation in lip shape as undivided to sinuately 3-lobed. Rupp (1937) extended this even more to include a variant from coastal New South Wales with a lip as strongly 3-lobed as any of the other Australian species. Although it is much less common and widespread than the entire or obscurely lobed variants, he was able to illustrate a complete range of variation from entire to strongly lobed lips.

Dockrill (1966, 1969) cited *C. gomphocarpum*, described by Fitzgerald in 1883, as the only synonym of *C. suave*. No locality information was given in the original description, and until Rupp

discovered an unpublished plate drawn by Fitzgerald in the Herbarium of the Royal Botanic Gardens, Sydney, little information was available about the species. The description and plate, subsequently published by Rupp (1939), are both undoubtedly of *C. suave*, differing from the typical species only in minor details. Fitzgerald stated that it differed from *C. suave* in having a club-shaped or almost terete capsule, but his plate shows that it is

Fig. 55 (left). *Cymbidium suave*. New South Wales, cult. Munich B.G. **Fig. 56** (right). New South Wales, Nowra.

Plate 7. *Cymbidium suave*. Australia, New South Wales. Del. George Bond for Alan Cunningham, 1823. Kew collection.

simply immature. Rupp stated that the lip also differed from *C. suave*, and that the scape was erect rather than arching. The plate does show a suberect scape, but it appears to be supported by the leaves, and older scapes are shown as arching. The lip is shown as strongly 3-lobed, and the mid-lobe appears to be truncated. The side-lobes and the base of the mid-lobe are finely dark-spotted. Although less common than plants with an entire or weakly lobed lip, specimens with a strongly 3-lobed lip are not outside of the range of variation in *C. suave* (see Rupp, 1937). Furthermore, the lip has not been flattened, and the scape has flowers with a lip shape similar to *C. suave*. The coloured markings on the lip therefore appear to be the only unusual character. This seems an insufficient basis on which to recognise it as a distinct taxon. The spotting may, perhaps, be indicative of hybridisation with *C. canaliculatum*, a possibility reinforced by the rare

Fig. 57. *Cymbidium suave*. New South Wales, cult. Sydney B.G.

colour variants of *C. suave* discussed previously (see Rupp, 1937) which are green with faint red-brown blotches on the sepals and petals that could also be explained by natural hybridisation. Tierney (1957) claimed that *C. gomphocarpum* had been re-collected at Harvey's Creek on the Atherton Tableland, but this record has not been confirmed. Therefore *C. gomphocarpum* is treated here as a synonym of *C. suave*.

Leaney (1966) included an interesting note on the vegetative growth of this species. He stated that *C. suave* can send rhizomes considerable distances inside the rotting trees in which it grows. These eventually reappear farther down the trunk from the main plant, and form a new young plant. He noted that careful splitting of the trunk has shown connections between apparently separate plants on the same tree, and suggested that the distance between connected plants may be as much as 2.5 m (8.2 ft).

8. Cymbidium suave R.Br.*, Prodr. Fl. Novae Holl.*: 331 (1810). Type: Australia, *R. Brown s.n.* (holotype BM!).

C. gomphocarpum Fitzg. in *J. Bot.* 21: 203–4 (1883). Type: Australia, *Fitzgerald s.n.* (holotype BM!).

A medium-sized, perennial, epiphytic *herb. Pseudobulbs* not strongly inflated, apparent only in young plants, developing in older specimens into an elongated stem up to about 50 cm long, which is covered by the fibrous remains of the sheathing leaf bases. Each shoot will grow and flower for many years before a new growth is produced near the base, the shoot extending only a few centimetres in length each year. Cataphylls present only in young plants. *Leaves* 6–11, carried apically on the shoot, distichous, up to 30–60 × 0.8–1.4 cm, ligulate, tapering, obtuse to subacute, oblique at the apex, thin and grass-like in texture, arching, eventually deciduous, articulated to a sheathing base 4–7 cm long with a narrow membranous margin. *Scape* usually 15–24(35) cm long, arching or pendulous, produced in the axils of the leaf bases just below the current leaves, often more than one per stem and often persistent on the stem for several years; peduncle 5–10 cm long, suberect at the base, arching above, covered in the basal half by 7–8 keeled, cymbiform, acute, spreading sheaths up to 3.5 cm long, exposed in the apical half, with a few small, sterile bracts; rhachis about 10–20 cm long with (10)20–40(50) closely spaced flowers; bracts 1–2 mm long, triangular, subacute. *Flowers* 1.5–2.5 cm across; strongly sweet-scented; rhachis often stained purplish, pedicel and ovary green; sepals and petals light green, sometimes yellow-green or olive-green, occasionally with pale reddish blotches; lip bright yellow to greenish; side-lobes stained orange-brown, or rarely green; disc dark red-brown in front, often paler behind; column pale yellow-green. *Pedicel and ovary* (8)12–18 mm long. *Dorsal sepal* 12–15 × 5–6 mm, oblong-elliptic to obovate, obtuse, mucronate, erect or porrect; lateral sepals similar, often slightly broader than the dorsal sepal. *Petals* shorter than the sepals, 10.5–13 × 4–5.5 mm, oblong-elliptic, slightly oblique, weakly porrect to almost parallel with the column. *Lip* 9.5–11 × 5–6 mm, almost entire to strongly 3-lobed, minutely papillose; side-lobes often weakly differentiated, erect, with rounded to acute apices; mid-lobe about 4 × 4 mm, ovate, obtuse to subacute, porrect, of thicker texture than the rest of the lip, margin entire; callus a raised, ligulate ridge that is glabrous and shiny, with a shallow depression in front which extends into the base of the mid-lobe. *Column* 6–8 mm long, winged towards the apex; pollinia 2, 1.3 mm long, ellipsoidal, deeply cleft, on a minute (0.5 mm), broadly crescent-shaped viscidium extending into 2, thread-like appendages at the tips. *Capsule* about 3 × 2 cm, sphaerical-ellipsoidal, on a 10–18 mm pedicel and with a 6-mm-long apical beak formed by the persistent column. Pl. 7, Figs. 48: 5a–f, 55–57.

DISTRIBUTION. Eastern Australia, from southern New South Wales to northern Queensland (Map 9).

This has the most restricted distribution of the Australian *Cymbidium* species. It is found from southern New South Wales, 14 km (9 miles) from the border with Victoria, northwards along the east coast to southern Cape York. It has the most southerly distribution of all the *Cymbidium* species, at about 36°S, almost equivalent to the most northerly latitudes at which the genus occurs, as represented by *C. goeringii* in Japan and Korea. *Cymbidium suave* is found from sea level up to about 1200 m (3940 ft), but does not occur on the drier western side of the Great Dividing Range.

This species prefers a drier habitat than *C. madidum*, and only occasionally occurs in cleared areas in the rainforest. Neither will it tolerate the dry conditions in which *C. canaliculatum* thrives, although it is more cold-tolerant than either of these species. It will also survive if its growths are burned in bush fires, perhaps because its roots are well protected.

HABITAT. In damp, open woodland, usually near the coast. *Cymbidium suave* is found in two main types of habitat, both of which provide some moisture and protection for the roots. In damp, open hardwood forest, particularly on *Eucalyptus*, it grows in hollows in branches and trunks of trees, forming a huge root system that may penetrate deep into the damp, rotting core of the tree, the roots sometimes extending 10 m (33 ft) from the plant. It is also found on tree stumps and even on fence posts made from *Eucalyptus*, if they have started to rot.

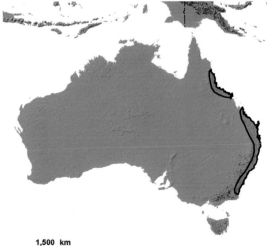

Its other common habitat is on *Melaleuca* trees that grow along water-courses or on swamp margins. The bark of this tree peels off like sheets of paper, and the roots of the orchid penetrate beneath the top layers, eventually forming an extensive network around the tree. When plants are taken from the wild into cultivation, so much of this root system is lost that the plants will often die. Rupp (1937) also reported this species growing on tree fern trunks.

It requires an open position where it receives strong sunlight. It flowers during August to October in northern Australia, and until December or January in the southerly temperate part of its range.

1,500 km

Map 9. Distribution of *C. suave*.

CONSERVATION STATUS. NT.

SECTION CYMBIDIUM

Section **Cymbidium** P.F. Hunt in *Kew Bull.* 24: 94 (1970). Type: *C. aloifolium* (L.) Sw.
Cymbidium section *Eucymbidium* Lindl., *Gen. Sp. Orch. Pl.*: 162 (1833).

The five species in this section are distinguished by their thick, often rigid, ligulate leaves with obtuse to emarginate, bilobed apices, and pendulous to arching or rarely suberect scapes with small, well-spaced flowers. The flowers are cream or greenish, with red or brownish markings. The leaf anatomy and leaf surface morphology are distinct and characteristic for this section (see Chapter 4). The mesophyll contains elongated, palisade-like cells in its upper layers, and the subepidermal fibre bundles are linked together by a complete subepidermal layer of lignified sclerenchyma (Fig. 5A). The characters of the epidermis are shared by section *Borneense*, both sections having elliptical stomatal coverings and narrow, slit-shaped apertures (Fig. 8).

9. CYMBIDIUM ALOIFOLIUM

Cymbidium aloifolium was the first species of the genus known in Europe and is the type of the genus. Linnaeus described it as *Epidendrum aloifolium* in his *Species Plantarum* in 1753, based on an illustration by Rheede of a plant found at the end of the 17th century in south-western peninsular India, growing on trees of *Strychnos nux-vomica*. It was published in 1703 in his *Hortus Malabaricus*. Olof Swartz transferred it to the genus *Cymbidium* in 1799.

Roxburgh described *Epidendrum pendulum* in 1795, from a plant collected in south-east India, and the illustration he published in his *Plants of the Coast of Coromandel* now serves as the type. It was also transferred to the genus *Cymbidium* by Swartz in 1799. Lindley (1833), King & Pantling (1898), Duthie (1906) and Rolfe (1917) upheld them as separate species, each of them providing different distinguishing characters and including different synonyms. However, Hooker (1891) considered them conspecific. From an examination of the type material and the original descriptions, it must be concluded that *C. pendulum* is conspecific with *C. aloifolium*, although the former name has often been wrongly used for specimens of *C. bicolor* from northern India.

Cymbidium erectum was described by Wight in 1851 based upon another specimen from peninsular India. His description indicated that the leaves were ligulate, and deeply and obliquely emarginate at the apex. The side-lobes of the lip were acute and, although the description states that the callus ridges were straight, the illustration shows them as interrupted and similar to those of *C. aloifolium*. Its main distinguishing feature was the erect scape. The description and drawing were probably prepared from a preserved specimen because all of the flowers are inverted, indicating that the scape was pendulous in life, as in *C. aloifolium*.

In 1917, Rolfe made a study of the *C. aloifolium* / *C. bicolor* complex, and recognised five separate species. One of those he described as a new species, *C. simulans*, distinguished by its broader leaves and striped lip with interrupted, curved callus ridges. These characteristics are typical of *C. aloifolium*, with which it is now considered to be conspecific. No type specimen was cited, but Seth (1982) selected a specimen collected by Pantling as the lectotype, since Rolfe included a reference to King & Pantling's work in his description. Recently, this name has been applied to pale-coloured variants of *C. aloifolium* (Menninger, 1961), but these do not warrant specific status.

Jones (1974) described *C. intermedium*, a plant with straight, entire callus ridges, reminiscent of those of *C. finlaysonianum*. Otherwise the plant does not differ from *C. aloifolium*. Although unusual, the shape of the callus ridges is variable in *C. aloifolium,* and such variants are occasionally found. This variation is mirrored in the closely related *C. bicolor*, in which almost straight callus ridges are commonly found in subsp. *obtusum*.

Although *C. aloifolium* has a wide distribution from India and southern China to the Malay Archipelago, it is much less variable than the closely related *C. bicolor*, and the variation is more-or-less continuous. Leaf breadth is variable, but except in the Malay Islands and in Indo-China where the leaves are narrower, the broadest leaves on the plant usually exceed 2.5 cm (1 in) and, in the extreme case of the specimens from Yunnan in south-west China, may reach 6.3 cm (2.5 in). Flower size is also variable, usually being 3–4.2 cm (1.2–1.6 in) in diameter, but some specimens may have flowers up to 5.5 cm (2.2 in) in diameter.

Cymbidium aloifolium can be most easily distinguished from *C. bicolor* by the markings on the lip, and by the length and shape of the side-lobes of the lip. The lip of *C. aloifolium* is always strongly veined with maroon, both on the mid- and side-lobes. The side-lobes have long, acute or acuminate apices which extend beyond the anther-cap, whereas in *C. bicolor* they are often blunter and never exceed the anther-cap.

Fig. 58. Habitat of *Cymbidium aloifolium*. China, SW Guangxi.

Fig. 59 (above left). *Cymbidium aloifolium*, China, SW Guangxi. **Fig. 60** (top right). *Cymbidium aloifolium*, S of Surin, Thailand. **Fig. 61** (bottom right). *Cymbidium aloifolium*, China, Guangxi.

Several other characters can usually be used to distinguish these species (Table 8), but there is a degree of overlap in some characters, and no single character can be relied upon. *Cymbidium aloifolium* is usually a more robust plant, producing larger pseudobulbs. Seth (1982) stated that the leaf of *C. aloifolium* (more than 2.5 cm (1 in) broad) is broader than that of *C. bicolor* (less than 2.5 cm, 1 in). This is often the case, but in Thailand they have leaves of a similar width, about 2.3–2.4 cm (0.9 in) (Seidenfaden, 1983). In Java, the breadth of the leaves of both species can be about 1.5 cm (0.6 in). Other useful vegetative characters include the leaf number, which is usually 4–5 per growth in *C. aloifolium* and 5–7 in *C. bicolor*, and in the shape of the leaf apex, which is unequally bilobed in both species, but obtuse in *C. bicolor*, and almost truncate or emarginate in *C. aloifolium*. The scape of the latter is nearly always pendulous, whereas that of *C. bicolor* may be arching, although it can also be pendulous, but is then always much shorter than that of *C. aloifolium*. Seth (1982) uses the shape of the side-lobes as a distinguishing character, stating that whereas those of *C. aloifolium* are acute, those of *C. bicolor* are obtuse. However, both *C. bicolor* subsp. *bicolor* and subsp. *pubescens* have acute side-lobe apices, although they are shorter than in *C. aloifolium*. The callus ridges are strongly sigmoid and often broken in *C. aloifolium*, whereas in *C. bicolor* they are often only weakly sigmoid or almost straight.

The difficulty of distinguishing these species in the herbarium has led to great confusion between them. Seth (1982) and Seth & Cribb (1984) distinguished them, and their treatment is followed here. However, the literature cited must be used with caution as the species have frequently been confused. The synonymy given by Seth is followed here, with the exception of *C. wallichii* Lindl., which has been transferred to the synonymy of *C. finlaysonianum*.

Fig. 62. *Cymbidium aloifolium* (Kew spirit no. 45061/46664). **1a.** Perianth, × 1. **1b.** Lip and column, × 1. **1c.** Pollinarium, × 3. **1d.** Pollinium (reverse), × 3. *C. bicolor* subsp. *bicolor* (Kew spirit no. 45048). **2a.** Perianth, × 1. **2b.** Lip and column, × 1. **2c.** Pollinarium, × 3. **2d.** Pollinium (reverse), × 3. *C. bicolor* subsp. *obtusum* (Kew spirit no. 48296). **3a.** Perianth, × 1. **3b.** Lip and column, × 1. **3c.** Pollinarium, × 3. **3d.** Pollinium (reverse), × 3. *C. rectum* (Kew spirit no. 48674). **4a.** Perianth, × 1. **4b.** Lip and column, × 1. **4c.** Pollinarium, × 3. **4d.** Pollinium (reverse), × 3. *C. bicolor* subsp. *pubescens* (Kew spirit no. 47389). **5a.** Perianth, × 1. **5b.** Lip and column, × 1. **5c.** Pollinarium, × 3. **5d.** Pollinium (reverse), × 3. *C. atropurpureum* (Kew spirit no. 42895). **6a.** Perianth, × 1. **6b.** Lip and column, × 1. **6c.** Pollinarium, × 3. **6d.** Pollinium (reverse), × 3. *C. finlaysonianum* (Kew spirit no. 48379). **7a.** Perianth, × 1. **7b.** Lip and column, × 1. **7c.** Pollinarium, × 3. **7d.** Pollinium (reverse), × 3. All drawn by Claire Smith.

9. Cymbidium aloifolium (L.) Sw. in *Nov. Act. Soc. Sci. Upsal.* 6: 73 (1799). Lectotype: Illustration in *Rheede, Hortus Indicus Malabaricus* 12(8): t. 8 (1703) (selected by Seth, 1982).

Epidendrum aloifolium L., *Sp. Pl.*: 953 (1753).

Epidendrum pendulum Roxb., *Pl. Coast Coromandel* 1: 35, t. 44 (1795). Type: Illustration in *Roxburgh, loc. cit.* (lectotype chosen by Seth, 1982).

Epidendrum aloides Curtis, *Bot. Mag.* 11: t. 387 (1797), *sphalm. pro E. aloifolium.*

C. pendulum (Roxb.) Sw., in *Nov. Act. Soc. Sci. Upsal.* 6: 73 (1799).

Aerides borassi Buch.-Ham. ex J.E. Sm. in Rees, *Cyclop.* 39: *Addend. Aerides* 8 (1819). Type: India, Mysore, *Buchanan-Hamilton* (holotype BM).

C. erectum Wight, *Icon. Pl. Ind. Or.* 5: 21, t. 753 (1851). Type: India, Iyamally Hills, Coimbatore, *Wight* (holotype BM?).

C. simulans Rolfe in *Orchid Rev.* 25: 175 (1917). Type: India, Sikkim, *Pantling 268* (lectotype K!, selected by Seth, 1982).

C. atropurpureum auct. non (Lindl.) Rolfe; Yen, *Icon. Cymbid. Amoyens*, A4 (1964).

C. intermedium H.G. Jones in *Reinwardtia* 9: 71 (1974). Type: India, Bombay, cult. *Jones C/85* (holotype not located).

A medium-sized, perennial, epiphytic *herb*. *Pseudobulbs* strongly inflated, 6–9 × 3–4 cm, ovoid, bilaterally flattened, enclosed in the persistent leaf bases and 6–7 cataphylls. *Leaves* 4–5(6) per pseudobulb, (27)40–100 × 1.5–4.5(6.3) cm, ligulate, obtuse to emarginate and unequally bilobed at the apex, coriaceous, rigid, arching, articulated to a 4–18 cm long, broadly sheathing base; cataphylls up to 14 cm long, becoming scarious and eventually fibrous with age. *Scape* 30–70(90) cm long, from within the cataphylls, pendulous to arching except at the base, with (14)20–45 flowers; peduncle 4–14 cm, covered basally by 5–7 overlapping, cymbiform, acute, spreading sheaths up to 3.5–7 cm long; rhachis 24–76 cm long; bracts 2–5 mm long, triangular. *Flowers* (3)3.5–4.2(5.5) cm across; lightly scented; rhachis, pedicel and ovary pale green, sometimes stained red-brown; sepals and petals pale yellow to cream with a broad, central, maroon-brown stripe, often with darker streaks; lip white or cream, side-lobes and mid-lobe veined maroon, mid-lobe yellow at the base; callus ridges and disc yellow; column dark maroon; anther-cap yellow. *Pedicel and ovary* 11–25 mm long. *Dorsal sepal* (17)19–24(28) × 5–85 mm, narrowly oblong to narrowly ligulate-elliptic, obtuse, mucronate, erect; lateral sepals similar, oblique, spreading. *Petals* (15.5)18–23(26) × 5–8.5 mm, narrowly elliptic, obtuse to acute, sometimes mucronate, porrect and almost parallel above the column. *Lip* 15–23 × 10–14 mm when flattened, 3-lobed, saccate at the base, minutely papillose to minutely pubescent; side-lobes erect, acute to acuminate, porrect, clasping the column, exceeding the column and the anther-cap; mid-lobe (7)8.4–12 × 6.5–9 mm, ovate, acute to obtuse, often mucronate, usually recurved, margin entire; callus usually in two sigmoid ridges often broken in the middle and inflated only towards the base and the apex. *Column* 10–12 mm long, arching, winged towards the apex, with a short column-foot; pollinia 2, triangular, deeply cleft, on a broadly triangular viscidium drawn into 2 short thread-like appendages at the tips. *Capsule* 4–7 × 2–3 cm, oblong-ellipsoidal with a short stalk and an apical beak about 1 cm long. Pl. 8, Figs. 58–62: 1a–d, 63.

DISTRIBUTION. Nepal, India, Sikkim, Andaman Islands, Bangladesh, Sri Lanka, southern China, Hong Kong, Burma, Thailand, Cambodia, Laos, Vietnam, western Malaysia, Sumatra and Java (Map 10); sea level–1500 m (4920 ft).

It occurs over much of peninsular India and South-east Asia. Although *C. aloifolium* extends to Java and Sumatra (Holttum, 1957; Backer & Bakhuizen, 1968; Comber, 1980, 2001), it has not been recorded from Borneo. Further collections may well extend the known range of *C. aloifolium* in Malesia.

Plate 8. *Cymbidium aloifolium*. *Seth* 154, cult. Kew, del. Claire Smith.

HABITAT. In the forks and hollows of large branches and tree trunks, usually in open forest in partial shade provided by the leaf canopy, also on limestone rocks; flowering (March)April–June(August).

It is commonly a lowland plant, and among the few epiphytic species able to survive on isolated trees that remain when land is cleared for agricultural use. It also appears to tolerate trampling when the trees are climbed to remove branches for firewood. Both it and the related *C. bicolor* are able to withstand dry, exposed conditions, such as on exposed karst limestone. Their leaves have several xerophytic adaptations, including a thick, succulent and coriaceous texture, the presence of a complete subepidermal layer of lignified cells, and stomata that are less densely spaced than in other species and have a characteristic slit-shaped, rather than rounded, aperture in the stomatal cover that can close completely, sealing the leaf surface in a continuous cuticular layer.

Cymbidium aloifolium and *C. bicolor* grow in similar habitats and occupy similar niches. Typically, they occur in forks or hollows in the trunks or larger branches of forest and riverine trees, usually preferring to grow close to the bole of the tree. This position utilises the partial shade that the foliage of the tree provides, although both species can tolerate full exposure to the sun as often occurs when the tree is deciduous. They often grow in rotting wood, and will form large clumps on dead trees where their extensive root system grows into the rotting wood. The roots form a dense, spongy mass, and *C. aloifolium* may produce erect, aerial roots that trap leaves and other plant debris. *Cymbidium aloifolium* also occurs on the trunks of palm trees, on rocks, especially limestone, and even on the brick walls of houses (Barretto & Youngsaye, 1980).

Although *C. aloifolium* and *C. bicolor* occur sympatrically over much of their range, they do not appear to hybridise. They seem to have slightly different flowering seasons, and this is particularly pronounced in northern Thailand where *C. bicolor* flowers during the hot, dry season, from December to March and *C. aloifolium* during the rainy season, from May to August, although specimens flowering during March are not uncommon, especially in peninsular Thailand (Seidenfaden, 1983). However, in northern India, the flowering times are reversed, *C. bicolor* flowering in May and June and *C. aloifolium* in April and May (King & Pantling, 1898; Pradhan, 1979).

Elevational range may also contribute to maintaining genetic isolation. In Thailand, *C. aloifolium* often occurs at slightly higher elevations in the deciduous hill forests rather than in evergreen lowland forest. In northern India the reverse is encountered, with *C. bicolor* growing along the Himalaya at slightly higher elevations than *C. aloifolium*.

CONSERVATION STATUS. NT.

Fig. 63. *Cymbidium aloifolium* and *C. bicolor* var. *pubescens* lips.

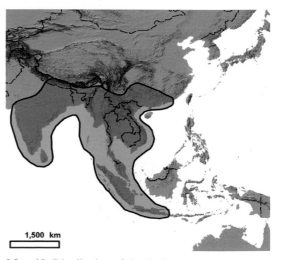

Map 10. Distribution of *C. aloifolium*.

124

10. CYMBIDIUM ATROPURPUREUM

Lindley (1854) originally described this species as a variety of his *C. pendulum*, a treatment which J.D. Hooker (1865) followed in *Curtis's Botanical Magazine*. Veitch (1894) included it as a variety of *C. finlaysonianum*. Rolfe (1903) was the first to treat it as a distinct species, and subsequent authors have accepted his treatment (Ames, 1908, 1925; Smith, 1933; Holttum, 1953; Seidenfaden, 1983). Seth (1982) included this species in section *Cymbidium*, and his treatment is followed here, with the exception of *C. pendulum sensu* Vidal, *non* (Roxb.) Sw., now considered a synonym of *C. finlaysonianum*. In the type description, Lindley cited two specimens, one cultivated by Knowles, the other collected by Cuming in the Philippines. The latter specimen seems to have been the one he used to prepare the description. However, it is not present in the Lindley herbarium, and has not been located elsewhere. Seth gave the type as "cult. Rollissons", a specimen which is said to have come from Java, although this locality is doubtful. This specimen is the plant figured in the *Botanical Magazine* (Hooker, 1865) and is also mentioned by Rolfe (1903).

Cymbidium atropurpureum varies over its extensive range principally in its leaf breadth, its scape length and its flower colour. The scape length varies in individual specimens, although apparently at random within populations. The most distinctively coloured variant of *C. atropurpureum* is found in the Philippines and lowland Sabah. It has deep wine-red sepals and petals, and narrow leaves up to 2.5 cm (1 in) broad. This was the variant described by Lindley (1854), Hooker (1865), Rolfe (1903) and Quisumbing (1940). However, plants from the more western part of its distribution (western Malaysia, southern Thailand, Sumatra) have greenish sepals, although the petals and the column remain deep maroon, and may be assignable to var. *olivaceum* described by Smith (1910) from Java. Du Puy & Lamb (1984) recorded a greenish-flowered specimen with purple petals in Sabah at higher elevations than the more common deep purple-flowered variant. This specimen also had broader leaves than the lowland variant (up to 4 cm (1.6 in) broad), indicating that further investigations might show that a variant with greenish sepal colour and broader leaves might be distinguished.

Holttum (1953) and Seidenfaden (1983) both state that this species can be distinguished from *C. finlaysonianum* by its narrower leaves. This is often the case, but some specimens have broad leaves, up to 4 cm (1.6 in), leading to confusion between the two taxa. The most reliable characters that distinguish them are the length and shape of the side-lobes of the lip, and the shape of the callus ridges In *C. atropurpureum* the side-lobes are much shorter than the column, and their apices are obtuse to rounded, appearing to be somewhat truncated (Fig. 62: 6a, b), whereas in *C. finlaysonianum* their apices are triangular, acute and strongly porrect, exceeding both the column and the anther-cap (Fig. 62: 7a & b). The callus ridges in *C. atropurpureum*, usually weakly defined, are rounded and confluent at their apices, merge into the base of the mid-lobe of the lip and are distinct from those of *C. finlaysonianum*, which are raised and terminate abruptly at the base of the mid-lobe.

10. Cymbidium atropurpureum (Lindl.) Rolfe in *Orchid Rev*. 11: 190–191 (1903). Type: ?Java, cult. Rollissons (neotype K!, selected by Seth (1982)).

C. pendulum (Roxb.) Sw. var. *atropurpureum* Lindl. in *Gard. Chron*. 1854: 287 (1854). Type: Philippines, *Cuming*, cult. Knowles (not located).
C. pendulum (Roxb.) Sw. var. *purpureum* W. Watson, *Orchids*: 151 (1890), *sphalm.* for var. *atropurpureum*.
C. finlaysonianum Wall. ex Lindl. var. *atropurpureum* (Lindl.) Veitch, *Man. Orchid. Pl.* 2: 16 (1894).
C. atropurpureum (Lindl.) Rolfe var. *olivaceum* J.J. Sm. in *Bull. Dép. Agric. Ind. Néerland*. 43: 60–62 (1910). Type: Java, Leomadjang, *Connell* (holotype L!).

A large, perennial, epiphytic or rarely lithophytic *herb*. *Pseudobulbs* up to 10 × 6 cm, ovoid, often obscurely and weakly inflated, bilaterally flattened, enclosed in the persistent leaf bases and about 4

Plate 9. *Cymbidium atropurpureum*. Curtis's Botanical Magazine t. 5710, del. W.H. Fitch.

scarious cataphylls. *Leaves* usually 7–9 per pseudobulb, up to 50–90(125) × 1.5–4 cm, ligulate, obtuse and unequally bilobed at the apex, coriaceous, rigid, arching, articulated to a broadly sheathing base up to 15–20 cm long, the shortest reduced to cataphylls with an abscission zone near the apex and a short lamina. *Scape* 28–75 cm long, from within the cataphylls, arching to pendulous, with (7)10–33 flowers; peduncle 5–16 cm, covered basally by 6–8 overlapping, cymbiform, acute, spreading sheaths up to 7 cm long; rhachis 20–55 cm long, pendulous; bracts 1–4 mm long, triangular. *Flowers* 3.5–4.5 cm across; usually coconut-scented; rhachis, pedicel and ovary pale green, often flushed with purple; sepals and petals deep maroon to dull yellow-green with strong maroon staining; lip white, becoming yellow with age, side-lobes stained maroon-purple, mid-lobe yellow in front of the callus ridges and blotched with maroon; callus ridges bright yellow in front, stained maroon behind; column deep maroon, sometimes paler in front, anther-cap white to pale yellow. *Pedicel and ovary* 15–26 mm long. *Dorsal sepal* 28–33 × 7–10 mm, narrowly ligulate-elliptic, obtuse, suberect, margins revolute; lateral sepals

Fig. 64. *Cymbidium atropurpureum*, N Sumatra.   **Fig. 65.** *Cymbidium atropurpureum*, Kalimantan.

Plate 10. *Cymbidium atropurpureum*. Sarawak, *Giles & Woolliams* s.n., cult. Kew, del. Claire Smith.

similar, falcate and oblique, pendulous and porrect. *Petals* 25–30 × 7.5–11 mm, narrowly elliptic, subacute, weakly porrect, margins sometimes revolute. *Lip* 21–25 × 13–15 mm when flattened, 3-lobed, usually broadest across the mid-lobe, minutely papillose to minutely pubescent, the longest hairs on the side-lobe tips; side-lobes erect, much shorter than the column, apices obtuse and appearing truncated; mid-lobe 11–13 × 13–14 mm, broadly ovate to rhomboid, obtuse to emarginate, weakly recurved, margin entire; callus of two sigmoid, raised ridges that are rounded and confluent at the apices and merge gradually with the base of the mid-lobe. *Column* 16–18 mm long, arching, winged; pollinia 2–2.5 mm long, triangular, deeply cleft, on a broadly triangular viscidium drawn into two acuminate tips. Pls. 9 & 10, Figs. 30, 62: 6a–d, 64–66.

DISTRIBUTION. Southern China (Hainan only), southern Vietnam, southern Thailand, western Malaysia, Sumatra, Java, Borneo and the Philippines (Map 11); sea level–2200 m (7220 ft). Smith (1910) reported a single collection of *C. atropurpureum* on Java, but Backer & Bakhuizen (1968) suggested that this may have been a cultivated plant. However, J. Comber (pers. comm.) found it in western Java. In Thailand, it is confined to the peninsula, near the western Malaysian border (Seidenfaden, 1983). Quisumbing (1940) gave its distribution in the Philippines as Luzon, Leyte and Mindanao, although he suggested that it may have been introduced into Luzon. Gloria Siu (pers. comm.) recently found it growing epiphytically in Hainan. It is a lowland forest species in the northern parts of its distribution, but it has been collected at 1200 m (3940 ft) in Sabah, and at 2200 m (7220 ft) in Sumatra.

HABITAT. In the forks of forest trees, and occasionally on rocks, usually in lowland evergreen and riverine forests and often near the sea; usually flowering between March and May, although it can flower at other times of the year.

CONSERVATION STATUS. NT.

Fig. 66. *Cymbidium atropurpureum*, Sumatra.

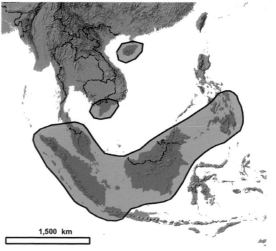

1,500 km

Map 11. Distribution of *C. atropurpureum*.

11. CYMBIDIUM BICOLOR

In his type description, Lindley (1833) stated that *C. bicolor* was related to *C. aloifolium*, but that it could be distinguished by the colour of the lip, and the saccate lip base. Although *C. aloifolium* also has a saccate lip base, further differences between these species have been documented, and they are maintained as distinct species in this study. The simplest method of distinguishing *C. bicolor* is by the colour of the lip, especially of the side-lobes, which are mottled with maroon or red-brown rather than veined and streaked with maroon. These characters are difficult to ascertain from herbarium material lacking flower colour notes, and the large variation in *C. bicolor*, in particular, has caused considerable taxonomic confusion. The misapplication of the name *C. pendulum* to the Himalayan specimens of *C. bicolor* is one prominent example of this. The literature and synonymy cited here is based mainly on the revision of section *Cymbidium* by Seth (1982) and Seth & Cribb (1984), with the exceptions of *C. pulchellum* Schltr., which is now considered to be conspecific with *C. chloranthum* Lindl. and *C. rectum* Ridl., which is a distinct species distinguished by its narrow, canaliculate leaves, suberect scape, short (8 mm, 0.3 in) column, connate pollinia, and the distinctive single maroon spot at the apex of the pale yellow, ligulate mid-lobe. The literature and specimens cited, and the distributions given by the various authors, are often confused and incorrect, and should be used with caution.

Cymbidium bicolor is a widespread and variable species, varying particularly in the breadth of its leaves, the angle at which its scape is held, length of its scape, length of its pedicel and ovary, size of its flowers, shape of its side-lobes apices, shape of its callus ridges and lip indumentum. Three more-or-less distinct, geographically separated taxa can be distinguished. Intermediates may occur where the distributions meet, and the differences are not sufficient to warrant the recognition of these taxa as separate species. However, as there is a geographically linked pattern in this variation, these taxa are recognised here at subspecific rank. The most useful distinguishing characters separating these three subspecies and *C. aloifolium* are summarised in Table 8.

11. Cymbidium bicolor Lindl., *Gen. Sp. Orchid. Pl.*: 164 (1833) & in *Bot. Reg.* 25: misc. 47 (1839). Type: Ceylon, [Sri Lanka], *Macrae 54* (holotype K!).

C. aloifolium Jayaw. in Dassanayake, M.D., ed., *Fl. Ceylon* 2: 183–185, f. 82 (1981), *pro parte, sensu* (L.) Sw. (1799).

A medium-sized, perennial, epiphytic *herb. Pseudobulbs* usually not inflated, up to 5 × 2.5 cm, elongate-ovoid, bilaterally flattened, enclosed in persistent leaf bases and 4–5 cataphylls. *Leaves* (4)5–7 per pseudobulb, up to 30–68(90) × (0.8)1.2–2.9(3.1) cm, narrowly ligulate, obtuse, unequally bilobed to oblique at the apex, cartilaginous, stiff, arching, articulated to a 3–12 cm long, broadly sheathing base; cataphylls up to 11 cm long, becoming scarious and eventually fibrous with age, the longest with an abscission zone near the apex and a short lamina. *Scape* 10–50(72) cm long, from within the cataphylls, arching to pendulous, with 5–26 flowers; peduncle 2–12(18) cm long,

Fig. 67. *Cymbidium bicolor* subsp. *bicolor*. Sri Lanka.

covered towards the base with about 5 overlapping, cymbiform, spreading sheaths up to 3.5–5.5 cm long; rhachis 8–33 cm long; bracts 1.5–4.5 mm long, triangular. *Flowers* 2.5–4.5 cm across; lightly fruit-scented; rhachis green, pedicel and ovary pale yellow to red-brown; sepals and petals pale yellow to cream, with a broad, weakly defined central stripe of maroon-brown; lip white or cream with a pale yellow patch at the base, the side-lobes finely mottled with maroon or purple-brown, the mid-lobe white to yellow at the base, variously spotted or blotched with maroon or purple-brown, with a narrow cream margin; callus ridges cream to yellow, often finely mottled behind with red-brown; column cream, pale green or yellow, variously shaded with maroon; anther-cap cream to pale yellow. *Pedicel and ovary* 9–12 mm long. *Dorsal sepal* 16.5–28 × (3.2)4–6.5 mm, narrowly oblong to narrowly obovate-ligulate, obtuse to subacute, often mucronate, erect; lateral sepals similar, oblique, spreading. *Petals* (14)15–21.5 × 4–6.2 mm, narrowly oblong to narrowly elliptic, obtuse to acute, porrect but not closely covering the column. *Lip* 12.5–18 × (8.5)9.5–15.5 mm when flattened, 3-lobed, saccate at the base, minutely papillose to shortly pubescent; side-lobes erect, weakly clasping the column, shorter than, or as long as the column, but not exceeding the anther-cap, acute to obtuse, porrect to recurved; mid-lobe 5.2–8.7 × 6–9.6 mm, elliptic to ovate, rounded to obtuse and often mucronate at the apex, usually recurved, minutely papillose to shortly glandular-pubescent, sometimes with an undulating margin; callus of two ridges that may be entire and almost parallel to sigmoid, or broken in the middle and inflated only towards the base and apex, minutely papillose to minutely pubescent. *Column* (8)9–12 mm long, arching, winged towards the apex, with a column-foot; pollinia 2, triangular, deeply cleft, on a triangular viscidium drawn into two thread-like appendages at the tips. *Capsule* 4–6 × 2–3 cm, oblong-ellipsoidal, with a stalk and an apical beak about 8 mm long.

DISTRIBUTION. India, Sri Lanka, southern China, Indo-China, western Malaysia, Java, Sumatra, Borneo, Sulawesi, the Philippines (Maps 12 & 13); sea level–1500 m (4920 ft).

HABITAT. In the forks and hollows of large branches and tree trunks, usually in open forest in partial shade provided by the leaf canopy; flowering (December) March–June. See *C. aloifolium* for a more detailed discussion of the habitats and ecology, and the individual subspecies for their essential characters and distributions.

CONSERVATION STATUS. NT.

Key to the subspecies of *C. bicolor*

1. Sepals 4–6.5 mm longer than the petals; petals usually porrect, connivent above the column ....
.... subsp. **bicolor**
Sepals as long as or up to 2 mm longer than the petals; petals weakly porrect and spreading **2**

2. Lip densely pubescent, usually with some glandular hairs especially near the base of the mid-lobe; side-lobes acute; scape strongly pendulous .... subsp. **pubescens**
Lip papillose or weakly pubescent on the side-lobe tips or the apices of the callus ridges; side-lobes obtuse to subacute; scape arching to pendulous .... subsp. **obtusum**

subsp. **bicolor**

This subspecies is characterised by an arching scape, a long pedicel and ovary, a dorsal sepal that exceeds the petals by 4–6.5 mm giving the flower a spidery appearance, petals that run parallel to each

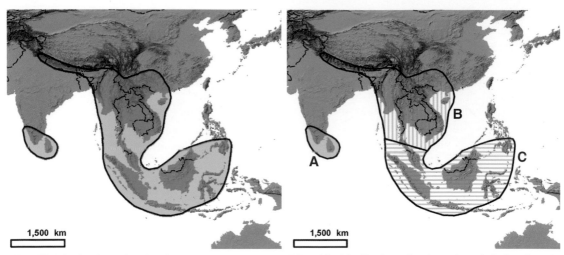

Map 12. Distribution of *C. bicolor*.

Map 13. Distribution of subspecies of *C. bicolor*. **A.** subsp. *bicolor*. **B.** subsp. *obtusum*. **C.** subsp. *pubescens*.

other or even cross above the column, a papillose lip with acute side-lobes (which do not, however, exceed the column) and a sigmoid or broken pair of callus ridges (Fig. 62: 2a & b). The lip is cream, with purple-mottled side-lobes, and a large, submarginal purple blotch towards the apex of the mid-lobe, and pale lemon yellow callus ridges, the colour of which extends into the base of the mid-lobe. Figs. 62: 2a–d; 67.

DISTRIBUTION. Sri Lanka, southern India (Map 13A).

Cymbidium bicolor was described by Lindley in 1833 from a specimen collected by Macrae in Ceylon (Sri Lanka). The typical subspecies is confined to Sri Lanka and southern and south-western peninsular India (Map 13), and its distribution appears not to overlap with that of subsp. *obtusum*.

subsp. **obtusum** Du Puy & P.J. Cribb, *The Genus Cymbidium*: 70 (1988). Type: Thailand, Uttaradit, *Menzies & Du Puy 120* (holotype K!).

C. crassifolium Wall., *Cat.* no. 7357, *nom. nud.*

C. mannii Rchb.f. in *Flora* 55: 274 (1872). Type: India, Assam, *Mann* (holotype W).

C. pendulum sensu King & Pantling. in *Ann. Roy. Bot. Gard. Calcutta* 8: 188, t. 251 (1898); Duthie, *Ann. Roy. Bot. Gard. Calcutta* 9: 136 (1906); Schltr. in *Fedde, Repert. Sp. Nov. Regni Veg. Beih.* 4: 271 (1919) *syn. excl.*, & *Fedde, Repert. Sp. Nov. Regni Veg.* 20: 104 (1924); Pradhan, *Indian Orchids* 2: 475 (1978); Hara, Stearn & Williams, *Enum. Flow. Pl. Nepal* 1: 38 (1978) *syn. excl.*; Y.S. Wu & S.C. Chen in *Acta Phytotax. Sin.* 18: 300 (1980), *non* (Roxb.) Sw.

Fig. 68. *Cymbidium bicolor* subsp. *obtusum*, *Cumberlege 712*, cult. Kew. **Fig. 69** (opposite). Vietnam, Kontum.

L. AVERYANOV

TABLE 8

A COMPARISON OF *C. ALOIFOLIUM*
AND THE THREE SUBSPECIES OF *C. BICOLOR*

Character	subsp. *bicolor*	*C. bicolor* subsp. *pubescens*	subsp. *obtusum*	*C. aloifolium*
Leaf breadth–max (mm)	(8)13–27	15–19(22)	13–29(31)	15–45(63)
Scape length (mm)	28–31 (approx)	10–39	18–50(72)	30–70(90)
Flower number	9–18	5–21	10–26	(14)20–45
Ovary length (mm)	18–38	9–21	9–42	11–25
Angle of the scape	arching	pendulous	arching or pendulous	pendulous
Dorsal sepal length (mm)	(18)20.5–28	17.5–22	16.5–21.5	17–28
Petal length (mm)	(14)15–21.5	17–21	15.5–19.5	15.5–26
Sepal minus petal length (mm)	4–6.5	0.5–1	1–2	1.5–2
Angle of the petals	connivent above the column	almost spreading	almost spreading	connivent above the column
Lip indumentum	papillose	pubescent, often with some glandular hairs	papillose or very weakly pubescent	papillose to weakly pubescent
Side-lobe apex	acute, long	acute, long	obtuse to subacute, short	acute to acuminate
Callus shape	sigmoid or broken	broken	sigmoid to straight	usually strongly sigmoid or broken
Side-lobe colour	mottled purple	mottled maroon	mottled maroon	striped maroon
Mid-lobe colour	lemon yellow at the base, with a large, submarginal purple blotch	pale yellow spotted and blotched with maroon	pale yellow or cream mottled and blotched with maroon or brown	striped maroon
Distribution	Sri Lanka and southern peninsular India	Malesia and the Philippines	Indo-China, to N. India, Nepal and S. China	Sri Lanka to N. India, China, Indo-China, W. Malaysia, Sumatra, Java

134

C. flaccidum Schltr. in *Fedde, Repert. Sp. Nov. Regni Veg.* 12: 109 (1913) & in *Beih.* 4: 267 (1919). Type: China, Kweichow [Guizhou], *Esquirol 2728* (holotype B†).

C. bicolor sensu Seidenf. in *Opera Bot.* 72: 81, f. 44, pl. 5b (1983), *syn. excl.*

C. paucifolium Z.J. Liu & S.C. Chen in *J. Wuhan Bot. Res.* 20 (5): 350 (2002), **syn. nov.** Type: China, Yunnan, Xishuanbanna area, *Z.J. Liu 21112* (holotype and isotype SZWN).

This subspecies often has broad leaves, up to about 25–30 mm wide. The scape is often longer than in the other subspecies, and varies from arching to pendulous. The petals are spreading, and the lip is papillose or weakly pubescent, especially at the tips of the side-lobes. The side-lobes of the lip are usually obtuse at the apex (Fig. 62: 3 a & b). The mid-lobe of the lip is characteristically broadly elliptic and rounded at the apex, appearing circular in outline. Its callus ridges are entire, varying in shape from sigmoid to almost straight and parallel with a furrow near the apex. The lip is cream, the side-lobes mottled with maroon, and the mid-lobe spotted and mottled with maroon, sometimes with a brighter yellow base. Figs. 62: 3a–d, 68 & 69.

DISTRIBUTION. Nepal, northern India (Sikkim, Assam, Arunachal Pradesh, Meghalaya), Bhutan, southern China, Burma, Indo-China (Map 13B).

Some differences can be seen between the Indo-Chinese specimens and those from China and the Himalaya. The latter have narrower leaves, a pendulous rather than arching inflorescence, shorter pedicel and ovary lengths, and the callus ridges are usually straight, not sigmoid. These are all variable characters with many intermediates, and the flowers appear to be almost identical in shape and colour. The decision to maintain all of this variation in a single subspecies is reinforced by the otherwise anomalous occurrence of plants resembling the northern Indian variant in southern China. Further investigation may indicate that these should be treated as distinct varieties, but lack of material makes this inadvisable at present.

King & Pantling (1898) misapplied the name *C. pendulum* (Roxb.) Sw. to specimens of *C. bicolor* from Sikkim. Subsequent authors followed them, including Duthie (1906), Schlechter (1919), Hara *et al.* (1978) and Pradhan (1979). *Cymbidium pendulum*, however, was described from a specimen collected in southern peninsular India, and is conspecific with *C. aloifolium*.

Cymbidium mannii was described by Reichenbach (1872) from northern India, and was considered by Seth (1982) to be a synonym of *C. bicolor*. However, Hooker (1891) and Seidenfaden (1983) included it as a synonym of *C. aloifolium*. The description is poor, but indicates an affinity with *C. bicolor* rather than *C. aloifolium*, especially in the shape of the callus ridges, which are described as straight and parallel or sigmoid, sometimes with a sulcate apex.

Cymbidium flaccidum was collected by Esquirol in China, the provenance being given originally as Sichuan but now thought to be from Guizhou, although the type specimen has since been destroyed. Schlechter (1913, 1919), who indicated its affinity with *C. bicolor*, described it as having leaves 2 cm (0.8 in) broad, the ovary short, side-lobes of the lip with obtuse apices, the mid-lobe of the lip broadly ovate and obtuse, and the callus ridges only slightly curved. It was considered as conspecific with *C. bicolor* by Seth (1982), and this treatment is followed here. The recently described *C. paucifolium* Z.J. Liu & S.C. Chen (2002), based upon a specimen collected in the Xishuanbanna, Yunnan, also belongs here. They compared it with *C. aloifolium* rather than *C. bicolor*.

Tso (1933) collected further specimens of this variant in Guangdong, his description closely matching the northern Indian specimens of *C. bicolor*. Further specimens have been collected in Hainan. Wu & Chen (1980) gave the distribution of *C. bicolor* in China as Guangdong, Guangxi, Guizhou, Hainan and Yunnan (Map 13B).

subsp. **pubescens** (Lindl.) Du Puy & P.J. Cribb, *The Genus Cymbidium*: 73 (1988). Type: Singapore, *Cuming* s.n., cult. Loddiges (holotype K!).

C. aloifolium sensu Blume, *Bijdr. Fl. Nederl. Indie*: 378, t.19 (1825), *non* (L.)Sw. (1799)
C. pubescens Lindl. in *Bot. Reg.* 26: misc. 75 (1840) & 27: t. 38 (1841).
C. aloifolium var. *pubescens* (Lindl.) Ridl. in *J. Roy. As. Soc. Str. Br.* 59: 196 (1911).
C. pubescens Lindl. var. *celebicum* Schltr. in *Fedde, Repert. Sp. Nov. Regni Veg.* 10: 190 (1911). Type: Sulawesi, Minahassa Peninsula, Lansot, *Schlechter 20627* (holotype B†).
C. celebicum (Schltr.) Schltr. in *Fedde, Repert. Sp. Nov. Regni Veg.* 21: 197 (1925).

This subspecies has leaves rarely more than 2 cm broad. It has a sharply pendulous, few-flowered scape, and a short pedicel and ovary. The petals are usually spreading. The lip is shortly pubescent; some of these hairs are glandular (swollen at the tips), especially towards the base of the mid-lobe. The side-lobe apices are usually acute, but shorter than the column, and the callus ridges are broken in the middle (Fig. 62: 5a & b). The lip is cream, mottled with maroon on the side-lobes, and spotted and blotched with maroon or red-brown on the mid-lobe, which is yellow towards the base. The callus ridges are also yellow. Pl. 11; Figs. 62: 5a–d, 70.

KATH BARRETT

DISTRIBUTION.
Western Malaysia, Java, Sumatra, Borneo, Celebes, Philippines (Mindanao) (Map 13C).

Cymbidium celebicum, described by Schlechter (1911, 1925) from Celebes, and tentatively included here as a synonym of subsp. *pubescens*, differs somewhat from the above description in that the inflorescence is arching, the side-lobes of the lip are blunt and the callus ridges are not broken. Specimens described by Ames & Quisumbing (1932) and Quisumbing (1940) from Mindanao have a large, red-brown, submarginal blotch on the mid-lobe. These are intermediate between subsp. *obtusum* and subsp. *bicolor*, respectively. Specimens from peninsular Thailand have the lip shape of subsp. *obtusum*, but are more strongly pubescent than the type.

Fig. 70. *Cymbidium bicolor* subsp. *pubescens*. Sarawak.

Plate 11. *Cymbidium bicolor* subsp. *pubescens*. Sarawak, del. Hugh Low, Kew Collection.

12. CYMBIDIUM FINLAYSONIANUM

John Lindley (1833) described *C. finlaysonianum* based upon a collection made by Finlayson in Vietnam (probably at Tourane Bay, now Da Nang). The holotype is in Lindley's herbarium, and an isotype is in the Wallich herbarium (no. 7358) at Kew. It had previously been described by Blume (1825) as *C. pendulum*, but this name is invalidated by its previous application by Swartz (1799) to a different species, now considered conspecific with *C. aloifolium*. We follow Seth (1982) in including it in section *Cymbidium*, and his synonymy is accepted with the addition of *C. wallichii* Lindl., which he included as a synonym of *C. aloifolium*, and of *C. pendulum sensu* Vidal, which he included as a synonym of *C. atropurpureum*.

Lindley (1833) cited three specimens when he described *C. wallichii*. They are all in the Wallich Herbarium at Kew (7352a – *Finlayson*; 7352b – Penang, *Porter*; 7352c – Attran R., *Wallich*). There is a fourth specimen in the Lindley herbarium that appears to be the same collection as 7352b. Of these, 7352c is a specimen of *C. aloifolium* (L.) Sw. Unfortunately, Seth (1982) selected 7352c as the lectotype of *C. wallichii*, and accordingly placed *C. wallichii* in the synonymy of *C. aloifolium*. The description by Lindley includes the phrase 'Lamellis continuis parallelis', and a drawing on his herbarium sheet of 7352b shows these two straight callus ridges. The specimen 7352c has the strongly sigmoid, interrupted keels characteristic of *C. aloifolium*, and therefore does not agree with the type description. Seth's lectotypification was therefore rejected in accordance with article 8 of the I.C.B.N.,

and the specimen in the Lindley herbarium (coll. Penang, Porter, *Wallich* 7352b) was selected instead as the lectotype of *C. wallichii (Du Puy & Cribb, 1988)*. This specimen is identical with *C. finlaysonianum*.

Lindley (1842) also described and illustrated *C. pendulum* var. *brevilabre* in the *Botanical Register*, distinguishing it by the broader, nearly round mid-lobe of its lip. The plate, type description and type specimen are indistinguishable from *C. finlaysonianum*, under which species this taxon is now included.

The leaf breadth and scape length vary in *C. finlaysonianum*, but this is often dependent on the age and general size of the specimen. It is a relatively uniform species, although some variation in flower colour and in flower size has been described. Du Puy & Lamb (1984) noted that two colour variants occurred in different habitats in Sabah, and that the differences were maintained when individuals were transplanted to the same environment. The variant that occurs as an epiphyte or lithophyte in lowland areas has greenish sepals and petals stained with red-brown, whereas the variant that occurs on rocks on the coast has golden to straw-coloured sepals and petals, with little red staining. Seidenfaden (1983) noted that in Thailand the specimens that occur towards the north of the distribution of this

P. CRIBB

Fig. 71. *Cymbidium finlaysonianum*, Sabah.

Plate 12. *Cymbidium finlaysonianum*. Sarawak, *Giles* 600, cult. Kew, del. Claire Smith.

J. COMBER

species have slightly smaller flowers (dorsal sepal about 25 mm (1 in) long) than those that grow in the more southerly, peninsular area.

Cymbidium finlaysonianum is closely related to *C. atropurpureum* (see Fig. 62: 6 & 7, and the discussion of the latter for the distinguishing characters), but it is more common, and has a wider distribution.

12. Cymbidium finlaysonianum Lindl., *Gen. Sp. Orchid. Pl.:* 164 (1833). Type: Cochinchina [Vietnam], Turon Bay [Tourane?], *Finlayson* in *Wallich 7358* (holotype K!).

C. *pendulum auct. non* (Roxb.) Sw.; Blume, *Bijdr.* 379 (1825); sensu Vidal, Phan. Cuming Philipp.: 150 (1885).
C. *wallichii* Lindl., *Gen. Sp. Orchid. Pl.:* 165 (1833). Type: Malaya, Penang, *Porter* (lectotype K!, selected by Du Puy & Cribb, 1988).
C. *pendulum* (Roxb.) Sw. var. *brevilabre* Lindl. in *Bot. Reg.* 30: t. 24 (1842). Type: Philippines, cult. Loddiges, *Cuming* s.n. (holotype K!).
C. *tricolor* Miq., *Choix Pl. Buitenz.* t.19 (1863). Type: Java, cult. Buitenzorg [Bogor] B.G. (holotype U).
C. *aloifolium auct. non* (L.) Sw.; Guillaumin in Lecompte, *Fl. Gen. Indo-Chine* 6: 415 (1932), *syn. excl.*

A perennial, epiphytic or lithophytic *herb*. *Pseudobulbs* up to 8 × 5 cm, ovoid, bilaterally flattened, usually obscure and little inflated, enclosed in the persistent leaf bases and 5–8 cataphylls that become scarious and eventually fibrous with age. *Leaves* distichous, 4–7 per pseudobulb, up to (36)50–85(100) × (2.7)3.2–6 cm, ligulate, obtuse to emarginate and unequally bilobed at the apex, coriaceous and rigid, almost erect, articulated to a broadly sheathing base up to 8–16 cm long, the shortest sometimes reduced to cataphylls with an abscission zone near the apex. *Scape* (20)30–115(140) cm long, from within the cataphylls, sharply pendulous, with (7)12–26 well-spaced flowers; peduncle 5–12 cm long, covered by 6–8 overlapping, cymbiform, acute, spreading sheaths up to 3.7–8 cm long; rhachis about 25–110 cm long, usually becoming slender and fractiflex towards the apex; bracts small, 1–3 mm long, triangular. *Flowers* 4–5.7 cm across; usually fruit-scented; rhachis, pedicel and ovary green, stained red-brown; sepals and petals dull green to straw-yellow, usually suffused with red-brown, especially towards the tips of the sepals and along the centre of the

Fig. 72. *Cymbidium finlaysonianum*, cult. ex Kalimantan.

petals; lip white, the side-lobes suffused and veined with purple-red, the mid-lobe yellow in front of the callus ridges and with a submarginal, U-shaped, purple-red blotch towards the apex and often some other reddish spotting; callus ridges bright yellow in front, stained purple-red behind; column deep purple at the tip, becoming yellowish towards the base; anther-cap cream. *Pedicel and ovary* 14–45 mm long. *Dorsal sepal* 25–33 × 7–11 mm, narrowly ligulate-elliptic, obtuse, erect, margins revolute; lateral sepals similar, oblique, spreading. *Petals* 24–30 × 7–11 mm, narrowly elliptic to ovate, obtuse to subacute, porrect, margins revolute, 7–9 veined. *Lip* 24–28 × 14–18 mm when flattened, 3-lobed, usually broadest across the side-lobes, papillose or with some minute hairs; side-lobes erect, longer than the column and usually also exceeding the anther-cap, upper margins involute, apices triangular, acute to acuminate, porrect; mid-lobe large, 11–14 × (9)11–14 mm, broadly elliptic, obtuse to emarginate, mucronate, recurved, margin undulate; callus of 2 parallel, raised and well-defined ridges that terminate abruptly at the base of the mid-lobe. *Column* 15–18 mm long, arching, winged; pollinia 2, about 2 mm long, triangular, deeply cleft, on a broadly triangular viscidium drawn into 2 short appendages at the tips. *Capsule* 5–10 × 3–4 cm, oblong-ellipsoidal, narrowing to a short pedicel and a 15–18 mm long apical beak formed by the persistent column. Pl. 12; Figs. 62: 7a–d, 71 & 72.

DISTRIBUTION. It occurs in southern Indo-China, Thailand, peninsular Malaysia, Sumatra, Borneo, widely in the Philippines (Quisumbing, 1940), Sulawesi and western Java where the climate is less seasonal (Backer & Bakhuizen, 1968; Comber, 1980) (Map 14); from sea level to 300 m (0–985 ft).

It is a lowland species, often growing near the sea and rarely above 300 m (985 ft), and is therefore more commonly encountered than the other tropical *Cymbidium* species. The robust habit of this species and its ability to form enormous clumps make it a conspicuous element in the lowland flora. In common with *C. aloifolium* and *C. tracyanum,* it has the ability to form erect roots that trap dead leaves and other plant debris. It often harbours biting ants, bees and even snakes. Its large size makes it suited to forks in the trunks and sturdy branches of trees in open forest, and its tolerance of exposed sites has made it common in secondary forest. This latter attribute has also allowed it to occur on economic tree crops such as rubber in Sabah (Du Puy & Lamb, 1984) and Java (Comber, 1980), and *Borassus* (Toddy) palms and other fruit trees in Thailand, where it can become a pest. It interferes with the morning tapping of rubber trees by retaining and dripping overnight rain. Its tolerance of exposed habitats is demonstrated by its ability to form large clumps on rocks on the coast in Sabah and western Malaysia, where it is exposed to full sun and salt-bearing winds. In Sarawak and Java, it has also been reported growing on trees in humid mangrove swamps (Backer & Bakhuizen, 1968). In Sabah, large bees have been seen pollinating it (A. Lamb, pers. comm.).

HABITAT. On trees in open lowland forest or secondary forest, usually near the coast, or on exposed coastal rocks, sometimes on rubber, oil-palm and other lowland tree crops. The flowering season varies throughout the range of this species, and seems to be dependent on climate fluctuations. In more northerly and seasonal regions of the Philippines it usually flowers between February and May, but in Indo-China it flowers from April to September. In the more equatorial regions, flowering occurs all year round. In cultivation in Europe and North America it is a summer- and autumn-flowering species, usually between June and September.

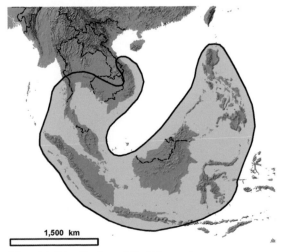

1,500 km

Map 14. Distribution of *C. finlaysonianum.*

CONSERVATION STATUS. NT.

13. CYMBIDIUM RECTUM

Cymbidium rectum was discovered in 1902, and had been in cultivation in the Singapore Botanic Garden for many years before Henry Ridley described it in 1920. It had been known in the gardens as *C.* '*erectum*' (a name already used by Wight (1852) for another species), the specific epithet referring to its upright scape, which is an unusual character in section *Cymbidium*. A watercolour sketch of *C. rectum* by Ridley in the collection at Kew, taken from the specimen that flowered in 1902 (*Ridley 11370*), is mistakenly labelled as 'near *C. acutum*'. He thought that the specimen had been collected in Perak, from the hills behind Taiping. It was not until the type specimen was collected by Genyns-Williams in 1916, near Siliau in the Negri Sembilan Hills, that its occurrence in Malaya could be confirmed. Holttum (1953) gave a second locality as Selangor, also in peninsular Malaya.

Subsequently, the species was lost to cultivation, and the name fell out of usage. Seth (1982) placed it in the synonymy of *C. bicolor*. However, Tony Lamb rediscovered it in Sabah and sent a living plant to Kew, where it flowered in September 1984. It corresponded closely with Ridley's sketch of 1902. Plants have also been seen by one of the authors (PC) in kerangas forest in eastern Kalimantan.

All of the specimens so far examined have an unusual pollinarium structure (Fig. 62:4c & d). The pollinia are similar to other species of section *Cymbidium* except that they are connate at the top, the two cleft pollinia being fused into a single structure. *Cymbidium rectum* may be further distinguished from allied species by its narrow, strongly channelled leaves (up to 1.4 cm (0.6 in) broad); suberect to

J. COMBER

Fig. 73. *Cymbidium rectum*, Kalimantan.

Fig. 74. *Cymbidium rectum.* Sabah, *Lamb* K12.　　**Fig. 75.** *Cymbidium rectum.* Sabah.

horizontal scape; and a lip mid-lobe that is ligulate, with an undulating margin and coloured cream with a primrose-yellow central stripe, a single apical maroon spot and usually some sparse maroon spots at the base. The callus ridges are almost straight and are unbroken, the column 8 mm (0.3 in) long, and the anther-cap about 1.5 mm (0.6 in) across (Fig. 62: 4a–d).

The general habit of the plant and the scape are similar to other species in section *Cymbidium*. The thick, fleshy, coriaceous leaf, with an unequally bilobed apex and a continuous subepidermal layer of sclerenchymatous cells linking the subepidermal fibre strands, is characteristic of this section. The leaf may be confused with narrow-leaved specimens of *C. bicolor* or *C. atropurpureum*, but it is normally narrower than in those species, and more strongly V-shaped in section. The flower shape and colour are also typical of section *Cymbidium*, closely resembling *C. aloifolium* and *C. bicolor*.

Cymbidium rectum is also, in some respects, similar to *C. chloranthum* in section *Floribundum*. They both have erect inflorescences, broad, elliptic floral segments, an undulating mid-lobe margin and a short column with a small anther.

13. Cymbidium rectum Ridl. in *J. Roy. Asiat. Soc. Str. Br.* 82: 198 (1920). Type: Malaysia, Negri Sembilan, *Genyns-Williams s.n.* (holotype SING!).

A perennial, epiphytic *herb. Pseudobulbs* not strongly inflated, about 5 × 2 cm, elongate-ovoid, enclosed in the sheathing leaf bases and 2–3 cataphylls. *Leaves* 7–9 per pseudobulb, up to 60 × 0.8–1.4 cm, narrowly ligulate, unequally bilobed to oblique at the tip, V-shaped in section, coriaceous, stiff, arching, articulated 1–8 cm from the pseudobulb. *Scape* up to 40 cm long, suberect to horizontal, often pendulous in fruit, with up to 17 flowers; peduncle up to 9 cm, covered towards the base by about 5 sheaths; sheaths up to 4 cm long, cymbiform, overlapping, spreading; bracts 2–3 mm long, triangular.

143

Pedicel and ovary 1.5–3 cm long. *Dorsal sepal* 17–20 × 7–8 mm, narrowly oblong to elliptic, obtuse or weakly mucronate, erect, margins revolute; lateral sepals similar, oblique, spreading. *Petals* narrower, 17–18 × 6 mm, narrowly elliptic, acute, porrect but not covering the column. *Lip* 15 mm long, 3-lobed; side-lobes erect, clasping the column, obtusely angled and appearing truncate at the apex, shorter than the column; mid-lobe 8–9 × 5–5.5 mm, ligulate, acute, recurved, minutely papillose, margin undulate towards the base; callus of two entire sigmoid ridges, less inflated in the centre than at either the base or apex. *Column* 8 mm long, weakly winged at the apex, with a short column-foot; pollinia 2, deeply cleft, connate at the top and forming one single pollinium, the viscidium crescent-shaped, drawn into two short, thread-like appendages. Figs. 62: 4a–d, 73–76.

DISTRIBUTION. Western Malaysia, Sabah and Kalimantan (Map 15); 450–800 m (1475–2625) ft. It has not been rediscovered in western Malaysia, and is currently known only from the Sook Plain in Sabah, at between 450–500 m (1475–1640 ft) and in a similar habitat in eastern Kalimantan. Both localities have been largely cleared, making *C. rectum* a species in imminent danger of extinction.

HABITAT. Epiphytic in full sun or light shade where it grows in open, damp *Baeckia frutescens* forest, on podsolic soils, and is usually seen on small, stunted trees less than 6 m (20 ft) above the ground. Its root system is full of biting ants that feed on the nectar exuded behind the sepals and at the base of the pedicel. In return, the ants protect the flowers from herbivorous insects. The beautiful forest where it grows also contains the lovely 'sealing wax' palm *Cyrtostachys renda* (syn. *C. lakka*). *Cymbidium rectum* also grows at slightly higher elevations in adjacent valleys that contain more swampy forest, where it appears to grow more luxuriantly. It flowers sporadically throughout the year but probably with a peak during September–December.

CONSERVATION STATUS. EN A1a-d; B1a,b(ii)(iii).

Fig. 76. *Cymbidium rectum*, Sabah.

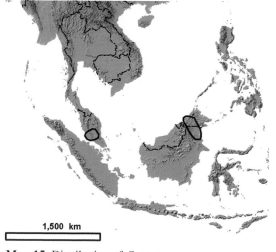

1,500 km

Map 15. Distribution of *C. rectum*.

SECTION BORNEENSE

Section **Borneense** Du Puy & P.J. Cribb, *The Genus Cymbidium*: 80 (1988). Type: *C. borneense* J.J. Wood.

Cymbidium borneense is an unusual terrestrial species characterised by its callus ridges reduced to two small swellings at the base of the mid-lobe of the lip, and the presence of four pollinia.

Studies of the leaf surface show that species in section *Jensoa* have strongly raised, hemispherical stomatal covers with a circular aperture and minutely papillose epidermal cells. *Cymbidium borneense* lacks these leaf surface features, having smooth epidermal cells and ellipsoidal stomatal covers with slit-shaped apertures, a pattern otherwise characteristic of species in section *Cymbidium*. The texture of its leaf is more coriaceous than that of the species in section *Jensoa*, more like that of section *Cymbidium*. However, the leaves of *C. borneense* lack the unequally bilobed apex of section *Cymbidium*. Transverse sections show that the leaves do not have a subepidermal layer of lignified cells, and that the leaves are thinner and lack the layer of palisade-like mesophyll cells of section *Cymbidium*. These comparisons are summarised in Table 9.

The number of pollinia is usually considered to be a conservative character, and has been used by Dressler (1981) to differentiate tribe Cymbidieae from the other vandoid orchids. All of the genera related to *Cymbidium* have two, usually cleft, pollinia. This indicates that the pollinial number (four) of species in sections *Jensoa* and *Pachyrhizanthe* is an advanced character in the tribe Cymbidieae, probably evolving from the two deeply cleft pollinia by the loss of the fusion along the inner margin. It seems possible that this step could have occurred twice in *Cymbidium*, giving rise to four pollinia in sections *Jensoa* and *Pachyrhizanthe*, and again separately in *C. borneense*. The leaf epidermis and stomatal characters indicate a link with section *Cymbidium* and the conclusion that the ancestor of *C. borneense* was a species with two deeply cleft pollinia and leaves similar in epidermal and stomatal morphology to those found in section *Cymbidium*. If it were proposed that *C. borneense* should be placed in section *Jensoa*, on the basis of pollinium number, the dissimilarities in stomatal cover shape, stomatal aperture shape and epidermal cell surface type would have to be accounted for (see Table 8). It would be necessary to postulate the separate evolution of this complex set of leaf characters in *C. borneense* and the loss of the characteristic leaf surface features found throughout section *Jensoa*. This alternative seems to be so unlikely that *C. borneense* is placed in its own section close to section *Cymbidium*.

Although *C. borneense* is similar in many respects to species of section *Cymbidium*, it has four pollinia and its callus is reduced. It has thinner leaves, and the apex is not strongly unequal or bilobed. It lacks the palisade-like mesophyll cells and the subepidermal sclerenchymatous layer in the leaves. It also has longer floral bracts. These differences were considered sufficient to justify a new section, *Borneense*, to accommodate it (Du Puy & Cribb, 1988). Yukawa *et al.* (2002) placed it sister to section *Cymbidium* in their molecular analysis of the genus.

Anatomical and DNA studies (Yukawa & Stern, 2002; Yukawa *et al.*, 2002) also showed that the little-known Philippines species *C. aliciae* is sister to *C. borneense*. The former, unlike *C. borneense*, has a well-developed two-ridged callus that is porate at the apex like that in species of section *Jensoa*. The affinities of section *Borneense* are more fully discussed in the phylogeny chapter (Chapter 11).

TABLE 9

A COMPARISON OF *C. BORNEENSE* WITH SECTION *JENSOA* AND SECTION *CYMBIDIUM*

Character	section *Jensoa*	*C. borneense*	section *Cymbidium*
Leaf texture	leaf not coriaceous	leaf coriaceous	leaf coriaceous
Stomatal cover shape	stomatal covers hemispherical	stomatal covers ellipsoid	stomatal covers ellipsoid
Stomatal aperture shape	stomatal aperture circular	stomatal aperture narrow, slit-shaped	stomatal aperture narrow, slit-shaped
Epidermal cell surface	epidermal cells papillose	epidermal cells smooth	epidermal cells smooth
Subepidermal strands	subepidermal sclerenchyma strands separate	subepidermal sclerenchyma strands separate	subepidermal sclerenchyma strands linked by a continuous subepidermal layer of sclerenchyma
Mesophyll	mesophyll cells all spherical	mesophyll cells all spherical	mesophyll cells in the upper layers elongated, palisade-like
Callus shape	callus in two ridges, convergent towards the tip, forming a short tube at the base of the mid-lobe	callus reduced to two small swellings	callus in two ridges, parallel, S-shaped or interrupted
Pollinia	four pollinia in two unequal pairs	four pollinia in two unequal pairs	two deeply cleft pollinia
Fragrance	often perfumed	coconut-scented	fruit-scented, except *C. atropurpureum* which is coconut-scented

14. CYMBIDIUM ALICIAE

The Philippino botanist Eduardo Quisumbing described *Cymbidium aliciae* in 1940, naming it for Mrs Alice Day who flowered the type material that had been collected in the Nova Vizcaya Mountains of Luzon. An excellent line drawing accompanied the description. It is apparently endemic to Central Luzon and Negros in the Philippines, where it grows as a terrestrial or occasionally an epiphytic species at from 300 to 2750 m (985–9020 ft) elevation. Cootes (2001) provided a photograph of the inflorescence.

 Cymbidium aliciae, like *C. faberi* and *C. cyperifolium*, has obscure pseudobulbs with 8–10 leaves and a lip callus in which the two ridges form an apical pore. These features led Du Puy and Cribb (1988) to consider it a synonym of *C. cyperifolium* and place it mistakenly in section *Maxillarianthe* (now included in section *Jensoa*). Superficially it appears to be closely allied to *C. cyperifolium,* but its floral bracts are much shorter, and its flowers differ in the narrow, linear-lanceolate, acuminate sepals (3–4 mm wide) and petals which are yellowish in colour with some purplish staining, and in the white, more triangular lip with truncate side-lobes, a longer tapering mid-lobe and yellow callus ridges.

Fig. 77. *Cymbidium aliciae*. Philippines. **Fig. 78.** *Cymbidium aliciae*. Philippines.

147

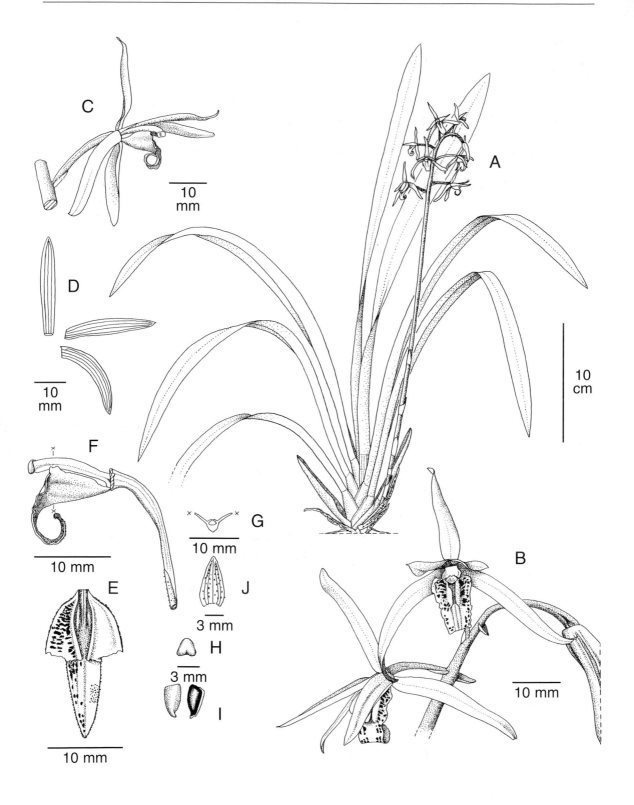

Fig. 79. *Cymbidium aliciae*. **A.** Habit. **B.** Part of inflorescence. **C.** Flower. **D.** Sepals and petal. **E.** Lip. **F.** Lip and column. **G.** Lip cross-section. **H.** Anther-cap. **I.** Pollinia, 4 in two pairs. **J.** Bract. After drawing from life by P. O'Byrne.

Both anatomical and DNA sequence data place it as sister to *C. borneense* with which it shares hypodermal fibre strands both in the abaxial and adaxial leaf surfaces (Yukawa & Stern, 2002; Yukawa *et al.*, 2002).

14. Cymbidium aliciae Quisumb. in *Philipp. J. Science* 72: 486 (1940). Type: Philippines, Luzon, Mts of Nueva Vizcaya, cult. Mrs. K.B. Day (holotype PNH†).

A terrestrial or occasionally an epiphytic *herb*, 40–60 cm tall. *Pseudobulbs* obscure. *Leaves* 7–10, subcoriaceous, linear-lorate, acute, 45–60 × 1.2–2 cm, prominently articulated near the base. *Inflorescence* erect, 4–10 flowered, 19–20 cm tall; peduncle bearing greenish sheaths 2.4–3 cm long; bracts subulate, 0.7–0.8 cm long, greenish. *Flowers* up to 5 cm tall, 3–3.5 cm across, fragrant, with cream to pale green sepals and petals, lip white with purplish red spots on the side- and mid-lobes; pedicel and ovary c. 2.5 cm long, greenish purple. *Sepals* spreading widely, linear, acute to acuminate; dorsal sepal erect, 2.8–3.5 × 0.3–0.6 cm; lateral sepals falcate, 2.5–3 × 0.35–0.6 cm. *Petals* porrect to suberect, lying horizontally above the column, narrowly linear-lanceolate, acute to acuminate, 2.3-3 × 0.45–0.5 cm. *Lip* obscurely 3-lobed, 2–2.4 × 1.1–1.6 cm; side-lobes erect, narrowly elliptic, truncate in front, 0.4 cm broad; mid-lobe recurved, triangular-oblong, obtuse, 0.7–1.1 × 0.5–0.55 cm, minutely papillose; callus of two connivent ridges forming a pore at the apex. *Column* clavate with a flattened lower surface, 1–1.1 cm long, white; pollinia 4, in two unequal pairs. Figs. 77–80.

DISTRIBUTION. Philippines only (Central Luzon, Negros) (Map 16).

HABITAT. Terrestrial or occasionally epiphytic in shaded places in montane forest; 300–2750 m (985–9020 ft).

CONSERVATION STATUS. EN A1cd; B1a,b(ii)(iii).

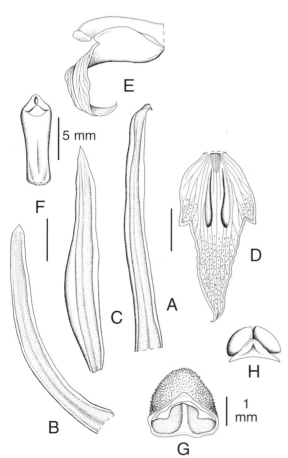

Fig. 80. *Cymbidium aliciae*. **A.** Dorsal sepal. **B.** Lateral sepal. **C.** Petal. **D.** Lip, flattened. **E.** Lip and column, side view. **F.** column, ventral view. **G.** Anther-cap. **H.** Pollinarum. Drawn from Kew Spirit Collection by Hazel Wilkes.

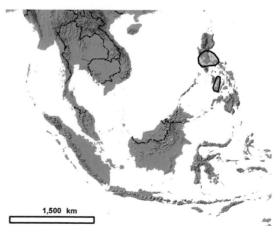

Map 16. Distribution of *C. aliciae*.

15. CYMBIDIUM BORNEENSE

Cymbidium borneense appears to be endemic to northern Borneo. Wood (1983) based his original description on material collected in northern Sarawak and it has since been found in central Sabah (Du Puy & Lamb, 1984). Specimens from Sabah flowered at Kew in March 1984.

The taxonomic position of *C. borneense* is complex. It has some characters that are normally diagnostic of section *Jensoa*, and some of section *Cymbidium*. Sections *Jensoa* and *Pachyrhizanthe* are usually distinguished by the presence of four pollinia in two pairs, whereas the other sections have two cleft pollinia. This would therefore place *C. borneense* in section *Jensoa*, close to *C. cyperifolium,* because of the numerous crowded, distichous leaves (Wood & Du Puy, 1984). However, the characteristically complex callus structure of that section is absent in *C. borneense*, its callus reduced to two small swellings. Its leaf micromorphology has also been demonstrated to share the elongated stomata with slit-shaped pores that are characteristic of section *Cymbidium*, and it lacks the suite of leaf surface characters found in section *Jensoa*. The molecular analyses of Yukawa *et al.* (2002) have confirmed its close relationship to section *Cymbidium*.

15. Cymbidium borneense J.J. Wood in *Kew Bull*. 38: 69–70, t. 1 (1983). Type: Borneo, Sarawak, *Lewis 314* (holotype K!, isotype SAR).

A perennial, terrestrial *herb*. *Pseudobulbs* 8 × 1.5 cm, fusiform, with 6–13 distichous leaves and covered by sheathing leaf bases with a membranous margin 2 mm broad, and occasionally 2–3 scarious or fibrous cataphylls on young specimens. *Leaves* (12)40–79 × 0.5–2.1 cm, linear-ligulate, acute, arching, coriaceous, articulated 8–12 cm from the base, not constricted into a petiole. *Scape* 16–18 cm

Fig. 81. *Cymbidium borneense*, Sabah.

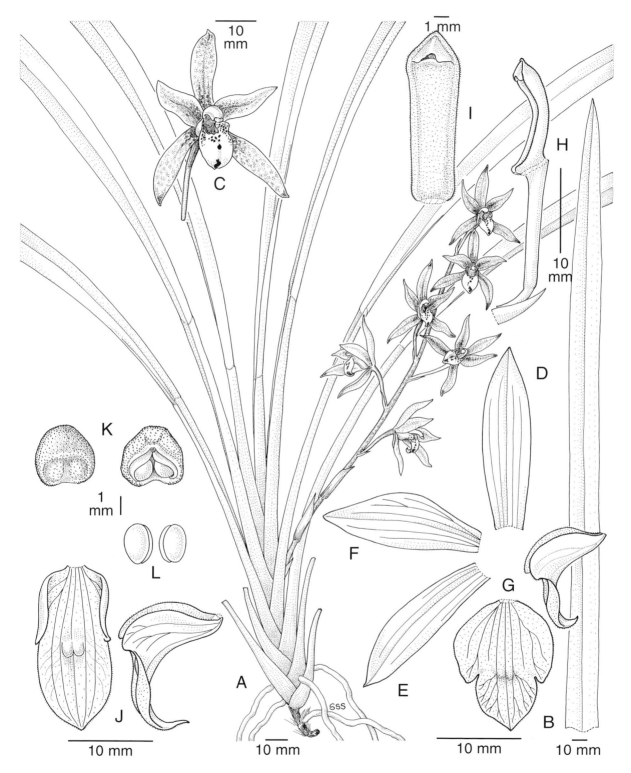

Fig. 82. *Cymbidium borneense*. **A.** Habit. **B.** Leaf. **C.** Flower. **D.** Dorsal sepal. **E.** Lateral sepal. **F.** Petal. **G.** Lip.
H. Column, ovary and bract. **I.** Column, ventral view. **J.** Lip, two views. **K.** Anther-cap, two views. **L.** Pollinaria,
4 in two pairs. Drawn from *Lewis et al.* 314; *Lamb* C18 and 7 by Susanna Stuart-Smith.

long, suberect, arising from within the lower leaf bases, with 3–5 flowers produced in the apical third of the scape and held below the leaves; peduncle arching upwards, with 3–5 distant, amplexicaul sheaths up to 1.5 cm long; bracts 0.5–1.5 cm long, narrowly ovate, acute. *Flowers* c. 4 cm across, coconut-scented; rhachis green, pedicel and ovary pale olive-green, stained with red-brown; sepals and petals cream with a narrow white margin, strongly stained and blotched with maroon-purple, especially in the centre; lip white, side-lobes speckled with maroon, mid-lobe with some maroon blotches and a pale yellow patch at the base; callus ridges pale yellow; column cream above, stained maroon-purple below; anther-cap pale yellow. *Pedicel and ovary* 2–3.5 cm long. *Dorsal sepal* 2.2–2.8 × 0.5–0.6 cm, narrowly oblong-elliptic, apiculate, erect, margins revolute; lateral sepals similar, subacute, curved, spreading to deflexed. *Petals* 2–2.4 × 0.6–0.8 cm, narrowly ovate, oblique, slightly broader than the sepals, porrect, but not forming a hood over the column. *Lip* about 1.5 cm long when flattened; side-lobes erect, rounded, subacute at the apex, minutely papillose; mid-lobe 7–8 × 7 mm, ovate to oblong, obtuse to subacute, recurved, minutely papillose, margin entire; callus reduced, composed of two small swellings at the base of the mid-lobe. *Column* 1.2–1.3 cm long, arching, winged, papillose; pollinia 4, ovate-elliptic, in two unequal pairs; viscidium triangular, with two short processes at the lower corners. Figs. 38: 2a–e, 81–83.

DISTRIBUTION. Eastern Malaysia (Sarawak, Sabah) (Map 17); 150–1300 m (490–4265 ft).

HABITAT. Terrestrial in rainforest, in deep shade and humus-rich soils over limestone or ultra-basic rocks, often near streams. In Sarawak, *C. borneense* is found on low limestone hills between 150 and 300 m (490–985 ft), but in Sabah it grows at higher elevations in lower montane ridge-top forest on ultra-basic rocks at about 1300 m (4265 ft) with *Gymnostoma* (*Casuarina*) *sumatrana*. It flowers in March–April in Sabah, and in October in Sarawak.

CONSERVATION STATUS. EN A1acd; B1a,b(ii)(iii).

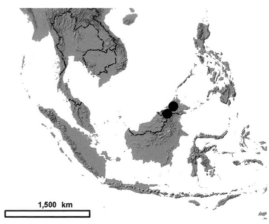

Fig. 83. *Cymbidium borneense*. Sepilok Forest Reserve, Sandakan, Sabah.

Map 17. Distribution of *C. borneense*.

1,500 km

152

SECTION BIGIBBARIUM

Section **Bigibbarium** Schltr. in *Fedde, Repert. Sp. Nov. Regni Veg.* 20: 105 (1924). Type: *C. devonianum* Paxt. (lectotype chosen by P.F. Hunt, 1970).

Schlechter (1924) established this section which contains a single, distinctive species characterised by its leaves narrowed to a slender petiole from a broad, elliptic lamina. The petals are rhombic in shape, and almost the entire lip has two small swellings replacing the callus ridges, and two large deep purple spots at the base of the mid-lobe. There is a short column-foot.

The two cleft, triangular pollinia and the pendulous scape indicate an affinity with section *Cymbidium*. The leaf margin is acuminate in transverse section, similar to that of section *Cyperorchis*, but the small flowers and the lack of a fusion between the base of the lip and the base of the column preclude its inclusion there. The anomalous placement of this section is reflected in both anatomical (Yukawa & Stern, 2002) and molecular analyses (van den Berg *et al.* 2002; Yukawa *et al.*, 2002). In none of the published cladistic analyses is the position of *C. devonianum* satisfactorily resolved (see Chapter 11), although a strict consensus tree based upon *matK* and ITS sequences (Yukawa *et al.*, 2002) placed it between a clade that included section *Floribundum,* the Asiatic species formerly placed in *Austrocymbidium* and sections *Jensoa* and *Pachyrhizanthe* on the one hand, and section *Cyperorchis* on the other (see Fig. 27). Van den Berg *et al.* (2002) included it in a clade with section *Jensoa*, but their ITS data, when analysed separately, combined it with section *Cyperorchis*.

16. CYMBIDIUM DEVONIANUM

Cymbidium devonianum is a distinctive species, both in its foliage and its flowers. Its leaves are superficially not unlike those of *C. lancifolium*, but are thicker in texture, oblique or sharply mucronate at the apex, and always have an entire margin. Moreover, *C. lancifolium* has a totally different growth habit, with cigar-shaped pseudobulbs, an erect scape and fewer, dissimilar, white flowers. The scape of *C. devonianum* is pendulous and usually bears 15–35 closely spaced, purplish flowers.

Although it is not a highly variable species, there is some variation in the flower colour, especially in the amount of red-brown on the sepals and petals. The type specimen had pale tepals and a lip with a pale margin. Variation is also found in the number of leaves produced per pseudobulb.

Several characters of *C. devonianum* are unusual, if not unique in the genus, and it is difficult to say to which species it is most closely related. The two pollinia and the pendulous scape place it alongside the species in section *Cymbidium*, but the differences from the other species are so great that it has been placed in a separate monotypic section.

It first flowered in Britain in 1843, in the Chatsworth collection of the Duke of Devonshire, a great orchid enthusiast and leading patron of horticulture of the day. It had been collected in 1837 by Gibson in the Khasia Hills in northern India. Paxton

Fig. 84. *Cymbidium devonianum*, Thailand, *Menzies & Du Puy* 492.

Plate 13. *Cymbidium devonianum*. Curtis's Botanical Magazine t. 9327, del. Lilian Snelling.

Fig. 85 (left). Habitat of *Cymbidium devonianum*, N Thailand. **Fig. 86** (right). *Cymbidium devonianum*, China, SE Yunnan.

(1843) described and figured this specimen in his *Magazine of Botany*. As with other species named by Paxton, no type specimen was preserved, so the illustration must serve as the type. Gamble and Pantling later collected it in Sikkim and Darjeeling, extending the known range.

Joseph Hooker collected an unusual flowering specimen, which lacked mature leaves, in May 1849, in the Lachen Valley in Sikkim at about 1700 m (5580 ft). He named and described *C. sikkimense* from this material in 1891, indicating that he considered it to be a deciduous species, which produced flowers followed later by the leaves. The type specimen consists of a flowering specimen of *C. devonianum* that lacks leaf laminas, which have broken off at the abscission zone towards the apex of the false petiole. An examination of the specimen shows that the growths are not immature, but rather that they are composed of several cataphylls that conceal the bases of 2–3 mature leaves. No new growths are visible, but this is normal in *C. devonianum,* which tends not to produce new growths until flowering has finished. The illustration by Hooker (1894) shows flowers that are identical with those of *C. devonianum*. As the specimen was collected in the same region as *C. devonianum*, at similar elevations and flowering at a similar time, and as the plant appears identical to this species in all characters except its lack of leaves, it must be assumed that this is in fact a specimen of *C. devonianum* which has, probably for some environmental reason, lost its leaves. Furthermore, this variant has never been found again, all other collections possessing leaves on the newer growths. Hooker also noted on the type specimen that the flowers and the scape appeared to belong to *C. devonianum*.

Plate 14. *Cymbidium devonianum*. NE India, cult. Kew, del. Claire Smith.

Liu and Chen (2000) recently described *C. rigidum* from Pingbian in Yunnan, stating that it was distinct in the genus. However, the type arrived with flowers that were over, and the drawing that accompanies the type description shows the spike mounted erect rather than pendent but with a right-angled bend near the base. The flowers and leaves are typical of *C. devonianum*, and it is undoubtedly conspecific with that species which is nowadays not uncommon in living collections of locally collected native orchids that we have seen in Kunming and Wenshan in Yunnan. To compound this problem, the name is a later homonym of *C. rigidum* Willd. (1804) and thus illegitimate.

There had hitherto been several dubious reports of this species in Laos, Cambodia and Vietnam by Guillaumin (1932, 1960, 1961a, 1961b). In 1932, he identified a leafless specimen collected by Contest-Lacour from Cambodia as *C. sikkimense*, but this has been shown by Seidenfaden (1983) to be *C. macrorhizon*. A leafless specimen (*d'Orleans* 395) from Laos and also mentioned by Guillaumin has now been identified by Seidenfaden as *C. aloifolium*. The three later references identify various specimens from near Dalat, Annam (Vietnam) as *C. devonianum*. Guillaumin (1961b) claimed that one of these specimens, collected by Tixier, corresponds exactly with the King and Pantling figure of *C. devonianum*. An examination of the flowers shows that this specimen, as represented in the Paris herbarium (flowers only), lacks the rhomboid petal and lip shape and strongly winged column apex of *C. devonianum*. Moreover, the callus is of two well-defined, sigmoid ridges, and the lip has well-defined side-lobes with obtuse apices and forms a sac towards the base, all features pointing to *C. bicolor*. Seidenfaden (1983) suggested that the flowers are close to *C. devonianum* as they have a distinct, short column-foot, but this feature is also found in *C. bicolor*. Therefore, all of the references to *C. devonianum* and *C. sikkimense* by Guillaumin can be rejected. Its presence in northern Thailand and Vietnam has since been confirmed.

16. Cymbidium devonianum Paxton, *Mag. Bot.* 10: 97–98, + fig. (1843). Type: Icon. in *Paxton*, loc. cit. (1843).

C. sikkimense Hook.f., *Fl. Brit. India* 6: 9 (1891) & *Icon. Plant.* 12: t. 2117 (1894). Type: Sikkim, Lachen Valley, *J.D. Hooker s.n.* (holotype K!).
C. rigidum Z.J. Liu & S.C. Chen in *Acta Phytotax. Sinica* 38 (6): 570 (2000), **nom. illegit.** Type: China, Yunnan, Pingbian, 14 May 2000, *G.N. Wang in S.C. Chen 2507* (holotype SZWN).

A perennial, lithophytic or epiphytic *herb*. *Pseudobulbs* represented by a swelling of the base of the shoot, about 3 × 2 cm, covered by 5–6 scarious cataphylls and the sheathing, persistent leaf bases; new growths bilaterally flattened, about 2–3 cm broad, composed of folded, flattened and keeled, distichous, purple cataphylls up to 10 cm long, from which the 2–4 true leaves emerge. *Leaves* suberect, 20–49 cm long including the channelled petiole; lamina 17–30 × 3.5–6.2 cm, elliptic, coriaceous, smooth, with a prominent mid-vein, obtuse to subacute, oblique, mucronate, margin entire; articulated 5–15 cm from the base. *Scape* (15)24–44 cm long, pendulous, with about 15–35 closely spaced flowers; peduncle 7–12 cm long, horizontal to pendulous, covered by 6–7 cymbiform, acute, spreading, purple sheaths up to 6 cm long; bracts 2–5 mm long, triangular, acute. *Flowers* 2.5–3.5 cm across, usually orientated towards one side of the scape; not scented; rhachis, pedicel and ovary purple to greenish; sepals and petals pale yellow to dull green, lightly to heavily mottled with purple-brown; lip purple, the side-lobes and the disc cream with maroon mottling, the mid-lobe maroon with two deep purple blotches at the base; column greenish, speckled red-brown above, and with a dark maroon spot at the base ventrally; anther-cap cream. *Pedicel and ovary* (8)12–24 cm long. *Dorsal sepal* 20–29 × 7–10 mm, elliptic, obtuse, erect, with recurved margins; lateral sepals similar, oblique, spreading, pendulous. *Petals* 16–22 × 6–10 mm, subrhombic, subacute to acute, oblique, spreading. *Lip* broad, 14–16(20) × 10–15 mm, rhombic, papillose, 3-lobed, attached to a column-foot; side-lobes obscure; mid-lobe about 8 × 8 mm, triangular-ovate, decurved to weakly recurved, obtuse to mucronate; callus reduced to two swellings at the base of the mid-lobe. *Column* 11–14 mm long, narrow at the base, broadening into a winged apex; pollinia 2, 1.5 mm long, triangular, deeply cleft, on a triangular viscidium. *Capsule* 3.5 × 2 cm, oblong-ellipsoidal, with an apical beak about 1 cm long formed by the persistent column. Pls. 13 & 14; Figs. 38: 6a–d, 84–86.

DISTRIBUTION. Nepal, north-east India (Sikkim, Meghalaya), Bhutan, ?Myanmar, northern Thailand, southern China (Yunnan) and possibly Laos and Vietnam (Map 18); 1200–2200 m (3936–7220 ft). *Cymbidium devonianum* is found between 1500–2200 m (4920–7220 ft) in northern India, from 1450 m (4760 ft) in northern Thailand and about 1200 m in Yunnan.

It has been collected most often in Sikkim and Meghalaya. Hara *et al.* (1978) included eastern Nepal and Bhutan in its distribution. Du Puy (1983, 1984) and Seidenfaden (1983) recorded its discovery in northern Thailand, and it has also recently been collected in Yunnan (Liu & Chen, 2000) and northern Vietnam (Averyanov *et al.*, 2003).

HABITAT. On mossy rocks and moss-covered trees where humus and leaf litter have accumulated, in broken shade; flowering April–June. It usually flowers during May and June in northern India, and was in flower in early April in Thailand. Some out-of-season flowering has been observed in cultivated specimens.

In northern Thailand, *C. devonianum* has been found growing in scrubland dominated by *Rhododendron lyi*, *Lyonia ovalifolia*, *Lithocarpus* and *Agapetes saxicola* at 1450–1500 m (4760–4920 ft) elevation (Du Puy 1983, 1984). It usually grows there on moss- and lichen-covered sandstone rocks in the broken shade cast by taller (3 m (9.8 ft)) shrubs, and in taller stands of vegetation surrounding the numerous creek beds. A layer of leaf litter and moss covered the rocks, and the orchid's roots spread widely through this moisture-retentive substrate, the inflorescence often lying horizontally on the mossy surface. Some specimens were also seen growing epiphytically on old, shaded, moss-covered branches, others in dry and exposed situations below small shrubs on almost bare rock. Although the plants were seen in flower in early April, during the dry season, the substrate retained enough moisture to maintain humidity around the roots of the orchid. New growth takes place during the rainy season from June to November. The pollination of this species seemed to be efficient, and almost all of the flowers had swollen ovaries. However, attempts to observe the pollinator were unsuccessful.

The first specimens of *C. devonianum* collected by Gibson were growing epiphytically on the trunks of decayed trees and in the forks of the branches of old trees where some humus had collected. J.D. Hooker, in his *Himalayan Journals* (2: 294–295, 1854) mentioned that it was growing lithophytically with other orchids on the top of Kollong Rock (Meghalaya). This is a red granite outcrop covered on top by mosses, lichens, club-mosses and ferns, among which, he reported, the orchids grew and flowered freely, even though they were exposed to the elements, including occasional frosts. It is sometimes found high up in the crevices of rocky cliffs in northern India. However, *C. Bailes* (pers. comm.) recently observed populations of this species in the Teesta Valley, in Sikkim (northern India), where it commonly occurs as an epiphyte on mossy branches.

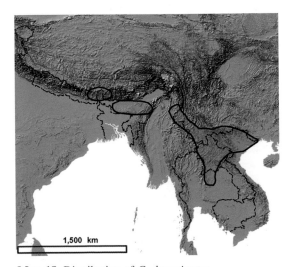

Map 18. Distribution of *C. devonianum*.

CONSERVATION STATUS. VU A1d; B1b(i)(iii).

SECTION HIMANTOPHYLLUM

Section **Himantophyllum** Schltr. in *Fedde, Repert. Sp. Nov. Regni Veg.*, 20: 103 (1924). Type: *C. dayanum* Rchb.f. (lectotype chosen by P. Hunt, 1970).

This section is related to section *Cyperorchis* and contains a single distinctive species, *C. dayanum*. It was placed in subgenus *Cymbidium* by Du Puy & Cribb (1988) as it has two cleft pollinia and lacks any fusion between the margins of the base of the lip and the base of the column. Results from DNA analyses place it as sister to, or nested in, section *Cyperorchis* and we follow that classification here (see Chapter 11). The cladogram (a strict concensus of ITS and *matK* sequences) published by Yukawa *et al.* (2002) places *C. dayanum*, along with *C. tigrinum*, as basal to the rest of the clade containing section *Cyperorchis* (Fig. 27). In the similar analysis of van den Berg *et al.* (2002) it was placed in a clade with *C. erythrostylum* (missing from Yukawa's analysis), basal to section *Cyperorchis* (Fig.26). Although there are similarities between these species, notably in the unusual flower shape with decurved lateral sepals and petals covering the column, we maintain these distinctive species in separate sections due to the notable morphological differences between them. Neither of the molecular systematic analyses included all three monotypic sections *Himantophyllum*, *Annamaea* and *Parishiella*, although they all appear to be basal to this clade.

Section *Himantophyllum* is characterised by its slender, acuminate leaves that resemble those of *C. ensifolium* (section *Jensoa*), but do not have the acuminate leaf margins in transverse section, observable as a narrow hyaline margin, which characterise the rest of the species in sections *Cyperorchis*, *Annamaea* and *Parishiella*. The maroon-and-white flowers are also distinctive, with sharply acute sepals and petals, the sepals decurved and the forward-pointing petals tending to form a hood over the column. The base of the lip and the column are not fused, and the lip has two strongly pubescent callus ridges, the hairs being glandular with swollen tips, a character that is unusual in the genus. These differences are sufficient to place *C. dayanum* in a section of its own.

Superficially, *C. dayanum* resembles some of the species in section *Jensoa*, especially in its vegetative habit and its slender, acute, arching leaves. It can be separated from those species by its epiphytic habit and rather thicker leaves, lacking protruding vascular bundles and with a slightly unequal tip. It can be easily distinguished from the species in section *Cymbidium* by its narrow, acute, thinner-textured leaves and its glandular-hairy callus ridges.

17. CYMBIDIUM DAYANUM

John Day of Tottenham, London, a well-known orchid grower, was the first person to flower this species in Europe in 1869, having imported it from Assam in 1865. Reichenbach described it in 1869, and John Day's drawing of the type specimen is preserved in one of his scrapbooks of orchid drawings at Kew (Cribb & Tibbs, 2004).

Cymbidium dayanum is a beautiful species with elegant, white and wine-red flowers on an arching to pendent inflorescence, and graceful, arching foliage. It can be easily recognised by its narrow, acute leaves, sharply pointed sepals, petals and lip, and well-defined, parallel, white callus ridges covered in short, white glandular hairs that extend in two lines beyond the callus tips into the mid-lobe of the lip.

King & Pantling (1895, 1898) described and illustrated *C. simonsianum* from material collected by Pantling in the Teesta Valley, Sikkim, although Pradhan (1979) states that the locality is actually Darjeeling. Neither the type specimen nor the description differs from *C. dayanum*, and it is included here as a synonym of *C. dayanum*.

In 1891, Sander named and described *C. pulcherrimum* based upon a cultivated specimen from Assam. Although a type specimen does not appear to have been preserved, the description is unquestionably of *C. dayanum*, and it is therefore reduced into synonymy.

Hooker (1891) included *C. dayanum* in his *Flora of British India* as a variety of *C. eburneum*. These two species are distinct and the confusion apparently arose through the use at that time of the name *C. eburneum* var. *dayi* Jennings (1875) for a colour variant of *C. eburneum* with a red-spotted lip. Hooker also noted that, at the time of writing, he had not seen a specimen of *C. dayanum*.

Although the first collections of *C. dayanum* were from northern India, it has a wide distribution, and it was not long before collections were made from other localities. The first collection from Taiwan, made by A. Corner and cultivated by Leach, was named by Reichenbach (1878) as *C. leachianum*. He distinguished this variant from *C. dayanum* by its broader leaf, less acute sepals, shorter mid-lobe of the lip and interrupted callus ridges. This description does appear to differ from *C. dayanum*, particularly in the shape of the callus ridges, but it is part of the population in Taiwan that has been studied by several later authors as *C. dayanum* var. *austro-japonicum* (Tuyama, 1941; Ohwi, 1965; Garay & Sweet, 1974; Lin, 1977), *C. alborubens* (Makino, 1902; Schlechter, 1919; Hu, 1973; Mark *et al.*, 1986) or *C. simonsianum* (Makino, 1912; Hayata, 1914). Wu & Chen (1980), Seidenfaden (1983), Seth & Cribb (1984) and Du Puy & Cribb (1988) have all included *C. leachianum* as a synonym of *C. dayanum*.

Makino (1902) described *C. alborubens* from a specimen collected in southern Japan. He did not compare it with either *C. dayanum* or *C. simonsianum*. Although Schlechter (1919) recognised this name, Hayata (1914) had previously included it in *C. simonsianum*. Its description does not differ from *C. dayanum*, and following the work of many authors on the floras of Japan and Taiwan, it is placed in the synonymy of *C. dayanum* (Tuyama, 1941; Ohwi, 1965; Garay & Sweet, 1974; Lin, 1977; Liu & Su, 1978; Wu & Chen, 1980). In 1914, Makino also published the name *C. simonsianum* f. *vernale*. This was a spring- rather than summer-flowering variant. Despite the observation by Nagano (1955) of these two variants in cultivation, it has proved impossible to distinguish them morphologically, and they are not given any formal taxonomic recognition here.

Fig. 87. *Cymbidium dayanum*, Borneo.

As noted previously, several authors accepted that there are differences between *C. dayanum* from northern India and the variant found in Japan and Taiwan. The most common classification for the latter is as *C. dayanum* var. *austro-japonicum*, named by Tuyama (1941) from material collected in the south of the island of Kyushu, Japan. He noted that it was morphologically indistinguishable from the specimens from Taiwan. He included *C. alborubens* Makino and *C. simonsianum sensu* Hayata, *non* King & Pantling as synonyms of var. *austro-japonicum,* which he distinguished from the type by its narrower leaves (1.1–1.3 cm (0.4–0.5 in) broad), and its narrower corolla lobes (sepal 7.5 mm (0.29 in) broad, petal 6 mm (0.24 in) broad, lip 21 × 15 mm, 0.8 × 0.6 in). These sizes are, in fact, typical for *C. dayanum*, and cannot be used to differentiate the taxa. The rest of the description is also in accord with that of *C. dayanum*. Of the other authors who have used this varietal name, only Lin (1977) provided any differences between it and the type variety. He stated that var. *austro-japonicum* differs in the apex of the mid-lobe of the lip, which is rounded rather than acute, and the inflorescence, which is pendulous rather than suberect. However, the inflorescence of *C. dayanum* is usually pendulous except where the crowded pseudobulbs and leaves artificially direct the scape upwards. Descriptions of *C. dayanum* from Taiwan and Japan vary. Garay & Sweet (1974) and Ohwi (1965) described the apex of the mid-lobe of the lip as cuspidate, Liu & Su (1978) as acute, Hayata (1914) as cuspidate-acuminate, Tuyama (1941) as acuminate to apiculate, and Seidenfaden (1983) as having an acute mid-lobe. This contradicts the statement by Lin concerning the lip shape, and there is no evidence that the mid-lobe is any different from the specimens collected from other parts of the distribution. Specimens from China also have an abruptly apiculate lip apex. There does not seem to be any consistent difference between the populations, and var. *austro-japonicum* is not therefore recognised here as distinct.

Shortly before his death, Rolfe examined Kerr's collections from Thailand, and had named and described, but not yet published, several new species. Downie (1925) subsequently published these descriptions. Rolfe distinguished his new species, *C. sutepense*, from *C. dayanum* by its larger flowers. However, the dimensions given fall within the range of *C. dayanum*. Seidenfaden (1983) concluded that *C. sutepense* cannot be maintained as a distinct taxon, a view which is followed here.

Fig. 88 (top right). *Cymbidium dayanum*, Sabah, *Bailes & Cribb* 751. **Fig. 89** (bottom). *Cymbidium dayanum*, Thailand, *Menzies & Du Puy* 74.

161

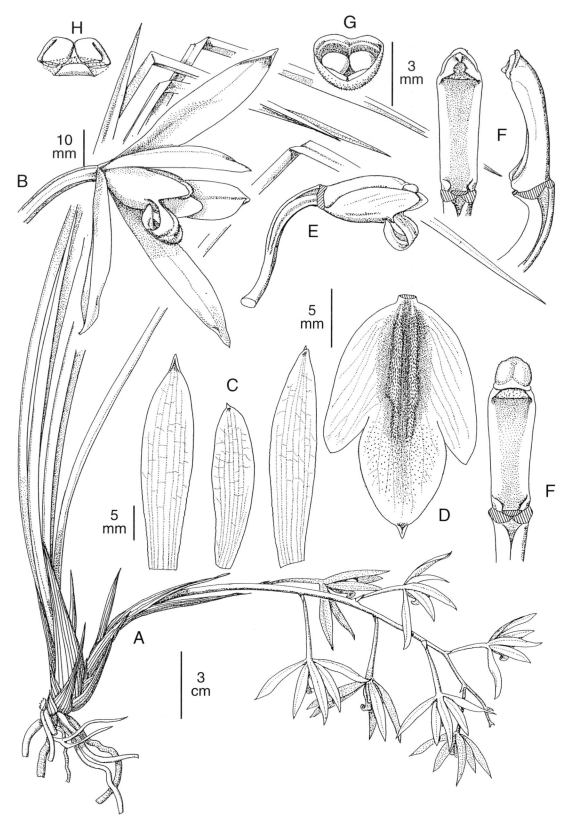

Fig. 90. *Cymbidium dayanum*. **A.** Habit. **B.** Flower. **C.** Dorsal sepal, petal and lateral sepal. **D.** Lip, flattened. **E.** Ovary, column and lip, side view. **F.** Column, three views. **G.** Anther. **H.** Pollinarium. A: specimen from Tanegashima Isl. (TI); B-H: cultivated specimen. All drawn by Mutsuko Nakajima.

Cymbidium aestivum, described by Z.J. Liu and S.C. Chen in 2004, like *C. sutepense*, might also fall within the circumscription of *C. dayanum*. The authors distinguish it from *C. dayanum* by its larger, purple flowers, longer floral bracts and mid-lobe three-fifths the length of the lip. Its habit does not differ otherwise from typical *C. dayanum*. In view of its large darker flowers and bracts, the possibility that a hybrid or polyploid origin, involving *C. dayanum*, needs investigation. Good photographs of the habit and flowers have been published by Liu *et al.* (2006)

Cymbidium poilanei was described by Gagnepain (1931) based on material from Cambodia. He considered it to be related to *C. ensifolium*, differing in having white flowers and prominent side-lobes to the lip. Seidenfaden (1983) recognised it as a separate species, and added that it could be further distinguished from the other species in section *Jensoa* by a row of hairs along the crest of the callus ridges. These characters are all strongly characteristic of *C. dayanum*, which also closely resembles *C. ensifolium* in its vegetative habit. Furthermore, Gagnepain noted that it had the same flower colour as *C. dayanum*. He distinguished *C. poilanei* from *C. dayanum* by its bilobed leaves, pendulous scapes, larger flowers and acuminate lip apex. Further differences evident in the description are the absence of callus ridges and the four pollinia of *C. poilanei*. However, Seidenfaden commented on the hairy callus of the type specimen, and the observed number of pollinia is uncertain, Gagnepain giving it tentatively as four in his description but following it with a question mark. The leaves of *C. dayanum* are not bilobed and appear shortly acuminate. Closer examination shows that they are in fact slightly oblique at the tip. The scape tends to be pendulous when possible, but may not be able to assume this habit because of the closely packed and congested pseudobulbs and leaves. The flowers are slightly smaller than would be expected in *C. dayanum*, but the difference is marginal. The description of shape of the lip of *C. poilanei* is incongruous, but further examination may show it to be apiculate. Furthermore, *C. poilanei*, like *C. dayanum*, is an epiphyte.

Guillaumin (1932) worked with Gagnepain, and they collaborated to produce the chapter on the Orchidaceae for Lecompte's *Indo-Chinese Flora*. The illustration given there of *C. poilanei* closely resembles *C. dayanum*, and the former is therefore treated here as a synonym of *C. dayanum*. Seidenfaden noted that later specimens identified by Guillaumin as *C. poilanei* should be placed in *C. ensifolium*.

Holttum (1953) described *C. dayanum* from Malaya as having either white or mauve petals and sepals with a median purple band. In Sabah, the extreme of this variation is found, the sepals and petals being deep wine-red in the centre and flushed with maroon to the margin. Although the leaves are narrow on this variant, they are not excluded from the range of expected variation in *C. dayanum*. It is otherwise identical in habit and floral morphology (Du Puy & Lamb, 1984). There appears to be a gradual darkening of the flower colour in a cline towards the southern and south-eastern extremes of the distribution of *C. dayanum*, with the darkest being found in Sabah. Specimens from Luzon (Philippines) have the white sepals and petals with a central stripe typical of *C. dayanum* from the more northerly parts of its distribution.

Cymbidium acutum, collected in Perak, Malaya, was described by Ridley as having petals and sepals 'nearly white ... with a medium band of purple'. In 1924 he included it in the synonymy of *C. dayanum*. This name has been used for the slightly darker variants found in Malaya and Sumatra.

Ames & Schweinfurth (1920) named dark red-flowered specimens, collected by Clemens on Mt Kinabalu, Sabah, as *C. angustifolium*. Although they did not compare them with *C. dayanum*, they differentiated them from *C. acutum* by the shorter leaves, the short, suberect scape and differently marked lip of the latter. The leaf length in *C. dayanum* is variable, and the variation includes the lengths of both. A suberect scape is not unusual for *C. dayanum*, as the closely set pseudobulbs with their dense covering of leaf bases and cataphylls often direct the scape upwards before it can adopt its usual pendulous habit. A shorter scape may become suberect under these circumstances. The lip of *C. acutum* is described as dark carmine with golden spots. The lip of specimens from all over the range of *C. dayanum* is distinctly coloured deep wine-red with whitish streaks and red veins on the side-lobes, a central yellow stripe on the mid-lobe and white callus ridges. The lip of the Sabah variant is almost identical in colour with that of specimens from Sumatra, Malaya and further north.

Plate 15. *Cymbidium dayanum*. Thailand, *Menzies & Du Puy* 72, cult. Kew, del. Claire Smith.

18. Cymbidium dayanum Rchb.f. in *Gard. Chron.*: 710 (1869). Type: Assam, cult. Day (holotype W, isotype K!).

C. leachianum Rchb.f. in *Gard. Chron.* n.s. 10: 106 (1878). Type: Taiwan, *Corner*, cult. Leach (holotype W).

C. pulcherrimum Sander in *Gard. Chron.* ser. 3, 10: 712 (1891). Type: cult. Sander (holotype not preserved).

C. eburneum var. *dayana* Hook.f., *Fl. Brit. India* 6: 12 (1891), *non* var. *dayi* Jennings, Orchids: t. 16 (1875) (= *C. eburneum*).

C. simonsianum King & Pantl. in *J. Asiatic Soc. Bengal* 64(2): 338–9 (1895) & in *Ann. Roy. Bot. Gard. Calcutta* 8: 188, t. 250 (1898). Type: Sikkim, Teesta Valley, *Pantling 51* (holotype K!, isotype P).

C. acutum Ridl. in *J. Linn. Soc.* 32: 334 (1896) & *Mat. Fl. Malay Penins.* 1: 140 (1907). Type: Malaya, Perak, Waterloo Estate, *Elphinstone* (holotype SING!, isotype K!).

C. alborubens Makino in *Bot. Mag. Tokyo* 16: 11 (1902). Type: Japan, Musashi, *Makino*, cult. Tokyo Bot. Gard. (holotype MAK).

C. simonsianum f. *vernale* Makino in *Iinuma, Somoku-Dzusetsu* ed. 3, 4(18): 1186 (1912).

C. angustifolium Ames & C. Schweinf. in Ames, *Orchid.* 6: 212–214 (1920). Types: Sabah, Kiau, *Clemens 74, 39, 82* (syntypes AMES).

C. sutepense Rolfe ex Downie in *Kew Bull.*: 382–383 (1925). Type: Thailand, Doi Suthep, *Kerr 113* (holotype K!).

C. poilanei Gagnep. in *Bull. Mus. Nat. Hist. Paris*, sér. 2, 3: 681 (1931). Type: Cambodia, Montagnes de l'Elephant, *Poilane 316* (holotype P).

C. dayanum Rchb.f. var. *austro-japonicum* Tuyama in Nakai, *Icon. Pl. As. Orient.* 4: 363–365, t. 118 (1941). Type: Japan, Kyushu, *Hurusawa*, cult. Koisikawa B. G. (not located).

C. eburneum var. *austro-japonicum* (Tuyama) Hiroe, *Orchid Fl.* 2: 96 (1971).

A perennial, epiphytic *herb*. *Pseudobulbs* about 4 × 2.5 cm, fusiform, slightly bilaterally compressed, usually covered by scarious remains of the persistent leaf bases and the cataphylls; cataphylls to 18 cm long, the longest becoming leaf-like with an abscission zone near the apex, becoming scarious and eventually fibrous with age. *Leaves* distichous, (4)5–8 per pseudobulb, up to (30)40–95(115) × 0.7–1.6(2.4) cm, linear-elliptic, acute to acuminate at the tip, with a slightly oblique apex, erect or arching towards the tip, thickened and coriaceous, dark green, the mid-vein prominent below, articulated 3–8 cm from the pseudobulb. *Scape* 18–30(35) cm long, arching to pendent; peduncle short, suberect to horizontal, covered in pink-veined sheaths up to 8 cm long, inflated and spreading in the apical half, cylindrical at the base; rachis with 5–15(20) flowers; bracts 2–10 mm long, triangular, acute, purplish at the base. *Flowers* 4–5 cm across; not usually scented; rhachis, pedicel and ovary pale green; sepals and petals white or cream with a central maroon stipe that does not reach the apex, or occasionally suffused wine-red with a deeper central stripe, and a narrow whitish margin that may be absent except towards the base; lip strongly marked with maroon, with an orange or yellow spot at the base; side-lobes white, veined maroon, with a maroon margin; mid-lobe deep maroon with a basal, triangular, pale yellow stripe that does not reach the apex; callus ridges white or cream; column dark maroon above and below; anther-cap pale yellow. *Pedicel and ovary* 2–3.5(4) cm long. *Dorsal sepal* (21)25–34 × (5)6–8 mm, narrowly elliptic to oblong-lanceolate, tapering to an acute or shortly acuminate apex, margin slightly erose towards the base, erect; lateral sepals similar, oblique and pendulous. *Petals* 18–28 × 5–7.5 mm, narrowly oblong to elliptic, obtuse to acute, apiculate, considerably shorter and narrower than the sepals, porrect and tending to cover the column. *Lip* 3-lobed, 15–22 × 10–15 mm when flattened; side-lobes well defined, erect and weakly clasping the column, as long as the column, with porrect, triangular tips subacute at the apex and minutely papillose; mid-lobe 8–12 × (6)7–9 mm, ovate, entire, recurved, apex subacute, mucronate to acuminate, minutely papillose, often with minute hairs towards the base and with two lines of longer, often glandular, hairs extending from the callus tips to the centre of the mid-lobe; callus ridges 2, well defined, parallel, from the base of the lip to the base of the mid-lobe, covered in glandular hairs about 0.2 mm long. *Column* 11–14 mm long, arching, winged; pollinia 2, triangular, about 1.2–1.5 mm long, deeply cleft, on an oblong viscidium usually with minute extensions at the lower corners. *Capsule* 4–6 × 1.5–2 cm, ellipsoid, tapering at the base to a stalk and at the apex to a short (1 cm) beak, formed by the persistent column. Pl. 15; Figs. 38: 3a–d, 87–90.

DISTRIBUTION. Northern India (Sikkim, Assam), Bhutan, Myanmar, southern China, Taiwan, Ryukyus, Japan, Philippines (Luzon), Thailand, Cambodia, western Malaysia, Sumatra, Borneo (Sabah) (Map 19); sea level–1800 m (5900 ft).

Cymbidium dayanum occurs in the Himalaya of northern India (Sikkim, Assam) and Bhutan. Its distribution extends east through Myanmar and southern China to Taiwan and southern Kyushu in Japan. Wu & Chen (1980) gave its distribution in China as the southernmost provinces: Fujian, Guangdong, Guangxi, Hainan, Yunnan and Taiwan. It has also been collected in the mountains of northern Luzon, in the Philippines. It occurs in Thailand, Cambodia and Vietnam, and probably also in Laos. Its most southerly localities are in peninsular Malaya, Sumatra and Sabah, but it has not been recorded from Java.

Holttum (1957) and Du Puy & Lamb (1984) noted that in the south of its distribution, it prefers the cooler, higher elevations of about 1000–1500 m (3280–4920 ft). Similarly, in Thailand and northern parts of western Malaysia it is often found at about 600–1200(1600) m (1970–3940(5250) ft) elevation, in China (Hainan) at 1700 m (5580 ft) and in Sikkim between 300–1300 m (985–4265 ft). In Taiwan it occurs in the mountains below 1000 m (3280 ft) (Lin, 1977).

HABITAT. The habitat of *C. dayanum* is almost exclusively in hollows on the limbs or trunks of tall trees or on rotten trees or logs in situations where the plant is in light shade, although it occasionally occurs on humus-covered rocks. In Thailand we have collected it from rotting, felled tree trunks in evergreen forest, where its roots formed a strong network between the bark and the damp, decaying core which some of the roots also penetrated (Du Puy, 1983). In Sabah, we have again noted it growing on rotten logs and dead trees, the roots penetrating into the rotting bark, in positions where it receives full sun for at least part of the day (Du Puy & Lamb, 1984). In Taiwan it is found as an epiphyte, often growing on ancient cypresses (Lin, 1977).

This species usually flowers between August and November (December), although out-of-season flowering is not uncommon in cultivation. In the more equatorial regions of its distribution (Sumatra, Sabah) it tends to flower during August and September, but due to the lack of strong seasonality in the climate it is not unusual for it to flower at other times of the year. There is a variant in cultivation in Japan that is reputed to flower in the early spring.

CONSERVATION STATUS. NT, but vulnerable in some parts of its range.

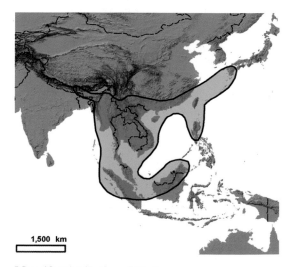

1,500 km

Map 19. Distribution of *C. dayanum*.

SECTION ANNAMAEA

Section **Annamaea** (Schltr.) P.F. Hunt in *Kew Bull.* 24: 94 (1970). Type: *C. erythrostylum* Rolfe.

Schlechter (1924) placed *C. erythrostylum* in its own section on the basis of its having only one leaf on the pseudobulb, a distinctive viscidium, broad petals and sepals and an unusual lip shape. It is now known that the pseudobulb has several leaves and that the hair-like processes on the viscidium and the broad sepals and petals are not unique in the genus. The shape of the lip, and especially the short, triangular mid-lobe are indeed distinctive. Several of its other characters are also unique among the large-flowered species of the genus: the scape produced on immature growths; long bracts; erect dorsal sepal; decurved lateral sepals; porrect petals, covering the column; side-lobes of the lip clasping the top of the column, and having long hairs mainly confined to the veins; short, triangular mid-lobe of the lip; column with a dense, ventral indumentum of long hairs; and deep purple-pink column. The wedge-shaped callus ridge structure is also unique, although reminiscent of that of *C. eburneum*.

In the molecular analysis of ITS and *matK* sequences by van den Berg *et al.* (2002), *C. erythrostylum* is sister to *C. dayanum* in a clade basal to section *Cyperorchis* (Fig 26). There is a resemblance in the flower shape and disposition of the sepals and petals in these two species: the overall flower shape of *C. erythrostylum* is similar to that of *C. dayanum*, although its flowers are much larger and its tepals much broader. However, they are very different in overall morphology and leaf anatomy. Moreover, *C. dayanum* lacks the fusion of the base of the lip and column, and has two separate, glandular-hairy callus ridges (not converging into a single structure as in *C. erythrostylum*). On this basis we are inclined to continue to recognise two separate sections for these two very distinctive species (see also the discussion under section *Himantophyllum*). The position of *C. wenshanense*, a recently described species resembling *C. erythrostylum*, has not been determined in molecular analyses, and may blur the differences between sections *Annamaea* and *Cyperorchis*.

18. CYMBIDIUM ERYTHROSTYLUM

Cymbidium erythrostylum was described by Rolfe in 1905 from a plant cultivated at the Royal Botanic Garden, Glasnevin, that had been collected by Wilhelm Micholitz in 1891 in Annam (Vietnam) for Messrs Sander & Sons of St Albans. It is one of the most attractive of the large-flowered species with its glistening, fragile-looking white petals and sepals, and its boldly red-marked lip and red column. Its early flowering period and long-lasting flowers have made it a useful species for hybridisation.

It is a distinctive species, readily recognised by its large white flower which is narrowly triangular in shape, taller and narrower than the other large-flowered species, with decurved lateral sepals and its petals covering the column, and by its short distinctively shaped and marked lip with a wedge-shaped callus.

The callus structure is reminiscent of that found in *C. eburneum*. Both have a broad, glabrous, ligulate basal portion, with several raised ridges that join at the apex into one inflated, wedge-shaped structure well behind the mid-lobe. However, in *C. erythrostylum* this callus apex is unique in being glabrous, and composed of three distinct lobes of which the central one is the largest, tapering well into the mid-lobe of the lip.

The recently described species *C. wenshanense* from Yunnan in south-western China has a similar flower shape to *C. erythrostylum*, with the lateral sepals decurved and the petals covering the column, and its column is also hairy, although the column of *C. wenshanense* is dorsally, not ventrally, hairy towards the base and is not so strongly coloured. It is undoubtedly closely related to *C. erythrostylum*, but its lip has two separate raised and hairy callus ridges, and the mid-lobe of its lip is broad, emarginate

Plate 16. *Cymbidium erythrostylum*. Curtis's Botanical Magazine t. 8131, del. Matilda Smith.

and glabrous. Its flowers are also somewhat larger, with longer sepals and petals, the dorsal sepal is not erect but arches over the column, and the floral bracts are much smaller.

18. Cymbidium erythrostylum Rolfe in *Gard. Chron.*, ser. 3, 38: 427 (1905), & ser. 3, 40: 265, 286, f. 115 (1906), in *Orchid Rev.* 14: 39 (1906) & in *Bot. Mag.* 133: t. 8131 (1907). Type: Annam [Vietnam], cult. R.B.G. Glasnevin, *Micholitz s.n.* (holotype K!).

Cyperorchis erythrostyla (Rolfe) Schltr. in *Fedde, Repert. Sp. Nov. Regni Veg.* 20: 427 (1924).
Cymbidium erythrostylum Rolfe var. *magnificum* Hort. in *Gard. Chron.*, ser. 3, 90: 81, + fig. (1931).
    Type: cult. *McBeans* (type not preserved).

A perennial, epiphytic, lithophytic or terrestrial *herb*. *Pseudobulbs* about 6 cm long, 2 cm in diameter, produced annually, narrowly ovoid, bilaterally flattened, with 6–8 distichous leaves, and 2–3 cataphylls that become scarious with age. *Leaves* up to 45(55) × 1.5(1.7) cm, narrowly linear-obovate, arching, apex unequal, apiculate, articulated 2–5 cm from the pseudobulb to a persistent, broadly sheathing, yellowish base with a 1–2 mm broad membranous margin. *Scape* 15–35 cm long, from within the axils of the cataphylls or lower leaves on immature growths; peduncle suberect, arching, covered by about 6 sheaths; sheaths overlapping, cymbiform, up to 8 cm long, the middle sheaths cylindrical near the base;

DAVID DU PUY

Fig. 91. *Cymbidium erythrostylum*, diploid (left flower) and tetraploid clones.

rhachis with (3)4–8(12) flowers; bracts (1.5)2.5–5 cm long, narrowly lanceolate, cymbiform. *Flower* about 6 cm across, appearing narrowly triangular, not scented; sepals and petals white, the petals pale pink along the mid-vein in the basal half, sometimes spotted with pink at the base, glistening as though covered in frost and thin in texture; lip yellow-white, darker yellow on the mid-lobe, veined with deep red, the veins becoming broken and spotted towards the margins of the side-lobes, and broader and blotched near the apex of the mid-lobe; callus cream, pink-mottled; column purple-pink above, paler below; anther-cap cream. *Pedicel and ovary* (2.5)4–5.5 cm long. *Dorsal sepal* 4.5–5.7 × 2.2–2.6 cm, obovate-elliptic, acuminate, concave, erect; lateral sepals similar, falcate, curved downwards. *Petals* broad, 4.1–4.9 × 1.7–2.2 cm, obovate-elliptic, acuminate, curved, porrect and covering the column. *Lip* 3-lobed, fused to the base of the column for about 3 mm; side-lobes 1.5–1.9 cm broad, clasping the column, with the hairs mostly confined to the veins, the hairs longest around the callus, apices broadly rounded; mid-lobe 1–1.2 × 1.3–1.5 cm, triangular, acuminate, recurved, hairy, margin undulating; callus glabrous, broad, with (3–)5 raised ridges behind, converging into three ridges in front that become strongly inflated and confluent to form a three-lobed, cuneate apex well behind the mid-lobe of the lip, the central lobe being the largest and tapering into the mid-lobe. *Column* about 3 cm long, arcuate, winged, hairy beneath; pollinia triangular, 2 mm across, on a rectangular viscidium with two long, hair-like processes from the lower corners. Pl. 16; Figs, 91, 92 & 95: 5a–d.

DISTRIBUTION. Vietnam (Map 20); about 1500 m (4920 ft). It appears to be a narrow endemic to a small region of southern Vietnam.

HABITAT. It grows epiphytically in open coniferous woodlands, which are the most widespread habitats in the northern part of south Vietnam, usually developing on degraded soils on granite and hill slopes between 1000–1800 m (3280–5905 ft) where *Pinus kesiya* is the commonest dominant tree species. Epiphytes are occasionally abundant and diverse in these open montane pine forests.

It flowers from May to July in the wild, but from October to December in cultivation in the Northern Hemisphere.

CONSERVATION STATUS. EN A1cd; B1a,b(ii)(iii).

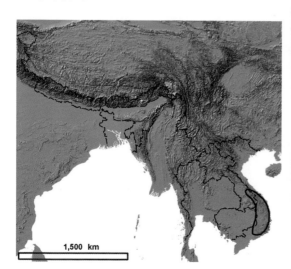

Map 20. Distribution of *C. erythrostylum*.

Fig. 92. *Cymbidium erythrostylum*. Cult. Kew.

SECTION CYPERORCHIS

Section **Cyperorchis** (Blume) P.F. Hunt in *Kew Bull*. 24: 94 (1970); Seth & Cribb in Arditti (ed.), *Orchid Biol., Rev. Persp*. 3: 303 (1984). Type: *C. elegans* Lindl.

Cyperorchis Blume in *Rumphia* 4: 47 (1848); Hook. f., *Fl. Brit. India* 6: 14 (1891).

Cyperorchis section *Eucyperorchis* Schltr. in *Repert. Sp. Nov. Regni Veg*. 20: 106 (1924).

Iridorchis Blume, *Orch. Arch. Ind. Jap*. 1: 90–92, t.26 (1858), *non Iridorkis* Thouars (1809). Type: *Iridorchis gigantea* (Wall. ex Lindl.) Blume (= *Cymbidium iridioides* D. Don).

Iridorchis Blume, *Orch. Arch. Ind. Jap*. 1: 90–92, t. 26 (1858). Type: *Iridorchis gigantea* (Wall. ex Lindl.) Blume (= *Cymbidium iridioides* D. Don).

Cyperorchis Blume in *Rumphia* 4: 47 (1848); Hook. f., *Fl. Brit. India* 6: 14 (1891). Type: *Cyperorchis elegans* (Lindl.) Blume (= *Cymbidium elegans* Lindl.).

Arethusantha Finet in *Bull. Soc. Bot. France* 44: 178–180, pl. 5 (1897). Type: *Arethusantha bletioides* Finet (= *Cymbidium elegans* Lindl.).

C. section *Iridorchis* (Blume) Schltr. in *Fedde, Repert. Sp. Nov. Regni Veg*. 20: 107 (1924).

C. section *Eucyperorchis* Schltr. in *Repert. Sp. Nov. Regni Veg*. 20: 106 (1924).

Cymbidium subgenus *Cyperorchis* (Blume) Seth & Cribb in Arditti (ed.), *Orchid Biol. Rev. Persp*. 3: 300 (1984).

C. section *Iridorchis* (Blume) P.F. Hunt in *Kew Bull*. 24: 94 (1970).

C. section *Eburnea* Seth & P.J. Cribb in Arditti (ed.), *Orchid Biol., Rev. Persp*. 3: 304 (1984). Type: *C. eburneum* Lindl.

Cyperorchis Blume (1848) was given sectional status within subgenus *Cyperorchis* following the treatment of Seth & Cribb (1984). Here we enlarge the concept to include all of the species formerly treated within subgenus *Cyperorchis* sensu Du Puy & Cribb (1988) except *C. erythrostylum* and *C. tigrinum*. Van den Berg *et al*. (2002) and Yukawa *et al*. (2002) published phylogenies of *Cymbidium* based on ITS of nuclear ribosomal DNA and plastid *matK* sequence data, which clearly show the close genetic relationships of the species involved and bring into question the desirability of subdividing *Cyperorchis*. These groupings may become more apparent if different gene sequences, which are more informative within this section, are compared in the future.

Although the molecular systematic studies failed to elucidate the relationships of the species within this large section, groups based upon macromorphological characters are still apparent. The sectional accounts given in Du Puy & Cribb (1988) discuss these groupings, and although it seems necessary to dissolve the section limits, and given the commercial importance of the species in this section, it seems appropriate and informative to record these species groups here (Table 10).

The section is characterised by having narrow, hyaline, incurved leaf margins that appear acuminate in transverse section, a character shared with *C. devonianum*. The flowers are relatively large for the genus with acute sepals and petals, and the dorsal sepal is porrect, tending to cover the column. The base of the lip is fused to the base of the column for about 2–6 mm. In the closely related *C. dayanum* the flowers have acute sepals and petals and lack the fusion of the base of the lip to the lower column margins. The pollinia of species in section. *Cyperorchis* are paired and are deeply cleft behind, but are variable from triangular to quadrangular or clavate, the shape often characteristic for the species groups (Table 10). These characters, however, do not merit its recognition as a distinct genus.

The species in this section, along with *C. erythrostylum*, have formed the basis for the breeding of modern hybrids. The flowers are large (about 7–15 cm in diameter), and mostly open widely except for the dorsal sepal, which is usually porrect and covers the column. The plants are robust, with long, acute leaves and often have large, more or less bilaterally compressed pseudobulbs.

TABLE 10

THE SPECIES GROUPS WITHIN SECTION *CYPERORCHIS*,
BASED ON MACROMORPHOLOGICAL CHARACTERS (AFTER DU PUY & CRIBB, 1988)

Pollinia shape	Growth habit	Flower colour	Other characters	Species
Pollinia clavate to subquadrangular	Pseudobulbs produced annually and flowering from the base	Flowers cream or dark green to purple-brown	Flowers not opening fully to hardly opening and pendulous	*C. elegans, C. cochleare, C. whiteae, C. sigmoideum*
Pollinia quadrangular	Stems not strongly pseudobulbous, growing indeterminately and flowering for several years from the leaf axils	Flowers white	Flowers not opening fully, the tepals not spreading widely Leaf apex finely and acutely bilobed, with a short mucro between the lobes	*C. eburneum, C. parishii, C. roseum, C. mastersii,* [*C. banaense*]
Pollinia triangular	Pseudobulbs produced annually and flowering from the base	Flowers not white, variously coloured	Flowers opening widely, the lateral sepals and petals spreading	*C. tracyanum, C. iridioides, C. erythraeum, C. wilsonii, C. lowianum, C. schroederi C. hookerianum*
		Flowers white		*C insigne, C. sanderae,* [*C. wenshanense*]

19. CYMBIDIUM BANAENSE

Cymbidium banaense was described by the French botanist, Gagnepain, based on a specimen collected at Bana [Ba Na] which is the hill station behind Da Nang in south-central Vietnam. It is possible that the material had been collected elsewhere and brought to Ba Na by an orchid grower. It has not been re-collected. Unfortunately, it was omitted from their monograph of the genus by Du Puy & Cribb (1988). Seidenfaden (1992) discussed it in his account of the orchids of Indochina. It appears to be closely related to *C. eburneum,* but lacks the wedge-shaped callus of that species, and also to *C. mastersii,* but perhaps most closely to *C. sanderae.* These three species all have finely bilobed leaf tips, but none have the broad, obtuse lobes of this species.

    Cymbidium banaense is distinctive, differing from all the other species with large white flowers by its leaves that are unequally roundly or obtusely bilobed at the apex, flowers that have a large and distinctively coloured lip with a mid-lobe much longer than wide, and a callus with two ridges dilated and free at the apex. In many ways its flowers seem closest to *C. sanderae,* but that has smaller flowers with a lip that is marked with purple and has a broader, shorter mid-lobe. Its suberect, few-flowered inflorescence is reminiscent of that of *C. mastersii,* but that has longer, acute leaves, smaller flowers and an ovate mid-lobe of the lip.

    The possibility exists that *C. banaense* is a natural hybrid with *C. eburneum, C. mastersii, C. sanderae* or one of the other white-flowered species as parents. However, the distinctive lip shape and colouring make it difficult to determine which species might have been involved. DNA sequence data might resolve the issue in the future.

19. Cymbidium banaense Gagnep. in *Bull. Mus. Hist. Nat. Paris,* sér. 2, 22: 626 (1950); Seidenfaden, *Contr. Orchid Fl. Cambodia, Laos & Vietnam*: 37 (1975) & in *Opera Bot.* 114: 343, fig. 230 (1992). Type: Vietnam, Bana [Da Nang], 1400 m, *Poilane 29022* (holotype P!).

An epiphytic *herb. Pseudobulbs* not seen. *Leaves* c. 20–25 × 2 cm, linear, unequally obtusely or roundly bilobed. *Inflorescence* arching-suberect, 3-flowered; bracts scarious, 1.2–1.3 cm long, ovate, acuminate. Flowers white with a rose-pink flush; lip with yellow side-lobes and a rose-pink tip to the mid-lobe; pedicel and ovary c. 3 cm long. *Dorsal sepal* 5.7 × 1.5 cm, lanceolate, acuminate, mucronate. *Lateral sepals* similar but falcate. *Petals* 5.7 × 1 cm, lanceolate, falcate, acuminate. *Lip* fused to column at base for 1 cm, 3-lobed above, 5 × 2 cm, papillose on the disc and the base of side-lobes; side-lobes erect, rounded in front, 0.5 cm wide; mid-lobe 1.5 × 0.6–0.8 cm, oblong-ovate, acute; callus of two ridges which are free and dilated at the apex. *Column* arcuate, clavate, 3.5 cm long. Fig. 93.

DISTRIBUTION. South Vietnam only (Map 21); 1400 m (*c.* 4590 ft).

HABITAT. Unknown.

CONSERVATION STATUS. EN A1cd; B1ab.

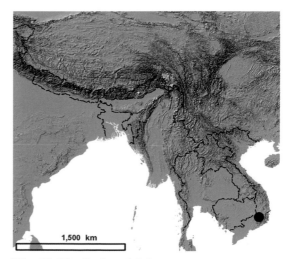

Map 21. Distribution of *C. banaense.*

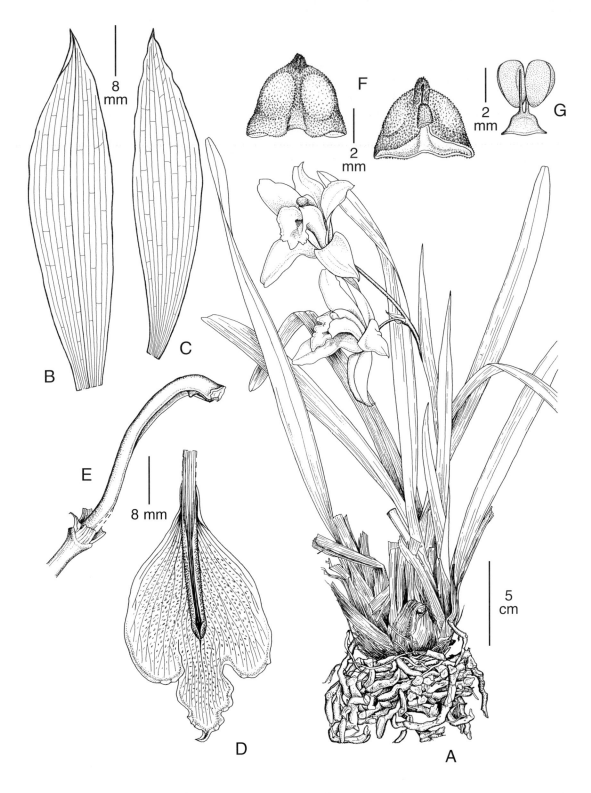

Fig. 93. *Cymbidium banaense.* **A.** Habit. **B.** Dorsal sepal. **C.** Petal. **D.** Lip. **E.** Column. **F.** Anther-cap. **G.** Pollinarium. Drawn from the type, *Poilane* 29022 (P), by Hazel Wilkes.

20. CYMBIDIUM COCHLEARE

With its long, slender leaves and flower spike, *C. cochleare* is an elegant species that resembles *C. elegans* but with greenish brown flowers. The rhachis of the scape is thin and wiry, purple-coloured, pendulous, and carries several bell-shaped, glossy, brown and olive-green flowers. The lip has broad side-lobes and is yellow with reddish mottling.

Cymbidium cochleare was described in 1858 by Lindley from a specimen collected by J.D. Hooker in Sikkim which had almost finished flowering and the ovaries were starting to swell. This poor material led Lindley mistakenly to describe the callus as spoon-shaped at the tip, and prevented description of the mid-lobe or flower colour. In 1879 H.G. Reichenbach was given fresh flowers from the nursery of Messrs Low, and was able to give a more complete description, indicating that the lip has two short but separate callus ridges. The Reichenbach plant had been collected by Boxall, reputedly from Burma (Myanmar), although this may not be the correct locality, there having been no further collections from that country. A cultivated specimen that was sent to Kew from Bangkok is likewise the only record for Thailand. It differs slightly in the shape of the side-lobe tips that are not as triangular and are more rounded than usual. There are many links between the floras of northern India and Thailand, and some other species have a similar disjunct distribution such as *C. mastersii* and *C. devonianum*, and *C. cochleare* may occur in the wild in Thailand.

Fig. 94. *Cymbidium cochleare*, Sikkim, *Bailes*, cult. Kew.

DAVID DU PUY

175

Fig. 95. *Cymbidium cochleare* (Kew spirit no. 12843). **1a.** Perianth, × 0.7. **1b.** Lip and column, × 0.7. **1c.** Pollinarium, × 3. **1d.** Pollinium (reverse), × 3. **1e.** Flower, × 0.7. **1f.** Flowering plant, × 0.1. *C. whiteae* (Kew spirit no. 49023). **2a.** Perianth, × 0.7. **2b.** Lip and column, × 0.7. **2c.** Pollinarium, × 3. **2d.** Pollinium (reverse), × 3. **2e.** Flower, × 0.7. **2f.** Flowering plant, × 0.1. *C. elegans* (Kew spirit no. 45555). **3a.** Perianth, × 0.7. **3b.** Lip and column, × 0.7. **3c.** Pollinarium, × 3. **3d.** Pollinium (reverse), × 3. *C. sigmoideum* (Java, *J.J. Smith* 150). **4a.** Perianth, × 0.7. **4b.** Lip and column, × 0.7. **4c.** Pollinarium, × 3. **4d.** Pollinium (reverse), × 3. **4e.** Flower, × 0.7. **4f.** Flowering plant, × 0.1. *C. erythrostylum* (Kew spirit no. 47871). **5a.** Perianth, × 0.7. **5b.** Lip and column, × 0.7. **5c.** Pollinarium, × 3. **5d.** Pollinium (reverse), × 3. *C. tigrinum* (Kew spirit no. 46422). **6a.** Perianth, × 0.7. **6b.** Lip and column, × 0.7. **6c.** Pollinarium, × 3. **6d.** Pollinium (reverse), × 3. All drawn by Claire Smith.

TABLE 11

A COMPARISON OF THE TWO RELATED SPECIES
C. COCHLEARE AND *C. ELEGANS*

Character	*C. elegans*	*C. cochleare*
Leaf width	up to 1.4(2.0) cm	up to 1.0 cm
Number of flowers	about 20–35	usually 7–20
Spacing of flowers	very dense	less dense
Flower colour	cream to yellow-green, with a cream to pale green lip, not or occasionally sparsely red-spotted	greenish-brown, the lip yellow to orange-yellow, with numerous red spots
Flower appearance	matt, not shiny; bell-shaped, full	waxy, lustrous; more slender, spidery
Side-lobe tip shape	porrect, tapering to acute or subacute apices	not porrect, front margins erect, with subacute apices
Mid-lobe shape	ligulate at the base, expanded into two lobes with an emarginate apex; appearing obcordate	cordate to elliptic, mucronate
Callus structure	callus long, tapering from an inflated apex to the base of the lip, usually with auricles near the base forming a trough-shaped depression	callus short, quickly tapering from an inflated apex and terminating about mid-way to the base of the lip

Cymbidium cochleare is closely allied to *C. elegans*. The distributions of these two species coincide, and both species flower in November. However, *C. elegans* is usually found in cooler forest at higher elevations. The two species can be distinguished by the characters listed in Table 11.

Despite the disjunction in distribution, *Cymbidium babae*, from Taiwan, appears to be conspecific with *C. cochleare*. Examination of several illustrations and photographs has failed to indicate any characters by which *C. babae* might be distinguished. Lin (1977) provided detailed dissections and good colour photographs of *C. babae*, in which the only obvious distinguishing character is the absence of callus ridges. However, even in *C. cochleare* the callus ridges are reduced.

Wu & Chen (1980) reduced *C. babae* to synonymy in *C. elegans* (as *C. longifolium*). However, the mid-lobe of the lip of *C. babae* is elliptical and lacks the obcordate shape of *C. elegans*, the leaves are narrower, the scape more slender, the rhachis darker coloured, the flowers less densely spaced, the segments narrower, the petals and sepals darker coloured and more lustrous, the lip darker with many fine red spots, the side-lobes of the lip less acute and the callus ridges less strongly expressed than in *C. elegans*.

20. Cymbidium cochleare Lindl. in *J. Linn. Soc.* 3: 28 (1858). Type: Sikkim, *J.D. Hooker 235* (holotype K!; isotypes K!).

Cyperorchis cochlearis (Lindl.) Benth. in *J. Linn. Soc.* 18: 317–318 (1881).
Cyperorchis babae Kudo ex Masam. in *J. Jap. Bot.* 8: 258–260, + figs. (1932). Type: Taiwan, *Kudo s.n.* (holotype TI).

Cymbidium babae (Kudo ex Masam.) Masam. in *Trop. Hort.* 3: 33 (1933).
C. kanran Makino var. *babae* (Kudo) S.S. Ying in *Chinese Flowers* 23: 7 (1976) & *Coll. Ill. Ind. Orch.
   Taiwan: 440 (1977).

A perennial, epiphytic *herb*, rarely terrestrial. *Pseudobulbs* up to 6 cm long, 2.5 cm in diameter, narrowly ovoid, bilaterally flattened, produced annually, with 9–14 distichous leaves and about 5 scarious cataphylls sheathing the new growths. *Leaves* up to 50–90 × 0.6–1 cm, linear, tapering to an acute apex, articulated 2.5–8 cm from the pseudobulb to a broadly sheathing base. *Scape* 30–65 cm long, wiry; peduncle suberect to horizontal or arching, covered by about 6 inflated, cymbiform sheaths up to 6 cm long; rhachis pendulous, with 7–20(30) flowers; bracts 1–5 mm long, triangular, scarious. *Flower* about 2.5 cm in diameter, bell-shaped, waxy, pendulous; not strongly scented; rhachis, pedicel and ovary purple; sepals and petals greenish-brown with a pale margin, glossy; lip yellow or orange-yellow, with numerous confluent red-brown spots on the side- and mid-lobes; callus yellow; column whitish-green, with some reddish spots ventrally; anther-cap cream. *Pedicel and ovary* 0.8–3.2 cm long. *Dorsal sepal* 3.9–4.5 × 0.6–0.9(1.1) cm, narrowly obovate, acute, porrect; lateral sepals similar, often mucronate, not spreading. *Petals* 3.8–4.4 × 0.4–0.6 cm, narrowly obovate or ligulate, subacute, curved, porrect, not spreading at least in the basal portion. *Lip* about 4–4.5 cm long, elongated, deltoid, 3-lobed, fused to the base of the column for about 2 mm; side-lobes 1–1.2 cm broad, triangular, subacute, erect, glabrous and glossy, front margin erect; mid-lobe 0.8–1 × 0.8–1 cm, cordate to elliptic, mucronate, porrect, the tip deflexed, with a dense patch of short hairs in the centre, margin minutely undulating; callus of two ridges, inflated and hairy in front, tapering quickly behind, reaching about half-way to the base of the lip. *Column* 3.2–3.6 cm long, deflexed at the tip, with short hairs and a shallow, elliptic, nectar-producing depression at the base; anther-cap and viscidium elongated into a distinct protruding rostellum; pollinia clavate, deeply cleft, about 1.5 mm long, on a rectangular viscidium without long hair-like processes from the corners. *Capsule* about 2.5 × 1.5 cm, broadly ellipsoidal, pointed, stalked, with a long beak about as long as the capsule. Figs. 94 & 95: 1a–f.

DISTRIBUTION. Northern India (Sikkim, Meghalaya), ?Myanmar (Burma), ?Thailand, Taiwan (Map 22); about 1500 m (4920 ft) (300–1000 m (985–3280 ft) in Taiwan).

HABITAT. In tropical valleys, in shade. Lin (1977) noted that this species is widespread but not common in Taiwan, and grows 'in the shaded and humid forests at altitudes from 300–1000 m'. It has also been reported growing as a terrestrial along streams in sand and as a semi-terrestrial in forest (Mark *et al.*, 1986). It is slightly fragrant, which explains its local name of 'Fragrant Sand Grass'. It should also be noted that Liu & Su (1978) do not accept this species as native to Taiwan, stating that it is probably an introduced, cultivated plant. This might explain the disjunct distribution of *C. cochleare*, but whether introduced or native, this species does now appear to be present in the wild in Taiwan. It flowers from November to January.

CONSERVATION STATUS. VU A1cd; B1ab.

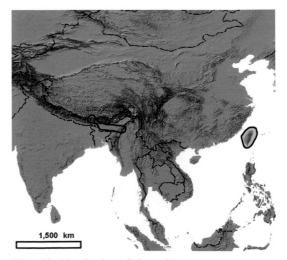

1,500 km

Map 22. Distribution of *C. cochleare*.

21. CYMBIDIUM EBURNEUM

Cymbidium eburneum was described by Lindley in 1847, from a specimen cultivated by Loddiges' nursery and said to be from the 'East Indies'. When, in 1851, Griffith described *C. syringodorum*, the lilac-scented *Cymbidium*, he gave the first precise locality information as the Khasia Hills in Assam. Lindley (1858) reduced this name to synonymy in *C. eburneum*. Later, Clarke and then King & Pantling collected it from Darjeeling and Sikkim. It is likely that the distribution extends along the Himalaya to include Nepal (Hara *et al.*, 1978) and Bhutan. Forrest collected a cultivated specimen of *C. eburneum* (wrongly identified as *C. hookerianum*) in western Yunnan where it is undoubtedly native. In 1939 a specimen from Myanmar (Burma), but without an exact locality, was sent to Kew, extending the known distribution of this species.

This species, along with *C. mastersii*, *C. parishii* and *C. roseum*, was placed in section *Eburnea* by Du Puy and Cribb (1988), which was characterised by slender and fusiform pseudobulbs that grow and flower indeterminately for two to many years. The leaf apex in these species tends to be acutely bilobed with a small mucro in the sinus. The few-flowered scape is produced from the axils of the leaves, not from the base of the pseudobulb. In common with section *Cyperorchis,* the flowers do not usually open fully, the hypochile is relatively long, and the rostellum is beaked. The quadrangular pollinia are placed on a rectangular viscidium with two long, hair-like processes from the lower corners (Table 10: Fig. 125: 3c).

Cymbidium eburneum has short inflorescences that usually carry a single, large, white flower, with a single, broad, yellow callus ridge. It can easily be distinguished from the other large, white-flowered species in this section by its characteristic callus ridges that are fused into one single, inflated, wedge-shaped structure at the apex. *Cymbidium mastersii* shares the tendency to grow indeterminately for several years before producing a new growth, the characteristic leaf tip shape, the elongated lip, and

Fig. 96. *Cymbidium eburneum*, Yunnan, Dali.

179

Plate 17. *Cymbidium eburneum*. Curtis's Botanical Magazine t. 5126, del. W.H. Fitch.

Fig. 97. *Cymbidium eburneum*, cult. Munich B.G.

similar colouring. Its flowers are much smaller and more numerous, the petals are much narrower, and it lacks the callus structure typical of *C. eburneum. Cymbidium insigne* has distinct swollen pseudobulbs which flower from the base, leaves which lack the complex leaf tip structure and much longer and more robust flower spikes, bearing many more flowers with two separate, hairy callus ridges.

The plants of *C. eburneum* seen in cultivation in the West until recently, usually imported from India or Thailand, had inflorescences bearing a single white flower, often pale pink-tinged, with a yellow callus. Pale pink spotting was occasionally apparent on the margin of the mid-lobe of the lip. We had the opportunity to examine a range of Chinese plants collected in the Gaoligong Shan Range, just east of the Nu Jiang (Salween River) in western Yunnan. They varied considerably in flower colour, particularly in the degree of pink on the sepals and petals and in the degree of spotting of the mid-lobe of the lip, and in the sepals, petal and lip shape. In lip colouring some approached the long-lost *C. parishii*, a Burmese species not currently in cultivation. Further work is needed to compare this Chinese material with Parish's type and his excellent drawing of the type, now at Kew. The possibility that there may be an intergradation between *C. eburneum* and *C. parishii* in flower coloration, that would see the latter disappear into synonymy, should be investigated.

Cymbidium eburneum was often well cultivated in the warm, humid stove greenhouses of the late 19th century, and large specimen plants were exhibited bearing upwards of 25 flowers. It is among the most beautiful of the *Cymbidium* species and has been used extensively in hybridisation. Its strong fragrance is usually not inherited when it is hybridised. It was one of the parents of the first artificial *Cymbidium* hybrid, *C.* × Eburneo-lowianum in 1889. The popularity of *C. eburneum* led to the naming of several cultivars. Reichenbach described two varieties based on variations in colour: the pure white-flowered var. *philbrickianum* and the pink-tinged var. *williamsianum*, both with a yellow callus and mid-lobe patch. A variant with some pale purple-pink spots on the mid-lobe of the lip was called var. *dayi* by Jennings in 1875.

A natural hybrid between *C. eburneum* and *C. mastersii*, *C. × ballianum*, originally imported from Burma (Myanmar) was recorded by Rolfe (1904). Specimens of a distinct species from Annam (Vietnam) have also been given this name (see under *C. sanderae*).

21. Cymbidium eburneum Lindl. in *Bot. Reg.* 33: t. 67 (1847) & in *Paxton, Mag. Bot.* 15: 145–146 + plate (1849). Type: India, Meghalaya (Khasia Hills), cult. Loddiges (holotype K!).
C. syringodorum Griff., *Notulae* 3: 338 (1851). Type: Northern India, Khasia Hills, Myrung, *Griffith 228* (holotype K!).
C. eburneum Lindl. var. *dayi* Jennings, *Orchids*: t. 16 (1875). Type: Illustration cited here (lectotype selected by Du Puy & Cribb, 1988).
C. eburneum Lindl. var. *williamsianum* Rchb.f. in *Gard. Chron.*, ser. 2, 15: 530 (1881). Type: cult. Williams (holotype W).
C. eburneum Lindl. var. *philbrickianum* Rchb.f. in *Gard. Chron.*, ser. 2, 25: 585 (1886). Type: cult. Philbrick (holotype W).
Cyperorchis eburnea (Lindl.) Schltr. in *Fedde, Repert. Sp. Nov. Regni Veg.* 20: 107 (1924).

A perennial, epiphytic *herb*. *Pseudobulbs* about 10 cm long, 3 cm in diameter, ovoid to fusiform, bilaterally flattened, not produced annually but growing in an indeterminate fashion for about three years before a new growth is produced, often covered in persistent leaf bases and bearing about 7 fresh leaves, each pseudobulb producing about 15–17 distichous leaves in total. *Leaves* up to 60 × 1.3(2) cm, narrowly ligulate, acute, with a finely unequally bilobed apex with a minute mucro in the sinus formed as an extension of the mid-vein; articulated to a broad, sheathing base 5–10 cm from the pseudobulb. *Scape* about 25(36) cm long, from within the axils of the leaves; peduncle erect or suberect, covered in about 8–21 sheaths up to 15 cm long; upper sheath cymbiform, middle sheaths mostly cylindrical below, inflated, cymbiform above; rhachis with 1–2(3) flowers; bracts 0.4–2 cm long, triangular, acute. *Flower* large, 8–12 cm across, not opening fully; sweetly lilac-scented; rhachis, pedicel and ovary bright green; petals and sepals white or faintly pink; lip white with a bright yellow central and basal patch on the mid-lobe and bright yellow callus, occasionally with some purple-pink spots on the mid-lobe; column white or flushed pale pink, sometimes spotted pink ventrally and with a small yellow patch at the base; anther-cap white. *Pedicel and ovary* 3.2–4.1 cm long. *Dorsal sepal* 5.6–7.6 × 1.8–2.9 cm, narrowly oblong-elliptic, acute, concave, porrect; lateral sepals similar, not fully spreading. *Petals* 5–7.3 × 1.5–2.2 cm, narrowly spathulate, slightly curved, the acute tips recurved and spreading. *Lip* 3-lobed, elongated, fused to the base of the column for 4–6 mm; side-lobes 1.2–1.8 cm broad, clasping the column, papillose or minutely hairy, apex broadly rounded, margin not fringed; mid-lobe 1.6–2 × 1.4–1.7 cm, ovate-triangular, rounded or mucronate, porrect or weakly recurved, minutely hairy with a dense central and basal patch of short hairs, margin entire and undulating; callus long, with 3 slightly raised ridges on a broad, glabrous disc behind, becoming confluent and strongly inflated and forming a cuneate apex terminating well behind the junction of the mid- and side-lobes, papillose or minutely hairy. *Column* (3.4)4.1–4.6 cm long, narrowly winged to the base, curved down near the apex,

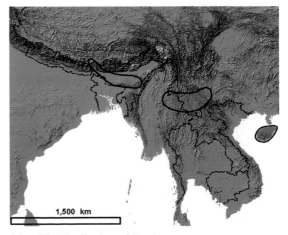

Map 23. Distribution of *C. eburneum*.

182

almost glabrous; pollinia quadrangular, about 4 mm long, on a rectangular viscidium with hair-like processes from the lower corners. *Capsule* about 7–10 cm long, ellipsoid, with an apical beak. Pl. 17; Figs. 96–98 & 125: 3a–e.

DISTRIBUTION. Northern India (Sikkim, Assam, Khasia Hills), Nepal, northern Myanmar, China (south-west Yunnan, Guangxi, Hainan) (Map 23); 300–1700 m (985–5580 ft).

HABITAT. On trees in warm, damp forest, and on rocks in ravines, in shade; flowering February –May (November–January in cultivation).

CONSERVATION STATUS. VU A1cd; B1ab.

Fig. 98. *Cymbidium eburneum*, W Yunnan.

22. CYMBIDIUM ELEGANS

This is one of the most striking and commonly cultivated of the *Cymbidium* species. Its large, densely crowded racemes of narrow funnel-shaped flowers with club-shaped pollinia are distinctive, and have led several authorities, including Blume, J.D. Hooker and Schlechter, to the conclusion that it, and *C. cochleare*, should be placed in a separate genus. This is discussed under the introduction to the taxonomy of the genus; molecular systematic studies do not support its separation even at sectional level (see Chapter 11), and the intermediate species *C. whiteae* and *C. sigmoideum* blur the distinctiveness of this group. This species is well known under the name *C. elegans* Lindl., described in 1833 from a specimen collected by Wallich at Gossaingsthan, probably now known as Gossainkunde, in Nepal.

Hara *et al.* (1978) used the name *C. longifolium* D. Don for this species in their *Enumeration of the Flowering Plants of Nepal*, and this usage was subsequently followed by Wu & Chen (1980). The name *C. longifolium* has, however, long been applied in the sense used by Lindley (1833) for another species that is also known in cultivation, to which Hara *et al.* have applied the name *C. erythraeum* Lindl.

Cribb & Du Puy (1983) proposed conserving the traditional usage of the names *C. elegans* and *C. longifolium*, citing specimens in the Wallich Herbarium, at Kew, as types. However, Hara (1985) has shown that Don based his description on specimens in the Lambert Herbarium that is now at the British Museum (Nat. Hist.). The specimens there support the usage of these names as Hara *et al.* have indicated. Although accepting that the correct name of Lindley's *C. longifolium* is *C. erythraeum*, we

P. CRIBB

Fig. 99. Habitat of *Cymbidium elegans* in W Bhutan.

Plate 18. *Cymbidium elegans*. N India, *Seth 109*, cult. Kew, del. Claire Smith.

Fig. 100 (top). *Cymbidium elegans* in Nepal. **Fig. 101** (above). Yunnan. **Fig. 102** (right). Cult. Munich B.G.

considered that the use of *C. longifolium* for the well-known *C. elegans* would create considerable confusion in the orchid world. We, therefore, applied for the name *C. longifolium* D. Don to be rejected under Article 69 of the *International Code of Botanical Nomenclature* to allow the retention of the name *C. elegans* for what is a widely grown and popular orchid (see Du Puy & Cribb – Proposal to reject the name *Cymbidium longifolium* D. Don (Orchidaceae). *Taxon* 37: 487–488, 1988).

Several varieties of *C. elegans* have been described. Variety *obcordatum* was differentiated by Reichenbach (1880) based upon its obcordate, emarginate mid-lobe of the lip and the colour of the auricles at the base of the callus. The former characters are in fact typical of this species, although the degree of their expression varies. The latter character is also highly variable within the species.

Variety *lutescens* Hook.f. (1891) was described (as a variety of *Cyperorchis elegans*) as 'a smaller plant, leaves 9 inches (23 cm), scape 7 inches (18 cm), densely clothed with imbricating sheaths 3 inches (7.5 cm) long; raceme suberect, secund, 5-flowered; flowers yellowish, $1^3/_4$ inches (4.5 cm) long', from a drawing in the Calcutta herbarium. This variety therefore differs from the type in its shorter leaves, and its shorter inflorescence with fewer flowers. The other characters in the description, and the drawing, fit the description of *C. elegans*. It appears that this variety is simply a smaller, poor specimen of *C. elegans*, and should not be recognised as a distinct variety.

Variety *blumei* Hort. (1907) was also described under *Cyperorchis elegans*, from a fine specimen shown at the Royal Horticultural Society, but neither the description nor the illustration shows any characters that could be used to maintain it as distinct.

In 1851, William Griffith described *C. densiflorum*, based on his own collection from Meghalaya (Khasia Hills). A second specimen described in the same account is attributed to *C. mastersii*. Lindley (1858) distinguished this variant from *C. elegans* by the lack of auricles at the base of the callus in *C. densiflorum*. However, the auricles are mentioned in the type description, and dissection of the type specimen has shown them to be present. The size of the auricles varies greatly between specimens, and they may sometimes be rather obscure. Reichenbach (1875) noted that examination of the flowers on the same plant in two consecutive years showed that the size of the auricles can vary from almost absent to conspicuous in a single specimen. He also noted that the colour of these auricles varied from orange to deep purple. The auricles may be cream, yellowish, orange, pink, reddish, purple or brown. The colour of the flower also shows some variation, from dull yellowish-green to clear cream or lightly pink-tinged. The number of flowers varies, but usually there are more than 20 in a single inflorescence.

Finet described *Arethusantha bletioides* (= *C. elegans*) in 1897 based on a specimen collected by the Prince of Orleans (*sic*) without any indication of where it was collected, or when. As both *Cyperorchis* (1848) and *Arethusantha* (1879) were based on the same species, and the former is in the earlier publication, the genus *Arethusantha* must therefore be considered synonymous with *Cyperorchis*.

Cymbidium elegans resembles *C. mastersii*, and both have auricles at the base of the callus. The key differences are outlined in the discussion of the latter. The differences with other allied species in section *Cyperorchis*, *C. cochleare*, *C. sigmoideum* and *C. whiteae*, are discussed under those species.

Hybridisation with *C. erythraeum* in Sikkim has led to hybrid swarms that display all stages of intermediacy between the two parents. This natural hybrid was first described by King & Pantling (1895, 1898) as *C.* × *gammieanum*. The natural populations are still present (C. Bailes, pers. comm.).

22. Cymbidium elegans Lindl., *Gen. Sp. Orchid. Pl.*: 163 (1833) & *Sert. Orch.*: t. 14 (1838). Type: Nepal, Gossaingsthan, *Wallich 7354* (holotype K!).

C. longifolium D. Don, *Prodr. Fl. Nepal.*: 36 (1825), **nom. rej.** Type: Nepal, Gossainkunde ('Gosaingsthan'), 1819, *Wallich s.n.* (lectotype BM).

Cyperorchis elegans (Lindl.) Blume, *Rumphia* 4: 47 (1848), in *Mus. Bot. Lugd. Bat.* 1: 48 (1849) & in *Orch. Archip. Ind.* 1: 93, t. 48c (1858).

Cymbidium densiflorum Griff., *Notulae* 3: 337 (1851). Type: Assam, Khasia, Myrung, *Griffith 229* (holotype K!).

Grammatophyllum elegans (Lindl.) Rchb.f. in *Walp. Ann. Bot.* 3: 1028 (1853).
C. elegans Lindl. var. *obcordatum* Rchb.f. in *Gard. Chron.*, n.s. 13: 41 (1880). Type: not located.
C. elegans Lindl. var. *lutescens* Hook.f., *Fl. Brit. India* 6: 15 (1891). Type: Illustration ex CAL (K!).
Arethusantha bletioides Finet, in *Bull. Soc. Bot. Fr.* 44: 179, t. 5 (1897). Type: without loc., *Prince
    d'Orleans s.n.* (holotype P).
Cyperorchis elegans (Lindl.) Blume var. *blumei* Hort. in *J. Hort.*, ser. 3, 54: 71 & fig. (1907). Type: cult.
    Colman (not located).

A perennial, epiphytic or lithophytic *herb. Pseudobulbs* up to 7 cm long, 4 cm in diameter, ovoid, bilaterally flattened, with 7–13 distichous leaves, produced annually. *Leaves* up to 65(80) × 1.4(2) cm, narrowly linear-elliptic, acute, hooded and unequally bilobed (but not forked) at the apex, articulated to a broadly sheathing base, 4–12 cm from the pseudobulb. *Scape* 30–60 cm long; peduncle suberect to horizontal, covered in sheaths up to 10–15 cm long; rhachis pendulous, usually with 20–35 closely spaced flowers; bracts triangular, 1–6 mm long. *Flowers* about 3 cm across; pendulous, bell-shaped, perianth not spreading, in a large, crowded, pendulous cluster; often lightly scented; rhachis, pedicel and ovary green; sepals and petals cream or pale straw-yellow, sometimes tinged pale pink, to pale yellow-green; lip cream to pale green, occasionally sparsely red-spotted, with two bright orange-yellow callus ridges and a brown, reddish or cream depression at the base; column pale green; anther-cap cream. *Pedicel and ovary* 0.8–2(3) cm long. *Dorsal sepal* 3.2–4.3 × 0.7–1.1 cm, narrowly obovate, acute, concave, porrect, closely covering the column; lateral sepals similar, often mucronate, not spreading. *Petals* 3.1–4.2 × 0.4–0.8 cm, narrow, ligulate or narrowly obovate, obtuse, sometimes curved, porrect. *Lip* deltoid, elongated, 3-lobed, fused to the base of the column for 2–3 mm; side-lobes 0.6–0.8 cm broad, triangular, erect and clasping the column, almost glabrous or minutely papillose, porrect, tapering to an acute or subacute apex, margin not fringed; mid-lobe small, 0.6–1 × 0.4–0.8 cm, base ligulate, expanded apically into two incurved lobes, emarginate, occasionally mucronate, porrect, with a dense patch of short hairs in the centre, margin often minutely undulating; callus ridges 2, converging towards their inflated, short-haired apices and usually with minute auricles at the base, forming a trough-shaped depression with some short marginal hairs. *Column* 2.8–3.4 cm long, sparsely hairy below, with a shallow, elliptical, nectar-producing depression at the base, narrowly winged, the wings inflated at the abruptly deflexed apex; anther-cap and viscidium elongated into a distinct protruding rostellum; pollinia clavate, deeply cleft, about 2 mm long, on a rectangular viscidium with long hair-like processes from the corners. *Capsule* 2–2.7 × 1.3–2 cm, broadly ellipsoidal, pointed, stalked, with a beak about as long as the capsule.
Pls. 1 & 18; Figs. 95: 3a–d, 99–102.

DISTRIBUTION. Nepal, north-east India (Sikkim, Darjeeling, Assam, Meghalaya, Naga Hills, Lushai Hills), Bhutan, northern Myanmar (Burma) (incl. Chin Hills), south-west China (south-west Sichuan, Yunnan, south-east Xizang) (Map 24); 1500–2800 m (4920–9185 ft).

HABITAT. On trees and rocks in damp, shady forest, sometimes on shaded rocks overhanging streams; flowering during October–November (December).

CONSERVATION STATUS. NT.

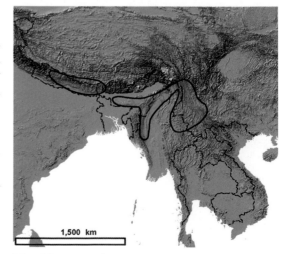

Map 24. Distribution of *C. elegans*.

23. CYMBIDIUM ERYTHRAEUM

Although still grown under the name *C. longifolium*, Hara *et al.* (1978) and Hara (1985) convincingly demonstrated that the type specimen of *C. longifolium*, collected by Nathaniel Wallich in Nepal, is referable to the species commonly known as *C. elegans*. Lindley (1833) misapplied both of these names in his *Genera & Species of Orchidaceous Plants*, and they have been incorrectly used since then. The correct name for this species is *C. erythraeum*, described by Lindley in 1859 from a collection made by J.D. Hooker in Sikkim. The application of the name *C. elegans* is discussed under that species.

Schlechter (1918) also described this species, as *C. hennisianum*, distinguishing it from *C. erythraeum* by its narrow leaves (3–5 mm (0.12–0.2 in) broad) and its fewer-flowered (4–7), more slender, shorter (about 40 cm, 16 in) spike. However, the other features mentioned are typical of *C. erythraeum*. The type specimen has been destroyed, but other specimens from Myanmar (Burma) agree well with *C. erythraeum*. It seems likely that the type specimen of *C. hennisianum* was a slender variant of *C. erythraeum* and should not be maintained as a separate taxon.

Based upon the published photographs and drawing that accompany the type description, we consider *C. flavum* Z.J. Liu & S.C. Chen to be an albinistic variant of *C. erythraeum*. Normal coloured plants of this species are found in the area of Yunnan where the type originated.

The flowers of *C. erythraeum* are heavily pigmented with brown on the sepals and petals, although a yellow-flowered variant, lacking red and brown pigments, has also been collected. They resemble those of *C. tracyanum* and *C. iridioides*, although the lip is less strongly marked. The petals are falcate as in *C. tracyanum* but are much narrower, giving the flower a spidery appearance, and the lip is much smaller and lacks the strongly undulating margin and cilia present in *C. tracyanum* and *C. iridioides*. Vegetatively, *C. erythraeum* is distinct in that its pseudobulbs are much smaller, and the leaves narrower than the other species in this section. Its flowers are borne on a slender rhachis and are smaller than in the other species.

DAVID DU PUY

Fig. 103. *Cymbidium erythraeum*, Nepal, C. Bailes *s.n.*, cult. Kew.

189

Cymbidium erythraeum has been found to be variable in the shape of the lip (Du Puy *et al.*, 1984). One variant, with a lip with truncated side-lobes and a cordate mid-lobe, is found in western China, whereas the other from Nepal, northern India and Bhutan has acute, porrect side-lobes and a reniform mid-lobe. Both of these vary greatly in the density of the indumentum on the lip and in the breadth of the leaf, although this latter is usually about a centimetre wide. These two major variants occur at opposite ends of the distribution of this species. This variation reflects its large east-west range.

23. Cymbidium erythraeum Lindl. in *J. Proc. Linn. Soc. Bot.* 3: 30 (1859). Type: Sikkim, *J.D. Hooker 229* (holotype K!).

C. longifolium auct. non D. Don (1825); Lindley, *Gen. Sp. Orch. Pl.*: 163 (1833); Hook.f., *Fl. Brit. India* 6: 13 (1891); King & Pantling in *Ann. Roy. Bot. Gard. Calcutta* 8: 191, t.254 (1898).
C. hennisianum Schltr. in *Orchis* 12: 46 (1918). Type: Burma, *Hennis* (holotype B†).
Cyperorchis longifolia (D. Don) Schltr. in *Fedde, Repert. Sp. Nov. Regni Veg.* 20: 108 (1924).
C. hennisiana (Schltr.) Schltr., *loc. cit.* 20: 107 (1924).
Cymbidium flavum Z.J. Liu & S.C. Chen in *Orchidee* 53 (1): 94 (2002), **syn. nov.** Type: China, Yunnan, Wen Shan Co., *Z.J. Liu 21128* (holotype PE; isotype SZWN).

A perennial, epiphytic or lithophytic *herb*. *Pseudobulbs* up to 5 cm long, 5 cm in diameter, ovoid, bilaterally flattened, with 5–9 distichous leaves. *Leaves* up to 90 × 1.6 cm, tapering gradually from the middle to a fine point, base broadly sheathing, narrowing to the abscission zone 3–6 cm from the base. *Scape* 25–75 cm long, suberect to horizontal, arching; peduncle covered in scarious sheaths up to 9.5 cm long; rhachis bearing 5–14 flowers; bracts triangular, up to 4 mm long. *Flowers* up to 8 cm across; scented; leaf bases and abscission zone purplish; rhachis, pedicel and ovary green; petals and sepals greenish, heavily spotted and irregularly striped red-brown; lip yellowish to white, side-lobes veined

Fig. 104. *Cymbidium erythraeum*. Yunnan.

dark red-brown; mid-lobe sparsely spotted red or red-brown, with a central stripe; callus cream or white; column pale yellow, shading to red-brown towards the tip, with red-brown dashes below; anther-cap cream. *Pedicel and ovary* 1.1–3.7 cm long. *Sepals* narrowly obovate, acute; dorsal sepal 3.7–4.1(5.3) × 0.75–1.1(1.25) cm, concave, porrect. *Petals* 3.6–4.3(4.9) × 0.5–0.8 cm, ligulate, acute, falcate, spreading. *Lip* 3-lobed, fused to the base of the column for 2–4 mm; side-lobes 0.7–1.2 cm broad, erect, acute and forward pointing or truncate, densely covered in short hairs at the front, papillose towards the base, margin sometimes fringed with short hairs; mid-lobe 0.9–1.1 × 1–1.4 cm, cordate to reniform, acute, recurved, papillose, sometimes with scattered short hairs, margin flat or undulating, callus tapering to the base of the lip, swollen at the tip, densely covered with short hairs, the indumentum not extending onto the mid-lobe. *Column* 2.3–2.9(3.3) cm long, sparsely hairy below, with wings narrowing strongly towards the base; pollinia obliquely triangular, 2.1–2.7 mm long. *Capsule* about 5–6 × 3 cm, fusiform-ellipsoidal, stalked, with a short (1.5 cm) apical beak. Pl. 2; Figs. 103–105, 106: 3a–g.

DISTRIBUTION. *Cymbidium erythraeum* grows in the forests of the Himalaya of northern India from Kumaon, through Nepal, Sikkim and Bhutan to Myanmar and the western provinces of China (Sichuan, Xichang, Yunnan) (Map 25); 1000–2800 m (3280–9185 ft). It has not been recorded in Meghalaya but probably also occurs there.

HABITAT. It is usually epiphytic on tree trunks in moist forest, but is also found growing on rocks, and there is one record of it on an overhanging grassy bank. Some plants may survive as terrestrials if they fall from overhanging branches. It appears to tolerate cool conditions, and in the Dafla Hills of north-east Assam it has been collected in forests which are regularly subjected to snow in December.

It flowers from August until November in the wild.

CONSERVATION STATUS. NT.

Fig. 105. *Cymbidium erythraeum*, Yunnan.

P. CRIBB

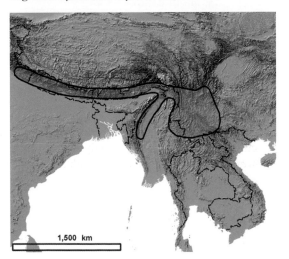

Map 25. Distribution of *C. erythraeum*.

191

24. CYMBIDIUM HOOKERIANUM

William Griffith was the first botanist to collect in Bhutan when he accompanied a British diplomatic mission to the country in 1848. He described this species in 1851 as *C. grandiflorum*, based on a specimen he had collected there. Swartz (1799) had, unfortunately, previously used this name for a distinct species now placed in *Pogonia*. The next available and legitimate name for the species is *C. hookerianum*, based on material collected by Thomas Lobb in the early 1850s. It flowered soon after its introduction, in the nursery of Messrs James Veitch & Sons, but not again until 1866, when Reichenbach described it. A flower from the same plant was figured in *Curtis's Botanical Magazine* (Bateman, 1866).

Variety *punctatum*, described by Cogniaux, based on a colour variant that has a few red-brown spots at the base of the sepals and petals, and more numerous spots on the mid-lobe of the lip, is otherwise identical to *C. hookerianum*. Two further varieties ascribed to *C. hookerianum*, var. *kalawense* T.A. Colyear and var. *lowianum* (Rchb.f.) Y.S. Wu & S.C. Chen, are discussed under *C. lowianum*.

Cymbidium hookerianum is similar in its large flower size and lip shape to *C. tracyanum* (see discussion of *C. tracyanum* for distinguishing characters), and is closely allied to *C. lowianum*, which also has clear green sepals and petals. The spots on the side-lobes of the lip are absent in *C. lowianum*, and the mid-lobe spots are replaced by a large, red, V-shaped patch. *Cymbidium hookerianum* can also be distinguished by its acute, porrect side-lobe apices, larger, more strongly recurved, ovate mid-lobe with a strongly undulating margin and more evenly distributed indumentum, and callus ridges that taper towards the base. *Cymbidium hookerianum* is also allied to *C. wilsonii* (see *C. wilsonii* for a discussion of their affinity and distinguishing characters).

24. Cymbidium hookerianum Rchb.f. in *Gard. Chron.*: 7 (1866). Type: cult. Veitch (holotype W).

C. grandiflorum Griff., *Notulae* 3: 342 (1851), *Itiner. Not.* 145 (1851) & *Icon. Pl. Asiat.* 3: t. 321 (1851); *non* Sw. (1799). Type: Bhutan, *Griffith 698* (holotype CAL).

C. giganteum Wall. ex Lindl. var. *hookerianum* (Rchb.f.) Bois., *Orch.*: 119 (1893), seen in *Lindenia* 9: 13 (1893).

C. grandiflorum Griffith var. *punctatum* Cogn. in *J. Orch.* 4: 76 (1893); L. Linden, *Lindenia* 9: 13–14, t. 389 (1893). Type: cult. Linden (holotype not located, but see the above reference for an illustration of the type).

Cyperorchis grandiflora (Griffith) Schltr. in *Fedde, Repert. Sp. Nov. Regni Veg.* 20: 107 (1924).

Cymbidium hookerianum Rchb.f. var. *hookerianum* Y.S. Wu & S.C. Chen in *Acta Phytotax. Sin.* 18: 302 (1980).

A perennial, epiphytic or lithophytic *herb*. *Pseudobulbs* 3–6 cm long, 1.5–3.5 cm in diameter, elongate-ovoid, bilaterally flattened, with 4–8 leaves. *Leaves* up to 80 × 1.4–2.1 cm, linear-elliptic, acute, articulated 4–10 cm from the pseudobulb to a broadly sheathing, strongly yellow- and green-striated base. *Scape* up to 70 cm long; peduncle suberect, loosely covered in scarious sheaths up to 12 cm long; rhachis arching to pendent, slightly flattened, bearing 6–15 flowers; bracts triangular, up to 4 mm long. *Flowers* up to 14 cm in diameter, with a strong fresh scent; rhachis, pedicel and ovary green; sepals and petals clear apple-green with some deep red spots towards the base, occasionally lightly shaded with red-brown; lip cream-coloured, becoming greenish at the margin, flushing strong purplish-pink after pollination, side-lobes spotted deep maroon, mid-lobe with a submarginal ring of red-brown blotches and spots and a broken central line of confluent red blotches, base of lip bright yellow with maroon spots; callus ridges cream with reddish spots; column cream, spotted maroon below, becoming green at the apex above, with a fine maroon line at the apex; anther-cap cream. *Pedicel and ovary* 3.5–6.0 cm long. *Dorsal sepal* 5.6–6.0 × 1.7–1.9 cm, narrowly obovate, acute, porrect; lateral sepals similar,

Plate 19. *Cymbidium hookerianum*. Curtis's Botanical Magazine t. 5574, del. W.H. Fitch.

Plate 20. *Cymbidium hookerianum.* Ex. *E. Young* s.n., cult. Kew, del. Claire Smith.

Fig. 106. *Cymbidium tracyanum* (Kew spirit no. 45390). **1a.** Perianth, × 0.7. **1b.** Lip and column, × 0.7. **1c.** Pollinarium, × 3. **1d.** Pollinium (reverse), × 3. *C. iridioides* (Kew spirit no. 45961). **2a.** Perianth, × 0.7. **2b.** Lip and column, × 0.7. **2c.** Pollinarium, × 3. **2d.** Pollinium (reverse), × 3. *C. erythraeum* (a–e, Sikkim, *Pantling* 8; f–g, Yunnan, *Henry* 11371). **3a.** Perianth, × 0.7. **3b.** Lip and column, × 3. **3c.** Pollinarium, × 3. **3d.** Pollinium (reverse), × 3. **3e.** Flowering plant, × 0.1. **3f.** Lip and column, × 0.7. **3g.** Lip (flattened), × 0.7. *C. hookerianum* (Kew spirit no. 47872). **4a.** Perianth, × 0.7. **4b.** Lip and column, × 0.7. **4c.** Pollinarium, × 3. **4d.** Pollinium (reverse), × 3. *C. wilsonii* (Kew spirit no. 37623). **5a.** Perianth, × 0.7. **5b.** Lip and column, × 0.7. **5c.** Pollinarium, × 3. **5d.** Pollinium (reverse), × 3. **5e.** Flower, × 0.7. All drawn by Claire Smith.

Fig. 107. *Cymbidium hookerianum*. Cult. Kew ex *E. Young*.

DAVID DU PUY

spreading. *Petals* 5.3–5.6 × 1.2–1.4 cm, ligulate or narrowly obovate, curved, spreading. *Lip* 3-lobed, fused to the column base for 4.5–5 mm; side-lobes 1.1–1.6 cm broad, triangular, acute and porrect, margins hairy (hairs over 1 mm long) or papillose; mid-lobe 1.7–2 × 2.7–2.9 cm, broadly ovate-cordate, mucronate, recurved, papillose, sometimes with scattered hairs, never with the central lines of hairs present in *C. tracyanum* or *C. iridioides*, margin erose and undulate; callus ridges 2, tapering to the cavity at the base of the lip, hairy along the top. *Column* 3.3–4 cm long, winged, with a few hairs or papillae ventrally towards the base; pollinia 3.2–4 mm long, triangular. *Capsule* up to 13 × 4 cm, fusiform-ellipsoid, stalked with a 2.5 cm apical beak. Pls. 19 & 20; Figs. 106: 4a–d & 107.

DISTRIBUTION. This fine orchid is found in the Himalaya of northern India (eastern Nepal, Sikkim Assam) and Myanmar, through to China where it occurs in northern Yunnan, south-west Sichuan and south-east Xizang (Map 26), 1500–2600 m (4920–8530 ft).

HABITAT. It grows as an epiphyte on trees in damp, shady forest or on steep banks or rocks, often where thick moss cover occurs. It is found at higher elevations than *C. iridioides* in regions where the distributions of these two species overlap. The tendency in cultivated specimens to drop their flower buds, or even to fail to form flower spikes, if the growing conditions are too warm, reflect the cooler climatic conditions that this species prefers; flowering January–March.

CONSERVATION STATUS. VU A1cd.

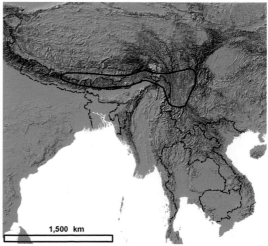

1,500 km

Map 26. Distribution of *C. hookerianum*.

25. CYMBIDIUM INSIGNE

Cymbidium insigne is easily distinguished from the other species in section *Cyperorchis* by its white or pink flowers with broad petals, broad side-lobes to the lip with rounded apices, tall, erect flower spike with flowers only in the apical region, longer bracts, more rounded pseudobulbs (in transverse section) and smaller, almost spherical capsules.

This species was first collected by Bronckart in 1901 in Annam, now Vietnam. He sent dried flowers and an inaccurate watercolour sketch to Kew (via Mr. Schneider), from which Robert Rolfe described it in 1904 as having rosy-lilac flowers. Plants exhibited by Bronckart in 1906 had paler flowers. Specimens collected by Micholitz in 1903, on the Lang Bian Plateau in south Vietnam, were exhibited by Sander and described by O'Brien in 1905 as *C. sanderi*. Subsequent comparison with Bronckart's specimens has shown the two collections to be conspecific. *Cymbidium insigne* has been awarded about ten times by the Royal Horticultural Society, and many unpublished cultivar names are therefore in circulation. It has been fundamental to the breeding of modern hybrids (see Chapter 8).

This species varies especially in the size, colour and shape of the mid-lobe of the lip. Specimens from Vietnam have a large, rounded mid-lobe, spotted with maroon-red. Specimens from northern Thailand have a smaller, triangular mid-lobe, more strongly veined and spotted maroon-red, or sometimes entirely lacking the red markings on the lip. They are described below as *C. insigne* subsp. *seidenfadenii*.

There is a natural hybrid between *C. insigne* and *C. schroederi*, *C. × cooperi*.

25. Cymbidium insigne Rolfe in *Gard. Chron.* ser. 3, 35: 387 (1904) & in *Bot. Mag.* 136: t. 8312 (1910). Type: Vietnam, *Bronckart 43* (holotype K!).

C. sanderi O'Brien in *Gard. Chron.*, ser. 3, 37: 115, t. 49 (1905). Type: Vietnam, cult. Sander, *Micholitz no. 1* (holotype K!).
C. insigne Rolfe var. *sanderi* (O'Brien) Hort. in *J. Hort.*, ser. 3, 58: 415, + fig. (1909).
Cyperorchis insignis (Rolfe) Schltr. in *Fedde, Repert. Sp. Nov. Regni Veg.* 20: 108 (1924).
Cymbidium insigne Rolfe var. *album* Hort. ex *Gard. Chron.*, ser. 3, 61: 101, f. 35 (1917). Type: cult.
    Armstrong & Brown (lectotype the above figure, selected by Du Puy & Cribb, 1988).

A perennial, terrestrial *herb*. *Pseudobulbs* up to 8 cm long, 5 cm in diameter, ovoid, lightly bilaterally flattened, with 6–10 leaves. *Leaves* up to 100 × 0.7–1.8 cm, narrowly linear-elliptic tapering to an acute apex, articulated to a broadly sheathing base 5–15 cm from the pseudobulb. *Scape* 100–150 cm long; peduncle erect (up to 120 cm long), robust, covered in about 13 scarious sheaths up to 23(33) cm long; rhachis 17–40 cm long, arching, becoming slender towards the tip, with up to 27 closely spaced flowers; bracts triangular, 3–15 mm long. *Flowers* 7–9 cm in diameter; not scented; rhachis, pedicel and ovary stained purple; sepals and petals white or pale pink, sometimes with some red spots at the base and over the mid-vein; lip white, becoming deep pink on pollination, side-lobes usually strongly veined and spotted maroon-red, mid-lobe yellow at the base and in the centre, usually streaked and spotted maroon-red, callus bright yellow at the tips, paler behind; column pale to deep purple-pink above, white, usually streaked with red below; anther-cap cream. *Pedicel and ovary* (1.5)2.3–4.5(6.5) cm long. *Dorsal sepal* 4.2–5.6 × 1.5–2 cm, obovate, acute, concave, erect or porrect; lateral sepals similar, spreading, slightly drooping. *Petals* 4–5.2 × 1.4–1.8 cm, narrowly obovate, acute, spreading. *Lip* 3-lobed, fused to the base of the column for 2–4 mm; side-lobes 1.2–2.2 cm broad, lightly clasping the column, papillose or minutely hairy, broadly rounded, margin not fringed; mid-lobe 0.9–2 × 1.2–2.4 cm, triangular to subcircular, acute, recurved, papillose, with a basal and central patch of short, dense hairs, margin entire and undulating; callus ridges 2, inflated at the apex, tapering to the base, densely hairy. *Column* 3–3.2 cm long, winged to the base, with some minute basal hairs ventrally; pollinia triangular to sub-quandrangular, 2–2.5 mm across. *Capsule* 4–5 cm long, round to oblong-fusiform, with an apical beak formed by the persistent column.

Plate 21. *Cymbidium insigne* subsp. *insigne*. Curtis's Botanical Magazine t. 8312, del. Matilda Smith.

subsp. **insigne**

Flowers pale to deep pink; lip marked with purple speckles and streaks; lip 3.4–3.7 × 3.8–4.2 cm; mid-lobe 1.6–2 × 2–2.4 cm, ovate to subcircular, not noticeably clawed. Pl. 21; Figs. 108, 109 & 130: 2a–d.

DISTRIBUTION. Southern Vietnam, China (Hainan) (Map 27); 1000–2500 m (3280–8200 ft).

HABITAT. In sandy soil and on sandstone rocks in low, open *Pinus*-Ericaceae woodland or in Ericaceae-*Arundinaria* associations; flowering (November) December until May. *Cymbidium insigne* is one of the few truly terrestrial orchids in section *Cyperorchis*. Bronckart stated that *C. insigne* was a terrestrial growing in sandy soil

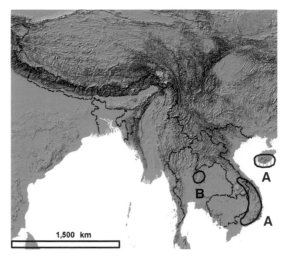

Map 27. Distribution of **A.** *C. insigne* subsp. *insigne*, **B.** subsp. *seidenfadenii*.

along ravines. Micholitz (1904) found it growing on steep banks among thick grass, usually in a stiff clay type of soil. Averyanov (pers. comm.) described it growing in southern Vietnam as terrestrial on ridge-tops at 1400–1500 m in open mixed forest of *Pinus khesya, P. dalatensis*, and *Lyonia ovalifolia* and at 2500 m on ridge-tops in Ericaceae-*Arundinaria* associations.

CONSERVATION STATUS. VU A1cd; B1ab.

Fig. 108. *Cymbidium insigne* subsp. *insigne*. Vietnam, cult. Munich B.G.

199

L. AVERYANOV

Fig. 109. *Cymbidium insigne* subsp. *insigne*. Vietnam, Kontum.

subsp. **seidenfadenii** *P.J. Cribb & Du Puy* **subsp. nov.** affinis subspeciei typica sed tepalis angustioribus, sepalis lateralibus deflexis, labello 2.7–2.9 × 2–2.3 cm valde rubro-striatis et lobo medio labelli minore, 0.9–1 × 1.2–1.3 cm, triangulari unguiculari satis differt. Typus: Thailand, Phu Luang, 1500 m, 11 April 1983, *Menzies & Du Puy 501* (holotypus K!).

This variety differs from the typical subspecies of *C. insigne* from Vietnam and Hainan in having more slender sepals and petals and a smaller lip, 2.7–2.9 × 2–2.3 cm, with a distinctively triangular, shortly clawed mid-lobe, 0.9–1 × 1.2–1.3 cm. The markings on the lip comprise bold red-purple stripes rather than dots and streaks, although a form lacking red markings occurs in ther wild. Pl. 22; Figs. 11, 17 110–111.

DISTRIBUTION. Northern Thailand (Map 27); 1400–1600 m (4590–5250 ft).

HABITAT. It grows in shallow, very sandy soil in low, open *Pinus*–Ericaceae woodland in the shade of low bushes in a region with many outcrops and boulders of sandstone that serve to break the vegetation cover, providing a diversity of habitats. Its long flowering spikes push through the twigs so that the flowers are held clear of the bush. Thus the flowers are at a similar height, and close to the *Rhododendron* flowers they mimic (see also Du Puy 1983, 1984 for more details of the habitat). It flowers from (November) December until May. Fig. 112.

In early April 1983, one of the authors (DD) discovered a population of what he identified as *C. insigne* in northern

DAVID DU PUY

Fig. 110. *Cymbidium insigne* subsp. *seidenfadenii*, N Thailand. *Menzies & Du Puy 501*.

Plate 22. *Cymbidium insigne* subsp. *seidenfadenii*. Thailand, *Menzies & Du Puy* 500 (right) and 501 (left), del. Claire Smith.

DAVID DU PUY

Thailand. The plants were terrestrial and had large white flowers, mostly marked with red stripes on the lip. Its distinction from *C. insigne* only became obvious when it was grown alongside true *C. insigne* in cultivation at Kew. The Vietnamese plants are often pale pink rather than white, have broader tepals, more spreading lateral sepals, a larger purple-speckled lip and a larger ovate mid-lobe with undulate margins.

Du Puy (1983, 1984) and Kjellsson *et al.* (1985) identified two colour variants in the natural populations in almost equal proportions. One of these had strong red spots and veins on the lip, whereas in the other any trace of this pigmentation was lacking, the lip being pure white or pinkish with a pale yellow patch on the centre of the mid-lobe and the apical region of the callus. The vegetation of the area was dominated by the white-flowered *Rhododendron lyi*, and another large-flowered white orchid, *Dendrobium infundibulum*, which was also common. All three were found to be pollinated by the same species of bumble-bee, *Bombus eximius*, the pollinaria of the two orchid species being placed on different regions of the thorax of the bee. The apparent absence of a food reward or other attractant in the orchid flowers, and their similar colouring, indicated that they were possibly both mimicking the *Rhododendron* flower (see Chapter 6). It is possible that the form with pure white flowers was being selected for, as the flowers more closely resembled those of *R. lyi*.

Fig. 111 (left). *Cymbidium insigne* subsp. *seidenfadenii*. N Thailand, *Menzies & Du Puy* 500. **Fig. 112** (below). Habitat of *Cymbidium insigne* subsp. *seidenfadenii*, N Thailand.

DAVID DU PUY

26. CYMBIDIUM IRIDIOIDES

Cymbidium iridioides was described by David Don in 1825 from material collected in Nepal by Nathaniel Wallich in 1821 and now in the Wallich herbarium. In 1832, Lindley published the name *C. giganteum*, also based on a specimen collected by Wallich. He noted the previous description by Don but questioned to which species the name referred. Don did not cite a collection number, and it was generally assumed that the name *C. iridioides* was applicable to a *Coelogyne* and, therefore, the name *C. giganteum* came into general usage. Don's description is brief, but the only characters that do not fit the present species are the leaves only 30 cm (12 in) long and the flower colour ('albi'). The Wallich specimens do indeed have short leaves and agree with Don's description. They also lack colour notes, and it is likely that Don did not know the colour of the living flowers. There is no reason why *C. iridioides* D. Don should not be accepted as the valid name for this species. Furthermore, Lindley's *C. giganteum* is a later homonym of *C. giganteum* Sw. (1800).

In 2003, Liu *et al.* described *C. gaoligongense* from the Gaoligong Mountains that lie parallel and close to the Myanmar border in western Yunnan. They compared it with *C. tracyanum* but the flowers are smaller and lack any of its distinctive red striping and spotting. It seems probable that it is no more than an albinistic form of *C. iridioides* which, from the description and accompanying illustration of *C. gaoligongense* provided by the authors, is similar and has previously been recorded from the region.

Cymbidium iridioides is one of the less showy of the large-flowered species in section *Cyperorchis*, with yellowish sepals and petals lined with red-brown. The flowers often do not open fully, hiding the relatively small, purple-spotted lip. It is similar in colouring, size and distribution to *C. erythraeum,* but the latter has much narrower floral parts, giving a spidery appearance, and the leaves are much narrower. Its whiter, less heavily spotted lip is also distinctive, also lacking the undulating margin and central lines of hairs present in *C. iridioides.*

Cymbidium tracyanum has a similar colouring, and has the lines of hairs on the mid-lobe of the lip, but it can be distinguished from *C. iridioides* by the larger size of its flower and especially of the mid-lobe of the lip, its strongly falcate petals, the hairs confined to the veins on the side-lobes and the fringe of long hairs on the margins of the side-lobes of the lip. *Cymbidium iridioides* is also allied to *C. wilsonii* (Du Puy *et al.*, 1984; see *C. wilsonii* for a discussion of their affinity and distinguishing characters).

26. Cymbidium iridioides D. Don, *Prodr. Fl. Nepal.*: 36 (1824). Type: Nepal, *Wallich s.n.* (holotype BM!).

C. giganteum Wall. ex Lindl., *Gen. Sp. Orchid. Pl.*: 163: (1833) & in *J. Proc. Linn. Soc. Bot.* 3: 29 (1859); *non* Sw. in *Schrad. J. Bot.* 2: 224 (1800) = *Cyrtopera gigantea.* Type: Nepal, *Wallich 7355* (lectotype K!, chosen by Du Puy & Cribb, 1988).
Cyperorchis gigantea (Wall. ex Lindl.) Schltr. in *Fedde, Repert. Sp. Nov. Regni Veg.* 20: 107 (1924).
Iridorchis gigantea (Wall. ex Lindl.) Blume, *Coll. Orchid Arch. Ind. Jap.* 1: 90, t. 26 (1858).
C. gaoligongense Z.J. Liu & J.Y. Zhang in *J. Wuhan Bot. Res.* 21, 4: 316 (2003), **syn. nov.** Type: China, W Yunnan, Gaoligong Mts, W of Baoshan, *Z.J. Liu 2582* (holotype SZWN No. 518114).

A perennial, epiphytic or lithophytic *herb. Pseudobulbs* 5–17 cm long, 2–6 cm in diameter, elongate-ovoid, bilaterally flattened, with about 10 leaves. *Leaves* up to 90 cm or more long, 2–4.2 cm broad, linear-elliptic, acute, mid-green, articulated 6–11 cm from the pseudobulb to a yellow-green, broadly sheathing base. *Scape* 45–85 cm long, suberect to horizontal; peduncle stiff, covered in scarious sheaths up to 11 cm long; rhachis 25–50 cm long, robust, tapering above, yellowish-green, bearing 7–20 flowers; bracts triangular, up to 2.5 mm long. *Flowers* up to 10 cm across, scented; rhachis, pedicel and ovary green; sepals and petals yellowish-green heavily stained with irregular veins and spots of red- or ginger-brown, with a narrow cream margin; lip yellowish, side-lobes dark red-veined, mid-lobe yellow

Plate 23. *Cymbidium iridioides*. Curtis's Botanical Magazine t. 4844, del. W.H. Fitch.

DAVID DU PUY

Fig. 113. *Cymbidium iridioides*, Nepal, cult. Kew 269-80-02505.

at the base, marked with a broad submarginal band of confluent deep red spots and blotches; callus ridges yellowish, spotted maroon in front, becoming muddy red behind; column yellowish-green, streaked red-brown beneath. *Pedicel and ovary* 2.2–4.2 cm long. *Dorsal sepal* 4.5–4.7 × 1.2–1.8 cm, narrowly obovate, acute, concave, porrect; lateral sepals similar, asymmetric and twisted forward giving the flower a half-open appearance. *Petals* 4.4–4.8 × 0.7–1 cm ligulate, curved, spreading. *Lip* 3-lobed, fused to the column base for 4–5 mm; side-lobes 1–1.2 cm broad, triangular, rounded at the apex, porrect, margin fringed with short hairs, the indumentum of short hairs evenly distributed; mid-lobe 1.2–1.6 × 1.4–1.8 cm, ovate, mucronate, recurved, sparsely hairy except in the centre where two or three lines of long hairs extend from the callus to beyond the centre of the mid-lobe, margin erose and strongly undulating, fringed with short hairs; callus ridges 2, short, reaching half way down the disc, dilated at the apex, tapering off rapidly below, covered in long hairs. *Column* 2.5–2.9 cm long, winged but narrowing at the base, giving a more slender appearance than in *C. tracyanum*, short hairs present ventrally near the base; pollinia 2.1–2.5 mm long, triangular. *Capsule* about 6–8 × 3–4 cm, fusiform-ellipsoidal, stalked, with the persistent column forming a short (1.5–2 cm), apical beak. Pls. 3 & 23; Figs. 106: 2a–d, 113 & 114.

DISTRIBUTION. Nepal, northern India (Kumaon, Assam, Sikkim, Meghalaya), Bhutan, Myanmar & south-west China (Yunnan, south-west Sichuan, south-east Xizang) (Map 28); 900–2800 m (2950–9185 ft), at higher elevations in its more southerly localities. In Kumaon, Nepal and Sikkim, the distribution of *C. iridioides* coincides with those of *C. hookerianum* and *C. erythraeum*, both of which, however, usually occur at higher elevations. It also occurs to the south, in Meghalaya, a region from which the latter two species are absent. The ranges of these three species also overlap in Yunnan, but *C. iridioides* extends farther south. It has been accidentally imported with collections of *C. lowianum*, a species endemic to the tropical forests of Thailand, Myanmar (Burma) and southern Yunnan; 900–2800 m (2950–9185 ft), highest in the more southerly localities.

HABITAT. It has usually been reported growing epiphytically on mossy trees in damp, shaded forest, the strongest specimens growing in hollows on rotting wood, or in tree hollows which have collected humus and leaf litter. It is also found on rocks in forests or on cliffs. Generally, the habitat of this species is frost-free, with some winter rain, but in regions where night frosts occur the winters are dry, as in Meghalaya. It flowers between August and December.

CONSERVATION STATUS. VU A1cd.

Fig. 114. *Cymbidium iridioides.* E. Bhutan.

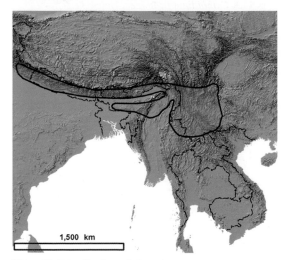

Map 28. Distribution of *C. iridioides.*

206

27. CYMBIDIUM LOWIANUM

Cymbidium lowianum was first collected in 1877 by William Boxall in Burma (Myanmar). He sent plants to Messrs Hugh Low & Co. of Upper Clapton, and the species was described in the same year by H.G. Reichenbach as a variety of *C. giganteum* (= *C. iridioides*). His description was based on a dried specimen, but he noted that it was possibly a distinct species, and when the imported plants flowered for Low, in 1879, he raised it to specific rank.

Wu & Chen (1980) considered *C. lowianum* to be a variety of *C. hookerianum*. Numerical taxonomic analyses (Du Puy *et al.*, 1984) demonstrated conclusively that these two taxa are distinct (see *C. hookerianum* for a discussion of their distinguishing characters).

Cymbidium lowianum is variable in flower size, shape, and colour and in the indumentum on the lip. A distinctive variant is recognised as var. *iansonii*, as indicated in the key below. There are several other variants of this species that have been recognised as distinct by some authors. Variety *superbissimum* L. Linden is a fine colour form with clear green sepals and petals and a strong red colour on the mid-lobe of the lip. Variety *concolor* Rolfe (Fig. 115) lacks red pigment in the flowers, giving clear green sepals and petals, and the red mark on the mid-lobe is replaced by yellow. Variety *viride* R. Warner & H. Williams and var. *flaveolum* L. Linden are the same. Variety *concolor* does not differ otherwise from the red-pigmented variants and does not form distinct wild populations. The most appropriate taxonomic treatment for this variant would appear to be as a cultivar.

27. C. lowianum (Rchb.f.) Rchb.f. in *Gard. Chron.* n.s. 11: 332, 404, t. 56 (1879). Type: Burma [Myanmar], *Boxall*, cult. Low (holotype W).

C. giganteum Wall. ex Lindl. var. *lowianum* Rchb.f. in *Gard. Chron.* n.s., 7: 685 (1877).

C. lowianum (Rchb.f.) Rchb.f. var. *concolor* Rolfe in *Gard. Chron.* ser. 3, 10: 187 (1891). Type: cult. Eastwood (holotype K!).

C. lowianum (Rchb.f.) Rchb.f. var. *superbissimum* L. Linden, *Lindenia* 9: 19, 20, t. 392 (1893). Type: Burma [Myanmar], cult. Linden (holotype the illustration in not located, but see the above reference for an illustration of the type).

C. lowianum (Rchb.f.) Rchb.f. var. *flaveolum* L. Linden, *loc. cit.* 12: 91–92, t. 572 (1896). Type: cult. *Linden* (holotype not located, but see the above reference for an illustration of the type).

C. lowianum (Rchb.f.) Rchb.f. var. *viride* Warner & Williams, *Orchid Album* 11: t. 527 (1897). Type: cult. Smee (lectotype the illustration in the above reference, selected here).

Cyperorchis lowiana (Rchb.f.) Schltr. in *Fedde, Repert. Sp. Nov. Regni Veg.* 20: 108 (1924).

Cymbidium hookerianum Rchb.f. var. *lowianum* (Rchb.f.) Y.S. Wu & S.C. Chen in *Acta Phytotax. Sin.* 18: 303 (1980).

A perennial, epiphytic or lithophytic *herb. Pseudobulbs* up to 13 cm long, 5 cm in diameter, bilaterally flattened, about 9-leaved. *Leaves* up to 90 × 3.5 cm, linear-elliptic, acute, articulated 6–10 cm from the pseudobulb to a yellow-green, broadly sheathing base. *Scape* up to 100 cm or more long, suberect to horizontal, arching; peduncle covered by sheaths up to 10 cm long; rhachis bearing 12–30(40) flowers; bracts triangular, 3 mm long. *Flowers* up to 10 cm in diameter; not scented; rhachis, pedicel and ovary green; sepals and petals bright apple-green to yellowish, sometimes lightly shaded with red-brown; lip yellowish to white, side-lobes not marked, the mid-lobe with a deep red, broad, V-shaped, submarginal patch, extending from near the tip to the angles between the mid-lobe and side-lobes, and a central red line to the callus tips, the base of the lip bright yellow or orange, spotted with red-brown (*C. lowianum* var. *iansonii* has light to heavy red-brown shading on the sepals and petals, and a light brown V-shaped mark on the mid-lobe); callus cream or white, occasionally with a few red spots; column green or yellowish above, tipped with red-brown, spotted red-brown below towards the base; anther-cap cream or white. *Pedicel and ovary* 3–5 cm long. *Dorsal sepal* 4.8–5.7 × 1.6–1.8 cm, narrowly obovate, acute, concave, porrect; lateral sepals similar, spreading. *Petals* 4.8–5.3 × 1.1–1.3 cm, ligulate to narrowly

Plate 24. *Cymbidium lowianum* var. *lowianum*. Cult. Kew, del. Claire Smith.

obovate, acute, curved, spreading. *Lip* 3-lobed, fused to the column base for 3–4.5 mm; side-lobes 1.4–1.6 cm broad, triangular, apex right-angled, margin not distinctly fringed, with an indumentum of velvety hairs especially at the front; mid-lobe 1.4–1.8 × 1.5–2.1 cm, cordate, mucronate, porrect to vertical (angled down at the base), with the indumentum in two zones comprising silky hairs in the centre and base, and very dense velvety hairs on the V-shaped, coloured region (var. *iansonii* has longer hairs on the lip and callus than var. *lowianum*), margin of the mid-lobe erose, minutely undulating; callus ridges 2, not reaching the base of the lip, spreading behind, dilated at the apex, finely hairy. *Column* 3–3.5 cm long, winged, with papillae or hairs present ventrally towards the base; pollinia 3–3.5 mm long, triangular. *Capsule* about 6–8 × 3–4 cm, fusiform-ellipsoidal, stalked, with a 1.5–2 cm apical beak.

DISTRIBUTION. *Cymbidium lowianum* occurs in northern Thailand, north-east and east Myanmar and south and west Yunnan (Map 29); 1200–2400 m (3940–7875 ft). Seidenfaden (1983) includes the north-west Himalaya in this range. Collections of this species from the wild occasionally included specimens of *C. iridioides* which generally has a more northerly distribution. The range of *C. lowianum* largely coincides with that of *C. tracyanum*, but it usually grows at higher altitudes. All of the larger flowered species in section *Cyperorchis* from northern Thailand, *C. insigne*, *C. lowianum* and *C. tracyanum*, have been under great collection pressure and are now rare and confined to a few of the more inaccessible mountains.

HABITAT. On trees in damp, shaded evergreen or mixed forest, often on *Schima wallichii*, or on shaded cliffs along valleys; flowering February–June.

CONSERVATION STATUS. VU A1cd.

Key to the varieties of *C. lowianum*

Scape arching, with a pendulous rhachis; flowers well spaced; sepals and petals light to apple green, sometimes slightly veined and shaded red-brown; lip with a dark red (or rarely yellow) V-shaped mark on the mid-lobe, and finely pubescent on the inner surface .... var. **lowianum**
Scape suberect, with a suberect rhachis; flowers closely spaced; sepals and petals lightly to strongly shaded and veined red-brown; lip with a light chestnut-brown V-shaped mark on the mid-lobe, and with a longer indumentum on the inner surface and especially on the callus .... var. **iansonii**

var. **lowianum**

The typical variety is highly distinctive and easily recognised by the short indumentum on its lip, as well as by its lip shape and markings. It is one of the most beautiful of the *Cymbidium* species, with its striking, large, apple-green flowers carried gracefully on a slender, arching scape. The mid-lobe is marked with a single, bold, V-shaped patch of maroon (occasionally yellow) that covers the entire apical and submarginal region. Only *C. schroederi* has a similar mid-lobe marking (see *C. schroederi* for a discussion of the relationship between these species and their distinguishing characters). It has frequently been used in hybridisation, the strong lip marking apparent in many modern hybrids. Pl. 24; Figs. 115–118 & 130: 4a–d.

Fig. 115. *Cymbidium lowianum* var. *lowianum* 'Concolor', cult. Kew.

DISTRIBUTION. Myanmar (Burma), south-west China (southern and western Yunnan), northern Thailand (Map 29).

Fig. 116. *Cymbidium lowianum* on a roof-top in W Yunnan.   **Fig. 117.** *Cymbidium lowianum* var. *lowianum*, cult. Kew.

Fig. 118. *Cymbidium lowianum* on a tree in western Yunnan.

var. **iansonii** (Rolfe) P.J. Cribb & Du Puy in *Kew Bull.* 40: 432 (1984). Type: Burma [Myanmar], cult. Low (holotype K!).

C. mandaianum Gower, cited in *Lindenia* 12: 91 (1896). Type: cult. *Manda* (holotype not located).
C. × iansonii Rolfe in *Orchid Rev.* 8: 191, 209, f. 34 (1900).
C. grandiflorum Griffith var. *kalawensis* T.A. Colyear in *Orchid Rev.* 42: 248 (1934) & 43: 165, 167 (1935). Type: Burma [Myanmar], *Colyear* (holotype CAL).

The flowers are slightly larger than in var. *lowianum*, and the lip and callus have longer hairs. The petals and sepals are marked with irregular red-brown stripes and shading as in *C. tracyanum*, and the V-shaped marking on the mid-lobe of the lip is light brown. Figs. 119 & 130: 5a–d.

DISTRIBUTION. Northern and eastern Myanmar (Burma), possibly western Yunnan.

Variety *iansonii* first flowered in the collection of Messrs Low in 1900, having originally been collected in northern Myanmar near Bhamo. Rolfe described it in 1900, considering it to be a hybrid between *C. lowianum* and *C. tracyanum*.

In 1912, a second plant was exhibited, under the name *C. mandaianum*, which Rolfe could not distinguish from the original specimen (Anon., 1912). In 1934, a specimen collected in Myanmar (Burma), near Kalaw, was described by Colyear as *C. grandiflorum* var. *kalawensis*. In 1935 it was figured in the *Orchid Review*. A plant of similar appearance is now in cultivation at Kew. These plants are intermediate between var. *lowianum* and the original plant of var. *iansonii*. The flowers are slightly larger than other specimens of *C. lowianum*, and the lip has a longer pubescence. The mid-lobe has a light-brown, V-shaped mark, but the sepals and petals are much less strongly marked with brown than in the type collection. This colour variation shows that a range of intermediates exists between the two varieties.

In a numerical taxonomic analyses of this group (Du Puy *et al.*, 1984), the specimens of *C. iansonii* linked closely with those of *C. lowianum*, although some distinction was apparent. No evidence could be found to suggest that hybridisation with any other species in the section was involved. *Cymbidium iansonii* has therefore been formally reduced to varietal rank within *C. lowianum*.

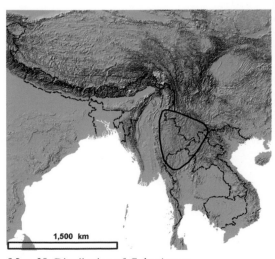

Map 29. Distribution of *C. lowianum*.

Fig. 119. *Cymbidium lowianum* var. *iansonii*. Cult. Munich B.G..

28. CYMBIDIUM MASTERSII

Cymbidium mastersii was first described by John Lindley in 1845 based upon a specimen flowered by Messrs. Loddiges in the previous December. He attributed the name to William Griffith, who had chosen it in honour of Dr. Masters of the Calcutta Botanical Garden. However, in 1851, Griffith published the name *C. affine* for the same species, but did not mention *C. mastersii*. In 1878, Reichenbach distinguished *C. affine* by the purple spotting towards the front of the lip, and by the slightly more hairy lip. This character is variable within the species, and the name *C. affine* has since been applied to specimens with the purple pigmentation. However, the type specimen of *C. mastersii*, illustrated in the *Botanical Register* of 1845, also had red-purple spots on the lip, and the two are undoubtedly conspecific.

This species varies in the degree of pigmentation of its lip and, in 1880, Reichenbach named the white-lipped variant var. *album*. At the same time he rescinded his previous differentiation of *C. mastersii* and *C. affine* on the basis of lip colour and instead proposed that they could be differentiated on the more upright flower spike and the lack of 'notches' (auricles) towards the base of the callus ridges in *C. affine*. However, the type specimen of *C. mastersii* also has both of these features.

Specimens of *C. mastersii* collected by one of the authors (DD) in Thailand have faint pinkish spots on the lip and are intermediate between the two major colour variants. Variation in the size of the auricles at the base of the lip is found in both *C. mastersii* and *C. elegans*, and seems to vary in both species from strongly expressed to absent. The angle at which the scape is held varies from erect to almost horizontal, depending on the robustness of the peduncle, and the weight of the flowers it carries.

In 1858, Lindley compounded the nomenclatural confusion by publishing *C. micromeson* based on a mixed collection by Griffith from the Khasia Hills. The type

Fig. 120 (top). *Cymbidium mastersii*, Thailand, *Menzies & Du Puy* 399. **Fig. 121** (right). Habitat of *Cymbidium mastersii* in northern Thailand.

212

Plate 25. *Cymbidium mastersii*. NE India, Khasia Hills, del. W. Griffith, used as a model by Sarah Drake for Edward's Botanical Register 31: t. 50 (1845).

DAVID DU PUY

Fig. 122. *Cymbidium mastersii*, Thailand, *Menzies & Du Puy* 399.

sheet comprises a shoot and flower spike of *C. mastersii*, a fruiting scape of *C. eburneum*, and a flattened lip of another species (probably in the genus *Coelogyne* according to J.D. Hooker on a note attached to the specimen). There is also a sketch of a lip from one of the flowers on the complete specimen which shows no obvious differences from *C. mastersii*, but is different from the flattened lip on the sheet. The description differs from *C. mastersii* in several characters that are attributable to this alien lip.

Cymbidium eburneum is closely related and shares many of the characters of *C. mastersii* but has a more inflated pseudobulb, which does not grow in an indeterminate manner for more than two or three years before producing a new growth. It can be distinguished by its usually single, large flower, and its confluent, cuneate callus tip (see also the discussion under *C. eburneum* and *C. whiteae*).

In 1996 F.Y. Liu described *C. maguanense* based upon her own collection from Maguan in southeast Yunnan. The type appears to be lost from the Kunming Herbarium. Although she compared it with *C. eburneum,* it is close to *C. mastersii* and, in our opinion, falls within the range of variation of that species. The latter is common in the area. It is possible, but unlikely, that it is a natural hybrid of *C. mastersii* with *C. eburneum*. Its flowers are similar in size and coloration to those of the former but are not unlike those of the latter in shape. *Cymbidium mastersii* is variable in its flower shape and coloration. However, the variability in these characters appears to us to be continuous.

Cymbidium mastersii can be distinguished from *C. elegans* by its conspicuous, stem-like pseudobulb and indeterminate growth habit with the scape arising within the axils of the upper leaves, not from near the base of a swollen pseudobulb. *Cymbidium mastersii* has distinctive, unequally bilobed, forked leaf apices, with a minute mucro in the sinus, and fewer, larger and more distant flowers that open more fully. The ovate mid-lobe of *C. mastersii* is distinct from the narrow mid-lobe expanding into two incurved apical lobes with a strongly emarginate apex typical of *C. elegans*. The pollinia of *C. elegans* are clavate rather than quadrangular. These two species often have auricles protruding at the base of the callus ridges.

Cymbidium erythraeum might also be confused with *C. mastersii*, but can be distinguished by the growth habit, pseudobulb and leaf characters, differently coloured and more distant flowers on a much longer rhachis that is exserted from the sheaths, spreading sepals and petals, more acute, porrect, hairy side-lobe tips, a cordate to reniform mid-lobe, and curved column not strongly deflexed near the apex.

Cymbidium mastersii produces nectar that often fills the pouch formed by the fusion of the base of the lip and the column.

There is a possible natural hybrid with *C. sanderae* or *C. insigne* collected in Annam (see the discussion of *C. sanderae*), and *C.* × *ballianum* is a natural hybrid with *C. eburneum*, imported from Burma [Myanmar] (*Orchid Rev.* 12: 85, 1904).

28. Cymbidium mastersii Griff. ex Lindl. in *Bot. Reg.* 31: t. 50 (1845). Type: cult. Loddiges (holotype K!).

C. affine Griff., *Notulae* 3: 336, t. 291, f. 3. (1851). Type: Assam, Khasia Hills, Churra, *Griffith s.n.* (holotype K!).

C. micromeson Lindl. in *J. Linn. Soc.* 3: 29 (1858). Type: Assam, Khasia Hills, Churra, *Griffith s.n.* (holotype K!).

C. mastersii Griff. ex Lindl. var. *album* Rchb.f. in *Gard. Chron.*, n.s., 13: 136 (1880). Type: not designated.

Cyperorchis mastersii (Griff. ex Lindl.) Benth. in *J. Linn. Soc.* 18: 318 (1881).

Cymbidium maguanense F.Y. Liu in *Acta Bot. Yunnanica* 18 (4): 412 (1996). Type: SE Yunnan, Maguan, *F.Y. Liu 88004* (holotype KUN!).

A perennial, epiphytic or lithophytic *herb*. *Pseudobulbs* not inflated, stem-like and inconspicuous within the numerous persistent leaf bases, growing indeterminately, forming cauline roots through the lower leaf bases and occasionally producing a new growth from near the base, with 6–17 apical leaves and numerous, distichous, alternate, sheathing, persistent leaf bases towards the base, the older growth

Fig. 123. *Cymbidium mastersii*. Vietnam, cult. Munich B.G.

215

appearing flattened and elongated. *Leaves* up to 64 × 1.8(2.5) cm, arching, narrowly ligulate, tapering to an acute, usually unequally bilobed, forked apex with a mucro in the sinus; articulated about 6–10 cm from the axis; leaf base with a 2–3 mm-wide membranous margin. *Scape* about 25–30 cm long, from the axils of the leaves; peduncle suberect, covered in 6–8 sheaths up to 16 cm long; upper sheath cymbiform, central sheaths cylindrical in the basal half, expanded, cymbiform in the apical half; rhachis 5–10 cm long, arcuate or pendulous, not strongly exserted from the sheaths, with (2)5–10(15) closely spaced flowers; bracts 2–7 mm long, triangular, acute. *Flower* about 6 cm across, not opening fully; almond-scented; rhachis, pedicel and ovary green; petals and sepals white or faintly pink; lip white, usually with a yellow central patch at the base of the mid-lobe, and bright yellow callus ridges, sometimes with some pale to strong purple-red spots and shading on the side- and mid-lobes; column white or pale green, sometimes spotted with pale red below, with a yellow patch at the base; anther-cap cream. *Pedicel and ovary* 1.6–2.5 cm long. *Dorsal sepal* 4.3–5.2 × 0.8–1.1 cm, narrowly oblong-elliptic to narrowly obovate, acute, concave, porrect but spreading towards the tips; lateral sepals similar, curved, porrect. *Petals* 3.9–4.9 × 0.5–0.7 cm, narrowly ligulate or narrowly obovate, slightly curved, porrect. *Lip* 3-lobed, elongated, fused to the base of the column for 3–5 mm; side-lobes 1–1.2 cm broad, lightly clasping the column, minutely hairy, apex broadly rounded to subacute and porrect, margin minutely fringed at the front; mid-lobe small, 1–1.3 × 1–1.3 cm, ovate, rounded or mucronate, porrect, minutely hairy with a dense central and basal patch of short hairs, margin undulating; callus long, with 2 slightly raised, well-separated ridges behind, sometimes minutely auriculate and forming a small trough-shaped depression at the base, inflated and converging towards the hairy apex at the base of the mid-lobe. *Column* 3.3–3.6(4) cm long, narrowly winged, curved down near the apex, glabrous or hairy below in an elliptical depression at the base; pollinia quadrangular, 2–3 mm long. *Capsule* broadly ellipsoidal, pointed, stalked, about 3 × 2.5 cm with a beak almost as long as the capsule. Pl. 25; Figs. 120–123 & 125: 2a–e.

DISTRIBUTION. Northern India (Sikkim, Meghalaya, Manipur), Myanmar, northern Thailand, southern China (north-west to south-east Yunnan), Laos and northern Vietnam (Map 30); 900–2200 m (2950–7220 ft).

HABITAT. On trees or rocks, in evergreen forest, often in deep shade, in humus, moss or on rotting wood. *Cymbidium mastersii* is found at about 1500 m (4920 ft) elevation in northern Thailand. Du Puy (1984) encountered it growing in the same region as *C. insigne*. Collections sent to Kew flowered in November 1984, corresponding to the end of the rainy season in Thailand. It is usually found growing epiphytically in the taller, denser, shaded and humid stands of forest surrounding creek beds. The crowns of older plants are held well clear of the branches and plant debris in which they grow, the long stems arching upwards and outwards from their place of anchorage. The largest specimens were found on moss-covered branches, although younger plants were also found on vertical tree trunks that were much less densely covered by vegetation. Occasional specimens were also found growing in humus and moss on boulders and cliffs. Kerr noted that on Doi Suthep it grew in 'shady, moist ravines in humus on dead tree trunks'. It flowers from (September) October until December (January).

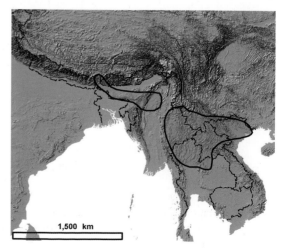

Map 30. Distribution of *C. mastersii*.

CONSERVATION STATUS. VU A1cd; B1ab.

29. CYMBIDIUM PARISHII

Cymbidium parishii was discovered in Burma (Myanmar), near Moulmein, on the border with Thailand in 1859 by the Rev. Charles Parish, who collected a number of plants that were lost in transit. In 1867 he sent two further plants to Messrs Low, and a dried flower to Hooker at Kew from the plant he had cultivated and painted. Hooker considered it to be a variety of *C. eburneum*, but did not publish this name until 1891, in his *Flora of British India*. Meanwhile, in 1872, Reichenbach examined Parish's dried material and named it *C. parishii*, publishing the name in 1874. Coincidentally, the two plants with Messrs Low were sold, one to John Day, the orchid enthusiast and artist, and another to the collection of Mr. Leech. Both of these plants flowered in June 1878, those grown by Mr. Swan, gardener for Mr. Leech, being the earlier. Leech's plant was subsequently sold to B.S. Williams for 100 guineas. Thus, 19 years after its discovery, living material was at last available to Reichenbach, enabling him to describe the species more fully, and to justify his claim that it was distinct from *C. eburneum* because the flower was smaller and the callus extended nearer to the base of the mid-lobe and lacked the middle velvet line of *C. eburneum*. He claimed that the viscidium had two long, spreading, hair-like appendages, which were said to be lacking in *C. eburneum*. Further investigation has shown the presence of these appendages in *C. eburneum* also.

Since then, *C. parishii* has been lost to cultivation. Little material has been preserved. The type specimen, *Parish 56*, consists of a 2-flowered scape and a leaf, and is the only specimen at Kew. Two unpublished paintings by Day and by Parish (Plate 26), both preserved at Kew, and Reichenbach's painting in *Xenia Orchidacea*, supply extra information about the colouring of the flower and, in particular, the habit of the plant. The paintings in R. Warner and B.S. Williams' *Orchid Album*, and L. Linden's *Lindenia* are both stylised, and depict an immature growth and old pseudobulb, and therefore they do not contribute much to the information available about this species.

Cymbidium parishii is closely related to both *C. eburneum* and *C. roseum*. Its tendency to grow indeterminately for more than one year, producing flowers over several seasons from the axils of the leaves towards the top of the growth, is similar to both of the above species and is typical of species in the former section *Eburneum*, the most extreme case of this being found in *C. mastersii*. The unequally bilobed, often split leaf tips, sometimes with a needle-like extension in the fork, are also typical of this section. It is similar to *C. eburneum*, differing in the characters listed in Table 12 but new collections of *C. eburneum* from Yunnan are closer in flower colour: callus structure differs in these species.

The differences between *C. parishii* and *C. roseum* are discussed under the latter species. *Cymbidium parishii* var. *sanderae* Rolfe is treated here as a distinct species, *C. sanderae*.

TABLE 12

A COMPARISON OF THE MORPHOLOGICAL DIFFERENCES
BETWEEN *C. PARISHII* AND *C. EBURNEUM*

Character	*C. parishii*	*C. eburneum*
Leaf width	1.8–3 cm	up to 1.3(2) cm
Number of flowers	2–3	1–2(3)
Dorsal sepal size (as a measure of flower size)	5.9 × 1.5 cm	5.6–7.6 × 1.8–2.9 cm
Lip colour	strong purple markings	no purple markings, or weakly spotted
Side-lobe apex shape	acute and porrect	broadly rounded
Mid-lobe shape	circular	ovate-triangular
Callus structure	two ridges, converging	one single cuneate ridge
Flowering time	June–July	March–May (November–January)

Plate 26. *Cymbidium parishii*. Myanmar [Burma], Megala Chyoung. Illustration of the type by Charles Parish, 1867, Kew collection.

29. Cymbidium parishii Rchb.f. in *Trans. Linn. Soc.* 30: 144 (1874), in *Gard. Chron.*, n.s. 1: 338, 566 (1874) and n.s., 10: 74 (1878). Type: Burma [Myanmar], *Parish 56* (holotype K!).

C. eburneum Lindl. var. *parishii* (Rchb.f.) Hook.f., *Fl. Brit. India* 6: 12 (1891).
Cyperorchis parishii (Rchb.f.) Schltr. in *Fedde, Repert. Sp. Nov. Regni Veg.* 20: 108 (1924).

A perennial, probably epiphytic *herb*, resembling *C. eburneum*. *Pseudobulbs* about 11.5 × 4 cm, fusiform, not produced annually but growing in an indeterminate fashion for several seasons, often covered in persistent, distichous leaf bases, with 11–14 apical leaves. *Leaves* 38–53 × 1.8–3 cm, ligulate, acute, with an unequally bilobed apex with a short mucro in the sinus, articulated to a broadly sheathing base. *Scape* about 25 cm long, from within the axils of the leaves; peduncle covered with several sheaths up to 15 cm long; upper sheath cymbiform, central sheaths cylindrical below, expanded and cymbiform above; rhachis with 2–3 flowers; bracts triangular, to 4 mm long. *Flower* smaller than that of *C. eburneum*, not opening fully; scented; rhachis, pedicel and ovary green; petals and sepals white; lip white with a yellow central and basal patch on the mid-lobe and orange-yellow callus ridges, with interrupted streaks of purple on the side-lobes, and a few purple spots on the mid-lobe except on the broad, white submarginal area; column white, yellowish and red-spotted below towards the base; anther-cap white. *Pedicel and ovary* 3–4 cm long. *Dorsal sepal* 5.9 × 1.5 cm, narrowly oblong-elliptic, subacute, concave, suberect to porrect; lateral sepals similar but shorter, 5 × 1.6 cm, curved. *Petals* 5.7 × 1.3 cm, oblong-spathulate, subacute, curved, not fully spreading but reflexed and spreading at the tips. *Lip* 3-lobed, elongated, fused to the base of the column for about 3 mm; side-lobes 1.5 cm broad, clasping the column, hairy, acute, porrect, margins not fringed; mid-lobe 1.6 × 1.7 cm, circular, rounded, mucronate, recurved, with a dense central patch of hairs extending from the front of the callus ridges, margin undulating; callus long, with 3 slightly raised ridges on a glabrous disc, the outer 2 becoming inflated and hairy towards the apex, converging but not confluent, terminating behind the junction of the mid- and side-lobes. *Column* 3.7 cm long, winged, curved down near the apex, almost glabrous. Pl. 26; Fig. 125: 4a–g.

DISTRIBUTION. Myanmar (Tenasserim; Map 31); 1500 m (4920 ft).

HABITAT. Montane forest. Little is known about the exact locality or habitat of this species. Parish, on a note with the type specimen, says he found it 'on the ascent of Nat-taung, near Toungoo (about 18°50'N, 96°50'E), at 5000 ft' [1650 m], in the same region as *Dendrobium crassinode*. In 1874, Reichenbach wrote about *C. parishii* at the request of Low, probably as a note had been received from his collector Mr Boxall, which Day quoted as saying 'I found 50 plants of this on the top of Moulle Tongue, one of the highest mountains in Burma, with *Dendrobium jamesonianum* and *Coelogyne reichenbachii*.' No records remain of the results of this collection. It flowers in June–July.

CONSERVATION STATUS. CR B1a.

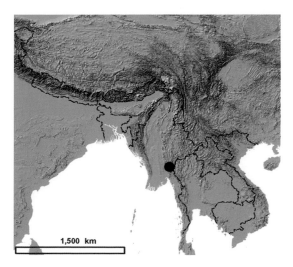

Map 31. Distribution of *C. parishii.*

219

30. CYMBIDIUM ROSEUM

Cymbidium roseum is a rare tropical species with a restricted distribution in the Malay Peninsula, Java and Sumatra. It resembles *C. eburneum* vegetatively but has 2–5 white or pink flowers, often with a boldly marked lip and a pair of yellow callus ridges.

It was described by J.J. Smith in 1905 based upon several collections from Java. Plants with pale pink-coloured flowers and purple markings on the lip, as described by J.J. Smith, are found throughout the range of the species. A yellowish white variant, lacking the spots on the lip, has been found in Sumatra and western Malaysia growing together with the pink form.

It is closely related to *C. parishii*, but the latter has more inflated pseudobulbs, longer leaves (about 38–53 cm (15–21 in) long), a larger flower (dorsal sepal 5.9 cm (2.3 in) in *C. parishii*, up to 4.8 cm (1.9 in) in *C. roseum*), a shorter, more sparse indumentum on the lip, acute side-lobes, a more strongly undulating mid-lobe and white rather than pinkish flowers. The flowers more closely resemble *C. mastersii*.

Cymbidium insigne is readily distinguished from *C. roseum* because it produces a conspicuous annual pseudobulb and longer, narrower leaves with acute, rather than bilobed, apices. Its flower spike is much longer, more robust, carries many more flowers and is produced from the base of the pseudobulb. Its lip is shorter but the mid-lobe is larger, and the side-lobe tips are more broadly rounded.

30. Cymbidium roseum J.J. Sm., *Orch. Java* 475 (1905) & in *Bull. Jard. Bot. Buitenzorg*, sér. 3, 6: t. 11 (1924). Type: Java, Malabar, *Bosscha s.n.*; Wanaredja, Desa Godong, *Ader s.n.*, Tijikora, *Kessler s.n.*, Goentoer, *Raciborski s.n.*; Slamat, *J.J. Smith s.n.* (syntypes BO!).

Cyperorchis rosea (J.J. Sm.) Schltr. in *Fedde, Repert. Sp. Nov. Regni Veg.* 20: 107 (1924).

Fig. 124. *Cymbidium roseum.* An albino variant from N Sumatra.

Fig. 125. *Cymbidium roseum* (Malaya, *Segerback* 2089, + photo). **1a.** Perianth, × 0.7. **1b.** Lip and column, × 0.7. **1c.** Pollinarium, × 3. **1d.** Pollinium (reverse), × 3. **1e.** Flower, × 0.7. **1f.** Flowering plant, × 0.1. **1g.** Leaf tip, × 0.5. *C. mastersii* (Kew spirit no. 14986/47814). **2a.** Perianth, × 0.7. **2b.** Lip and column, × 0.7. **2c.** Pollinarium, × 3. **2d.** Pollinium (reverse), × 3. **2e.** Leaf tip, × 0.5. *C. eburneum* (Kew spirit no. 29007/33346). **3a.** Perianth, × 0.7. **3b.** Lip and column, × 0.7. **3c.** Pollinarium, × 3. **3d.** Pollinium (reverse), × 3. **3e.** Leaf tip, × 0.5. *C. parishii* (Burma, *Parish* 56). **4a.** Perianth, × 0.7. **4b.** Lip and column, × 0.7. **4c.** Pollinarium, × 3. **4d.** Pollinium (reverse), × 3. **4e.** Flower, × 0.7. **4f.** Flowering plant, × 0.1. **4g.** Leaf tip, × 0.5. All drawn by Claire Smith.

A perennial, terrestrial, lithophytic or epiphytic *herb*. *Pseudobulbs* about 7.5 cm long, 2 cm in diameter, not strongly inflated, stem-like and inconspicuous within the leaf bases, growing and flowering in an indeterminate fashion for 2–3 years before a new growth is produced, usually with about 7 fresh leaves and numerous, persistent, sheathing leaf bases that eventually become fibrous, each pseudobulb producing about 13 distichous leaves in total. *Leaves* 20–40 × 2.2–2.7 cm, ligulate, obtuse, unequally bilobed at the apex, sometimes with a mucro in the sinus, coriaceous, articulated 4–6 cm from the pseudobulb. *Scape* (13)19–30 cm long, from the axils of the leaves; peduncle erect or suberect, covered in about 9 sheaths up to 16 cm long; central sheaths mostly cylindrical in the basal half, expanded and cymbiform in the apical half; rhachis with (1)2–5 flowers; bracts triangular, acute, up to 1 cm long. *Flower* 5–6 cm in diameter, not opening fully; faintly scented; rhachis, pedicel and ovary green; sepals and petals pale pink, speckled white, becoming darker or faintly pink-brown with age, sometimes white; lip pale pink or white with a bright yellow patch on the base of the mid-lobe, and bright yellow callus, sometimes with spots and interrupted streaks of purple on the side-lobes and callus ridges and a few purple spots on the mid-lobe; column purple-pink, yellowish or white, sometimes streaked purple-red beneath, yellow at the base; anther-cap cream. *Pedicel and ovary* 2.8–4.7 cm long. *Dorsal sepal* 4.4–4.8 × 1.3–1.6 cm, narrowly elliptic, acute, concave, porrect; lateral sepals similar, not fully spreading. *Petals* 4–4.7 × 1.1–1.3 cm, narrowly elliptic, curved, acute, tips recurved and spreading. *Lip* 3-lobed, elongated, fused to the base of the column for about 3.5 mm; side-lobes 1.1 cm broad, clasping the column, subacute to rounded and porrect, hairy, margin not fringed; mid-lobe 1.4–1.9 × 1.3–1.5 cm, broadly ovate, rounded, porrect, hairy with two lines of hairs at the base in front of the callus ridges, and a dense patch of hairs in the centre,

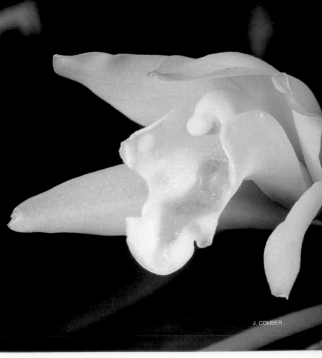

Fig. 126 (top). *Cymbidium roseum*, N Sumatra. **Fig. 127** (bottom). *Cymbidium roseum*, N Sumatra; a variant lacking spotting on the lip.

Fig. 128. *Cymbidium roseum*. An albino variant from N Sumatra.

margin entire and undulating; callus ridges 2, long, tapering to near the base of the lip from inflated, converging apices, terminating well behind the base of the mid-lobe, densely hairy. *Column* 2.9–3.5 cm long, narrowly winged to the base, curved down near the apex, with some hairs on the margin towards the base; pollinia quadrangular. *Capsule* 3.5–6 × 1.5–2.5 cm, fusiform, stalked, apex with a beak formed by the column. Figs. 124, 125: 1a–g, 126–128.

DISTRIBUTION. West Malaysia, Java, Sumatra (Map 32); 1500–2100 m (4920–6890 ft).

HABITAT. Holttum (1953) found it in the high mountains of West Malaysia, and reported that it grew on rocks or low down on trees in exposed places. In Java and Sumatra it is found growing as a lithophyte or terrestrial in full sun. It flowers from August until December.

CONSERVATION STATUS. VU A1cd; B1ab.

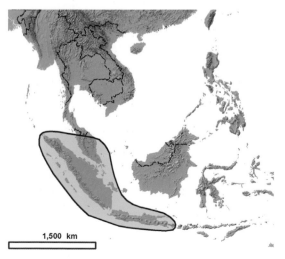

Map 32. Distribution of *C. roseum*.

223

31. CYMBIDIUM SANDERAE

This attractive orchid was discovered by Wilhelm Micholitz in 1904 on the Lang Bian Plateau in Vietnam (Annam), in the same locality as *C. insigne*. The specimens of *C. insigne* that he sent to Messrs. Sander's nursery were given the name *C. sanderi*, whereas those of this species were named after Sander's wife as *C. sanderae*. This latter name was not validly published, and Rolfe subsequently published it at varietal rank within *C. parishii*. That decision was based on the similar flower colour of the two species, but further study has shown that the two are distinct and that *C. sanderae* is more closely related to *C. insigne* than to either *C. parishii* or *C. eburneum*. The *International Code of Botanical Nomenclature* allows *C. sanderae* to be used, despite *C. sanderi* having been applied to a distinct taxon, since they are not homonyms (R.K. Brummitt, pers. comm.).

Cymbidium sanderae has a different growth habit from *C. parishii*. It has ovoid, well-developed pseudobulbs that are produced annually rather than fusiform pseudobulbs that grow and flower for several seasons before producing new growths; its leaves are acute rather than unequally bilobed or forked at the apex; the flower spike is produced from the base of the pseudobulb, not from the leaf axils towards the centre or apex of the pseudobulb; the spike is more robust and produces more (up to 15) flowers. The flower is also distinct, opening more fully, and with elliptic rather than oblong sepals and petals. The lip is not elongated and has a relatively large mid-lobe and rounded side-lobes that are not acute at the apex and has callus ridges that extend nearer to the base of the mid-lobe.

P. CRIBB

All of these characters are shared by *C. insigne*, indicating that *C. sanderae* should be regarded as closely allied to that species. Vegetatively these two species are similar. Their pseudobulbs are of a similar size and are ovoid with little bilateral flattening. The leaves are similar in shape, including the acute apices, but the leaves of *C. sanderae* are broader. Both species produce new pseudobulbs annually, and the flower spike grows from the base of the pseudobulb. *Cymbidium sanderae* has a much shorter flower spike usually with fewer flowers but is otherwise similar. The flowers of these two species are similar in shape, except that the mid-lobe of the lip of *C. sanderae* is rounded at the apex, not acute, and the markings on the lip are stronger. *Cymbidium insigne* is a terrestrial species, but Micholitz (1904) reported that *C. sanderae* usually grew epiphytically, in clumps of *Polypodium* fern and that the two species could be separated in the field.

Cymbidium sanderae is uncommon in cultivation, and was believed to have been lost until, in 1961, Emma Menninger uncovered a single specimen in the nursery of Armacost and Royston, California. This plant flowered in 1963 (Menninger, 1965). The named cultivar 'Emma Menninger' is a tetraploid plant converted from the diploid by Don Wimber. The measurements in brackets in the description usually

Fig. 129. *Cymbidium sanderae*, cult. Los Angeles Arboretum.

Plate 27. *Cymbidium sanderae*. *Seth* 87, cult. Kew, del. Claire Smith.

Fig. 130. *Cymbidium sanderae* (Kew spirit no. 46238). **1a.** Perianth, × 0.7. **1b.** Lip and column, × 0.7. **1c.** Pollinarium, × 3. **1d.** Pollinium (reverse), × 3. *C. insigne* subsp. *insigne* (Kew spirit no. 48010/22593). **2a.** Perianth, × 0.7. **2b.** Lip and column, × 0.7. **2c.** Pollinarium, × 3. **2d.** Pollinium (reverse), × 3. *C. schroederi* (Kew spirit no. 14989/14553). **3a.** Perianth, × 0.7. **3b.** Lip and column, × 0.7. **3c.** Pollinarium, × 3. **3d.** Pollinium (reverse), × 3. **3e.** Flower, × 0.7. **3f.** Flowering plant, × 0.1. *C. lowianum* var. *lowianum* (Kew spirit no. 45530/14543). **4a.** Perianth, × 0.7. **4b.** Lip and column, × 0.7. **4c.** Pollinarium, × 3. **4d.** Pollinium (reverse), × 3. *C. lowianum* var. *iansonii* (Kew spirit no. 37656). **5a.** Perianth, × 0.7. **5b.** Lip and column, × 0.7. **5c.** Pollinarium, × 3. **5d.** Pollinium (reverse), × 3. All drawn by Claire Smith.

F. HOECK

Fig. 131. *Cymbidium sanderae*, diploid clone. Cult. Munich B.G.

refer to this plant. The increase in the size of the flower, especially in the width of the sepals and petals, and the size of the lip, can be directly attributed to the increased ploidy level (Wimber & Wimber, 1968).

The type specimen of *C. sanderae* is supplemented by two further specimens imported by Sander from the Lang Bian plateau. One is similar to the type, and a note is attached saying 'only 2 or 3 plants flowered with a dark spotted lip. All the others turned out an inferior form which was named var. *Ballianum*—without purple markings on the lip—and sepals and petals waxy white.' The second sheet has leaves and flower spikes of this white-flowered variant. This latter is similar in appearance, but dissection of the flowers reveals a similarity to *C. mastersii* in the longer lip shape and the narrower callus ridges. The variations in the flower colour and lip shape indicate that the plants are part of a hybrid swarm, with *C. mastersii* as one parent, and either *C. insigne* or *C. sanderae* as the other. The possibility therefore exists that *C. sanderae* might be a natural hybrid. However, it would be difficult to account for the heavy marking on the lip and the leaves that are broader than either of the proposed parents. Furthermore, *C. mastersii* has not been collected in this region. Colour variation of this nature (presence or absence of red spots on the lip) is encountered in many related species including *C. insigne, C. roseum, C. mastersii* and *C. eburneum*, and might therefore also be expected in *C. sanderae*.

Sander's *Orchid Guide* (1927) mentions *C.* × *ballianum* from two separate sources. The first importation was from Burma, and was a natural hybrid between *C. eburneum* and *C. mastersii*. The use of this name for the collections made in Annam is therefore inadmissible, as they are certainly not hybrids of that parentage. Sander notes that 'the typical (Burmese) form has flowers slightly larger and of a more pure white than the Annamese varieties'.

227

31. Cymbidium sanderae (Rolfe) Du Puy & P.J. Cribb, *The Genus Cymbidium*: 131 (1988). Type: Vietnam, *Micholitz s.n.*, cult. Sander (holotype K!).

C. parishii Rchb.f. var. *sanderae* Rolfe in *Gard. Chron.*, ser. 3, 35: 338–339, f. 146 (1904) & in *Orchid Rev.* 12: 163–164, 279 (1904).

A perennial epiphytic *herb. Pseudobulbs* up to 6 cm long, 4 cm in diameter, ovoid, lightly bilaterally compressed, produced annually, with about 10 leaves. *Leaves* up to 50 × 2.5 cm, linear-elliptic, tapering to an entire, acute apex; articulated to a broadly sheathing base about 9 cm from the pseudobulb. *Scape* 30–50 cm long; peduncle erect to suberect, robust, covered in numerous cymbiform sheaths up to 19 cm long; rhachis about 10 cm long, suberect, becoming slender towards the tip, with 3–15 flowers; bracts 5–12 mm long, triangular. *Flowers* 8 cm across; scented; rhachis green, pedicel and ovary lightly stained with purple; petals and sepals white, usually flushed pink on the reverse with a few purple spots at the base of the petals; lip cream, usually heavily marked with maroon, side-lobes usually blotched and stained with maroon except towards the margins, mid-lobe yellow in the centre and at the base, with a cream margin 2 mm wide and usually marked with a submarginal band of confluent maroon blotches and random spots, callus ridges bright orange-yellow at the front; column cream above, usually streaked and stained maroon below and flushed yellow towards the base; anther-cap cream. *Pedicel and ovary* 2.5–4.3 cm long. *Dorsal sepal* 4.5–4.6 (5.7) × 1.4(2) cm, narrowly obovate to elliptic, acute, concave, erect or porrect; lateral sepals similar, spreading, curved. *Petals* 4.4–4.5(5.4) × 1.1–1.2(1.6) cm, narrowly obovate to elliptic, acute, spreading. *Lip* 3-lobed, 3–4 × 2–3 cm, fused to the base of the column for 4–5 mm; side-lobes 1.2–1.4(1.6) cm broad, lightly clasping the column, minutely hairy, apex broadly rounded to obtuse, margin not fringed; mid-lobe 1.2–1.5(1.9) × 1.1–1.5(2.4) cm, ovate, rounded, obtuse or mucronate, recurved, papillose with a dense basal and central patch of short hairs, margin entire and weakly undulating; callus ridges 2, inflated at the apices, tapering to near the base of the lip, hairy. *Column* 3.2–3.5 cm, winged, with some hairs ventrally towards the base; pollinia subquadrangular, about 2.5 mm long. Pl. 27; Figs. 129, 130: 1a–d, 131 & 132.

P. CRIBB

Fig. 132. *Cymbidium sanderae*, tetraploid clone. Cult. J. Coker.

DISTRIBUTION. Vietnam (Map 33); 1400–1500 m (4595–4920 ft).

HABITAT. It is found on trees, frequently in association with a *Polypodium* fern, in open, humid, evergreen, montane forest growing on acidic sandstone rocks. The canopy of these forests is dominated by cool-loving evergreen broad-leaved and coniferous trees; between 1300 and 2000 m elevation; flowering January–March (May).

CONSERVATION STATUS. EN A1cd; B1ab.

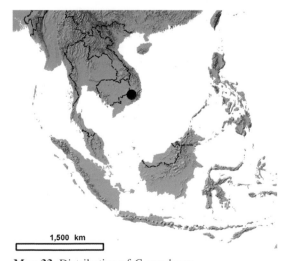

1,500 km

Map 33. Distribution of *C. sanderae*.

32. CYMBIDIUM SCHROEDERI

Cymbidium schroederi is similar to *C. lowianum* in habit, but its flowers are slightly smaller. Their colour combination of heavily red-brown marked sepals and petals, red-brown veined side-lobes, and the brownish, submarginal, V-shaped patch on the mid-lobe distinguish *C. schroederi* from the other species in section *Cyperorchis* with a similarly marked lip. It can also be recognised by its hairy lip, the abruptly terminating side-lobe tips, and callus ridges that taper towards the base.

This species was first collected at Trung Phan in central Vietnam (formerly Annam), and was imported by Messrs Sander. It flowered in 1905 in the collection of Baron Schroeder at the Dell, Englefield Green, Surrey. In the same year, Rolfe based his description of it on this material. Unfortunately, the type specimen has not been located. It may have been placed in Rolfe's collection of spirit-preserved material, which has not survived. Rolfe compared the plant with *C. iridioides* and *C. lowianum*, concluding that it was most closely related to the former, whereas Summerhayes (1942) came to the opposite conclusion. Table 13 illustrates the intermediacy of this species.

Numerical taxonomic analyses (Du Puy *et al.*, 1984) produced scatter diagrams that further confirm the intermediacy of *C. schroederi* between *C. iridioides* and *C. lowianum*. Dendrograms based on morphological similarity have shown *C. schroederi* to be closer to *C. lowianum*. The geographical distribution of *C. schroederi* is well to the south of either of the other species. However, *C. lowianum* and *C. iridioides* have been collected from the same locality in Yunnan. This indicates that *C. schroederi* might have originated from the natural hybridisation of these two species and probably following some introgression with *C. lowianum* has stabilised to form a distinct species occupying a distribution to the south of either of the proposed progenitors.

Fig. 133. *Cymbidium schroederi*. Vietnam, Kontum.

L. AVERYANOV

Fig. 134. *Cymbidium schroederi*. Vietnam, Kontum.

TABLE 13

A COMPARISON OF *C. LOWIANUM*, *C. IRIDIOIDES* AND *C. SCHROEDERI*, SHOWING THE INTERMEDIACY OF *C. SCHROEDERI*

Character	C. lowianum	C. schroederi	C. iridioides
Spike length	up to 100 cm	44–65 cm	45–85 cm
Spike strength	slender	slender	more robust
Number of flowers	12–30(40)	14–23	7–20
Flower shape	petals spreading, dorsal sepal porrect. Flower open	petals spreading, dorsal sepal porrect. Flower open	petals not fully spreading, dorsal sepal porrect. Flower not opening fully
Petal and sepal colour	clear green	greenish, stained with red-brown	greenish, heavily stained with red-brown
Length of fusion of lip and column	3–4.5 mm	3 mm	4–5 mm
Lip indumentum	short	long	long
Lip margin	fringe absent	fringed with short hairs	fringed with short hairs
Side-lobe tip shape	right angled, not porrect	right angled, not porrect	acute, porrect
Side-lobe markings	no markings	veined red-brown	veined red-brown
Mid-lobe shape	cordate	ovate-cordate	ovate
Mid-lobe recurvature	not recurved	slightly recurved	strongly recurved
Mid-lobe margin	minutely undulating	weakly undulating	strongly undulating
Mid-lobe indumentum	in two distinct regions, corresponding with V-shaped markings	in two distinct regions corresponding with V-shaped markings	shortly hairy, not in two regions. Lines of long hairs present from the base to the centre
Mid-lobe markings	submarginal V-shaped patch of deep maroon	submarginal V-shaped patch of red-brown	yellowish in the centre, with a submarginal ring of confluent red-brown spots and blotches
Callus shape	spreading towards the base	tapering towards the base	tapering towards the base
Callus indumentum	short hairs	long hairs	long hairs
Column length	3.0–3.5 cm	2.8 cm	2.5–2.9 cm
Column colour	spotted red-brown below	streaked red-brown below	streaked red-brown below towards the base
Scent	none noticeable	none noticeable	sweet-scented
Flowering	February–June	March–June	August–December

Seidenfaden (1983), in his account of the genus in Thailand, tentatively refers one specimen to *C. schroederi*. This specimen is, however, a small-flowered *C. tracyanum*, having narrow, curved petals, a hairy lip with the indumentum on the side-lobes confined to the veins, lines of hairs extending from the callus well into the mid-lobe and acute, porrect side-lobe tips.

A natural hybrid with the sympatric *C. insigne* has been recorded and named *C. × cooperi* Rolfe (1914). This cross has also been made artificially and named *C.* J. Davis.

32. Cymbidium schroederi Rolfe in *Gard. Chron.*, ser. 3, 37: 243 (1905) & in *Orchid Rev*. 14: 39 (1906) & 25: 101 (1917). Type: Vietnam, cult. Schroeder (not located).

Cyperorchis schroederi (Rolfe) Schltr. in *Fedde, Repert. Sp. Nov. Regni Veg.* 20: 108 (1924).

A perennial, epiphytic *herb. Pseudobulbs* up to 15 cm long, 4 cm in diameter, bilaterally flattened, about 6-leaved. *Leaves* up to 60 × 2.5 cm, linear-elliptic, acute, articulated to a broadly sheathing base. *Scape* 44–65 cm long, suberect, arching; peduncle covered in sheaths up to 14 cm long; rhachis robust at the base, becoming more slender towards the apex, bearing 14–23 flowers; bracts triangular, 3 mm long. *Flowers* 8–9 cm across; not scented; rhachis, pedicel and ovary bright green; sepals and petals green to yellowish-green, with dull, irregular, light brown veins and spots; lip pale yellow, side-lobes veined red-brown, mid-lobe marked with a large, red-brown, submarginal V-shaped patch and median line to the front of the callus; callus whitish, sparsely spotted red-brown; column greenish-yellow shading to dark purple around the anther above, cream with red dashes below; anther-cap cream. *Pedicel and ovary* 2–4.5 cm long. *Dorsal sepal* 4.5–4.9 × 1.4–1.6 cm, narrowly obovate, acute, porrect;

Plate 28. *Cymbidium schroederi*. Curtis's Botanical Magazine t. 9637, del. Lilian Snelling.

lateral sepals similar, spreading. *Petals* 4.2–4.5 × 0.8–1.2 cm, narrowly obovate, acute, curved, spreading. *Lip* 3-lobed, fused to the column base for 3 mm; side-lobes 1.1–1.2 cm broad, triangular, densely hairy with hairs towards the apex, apex right-angled to rounded, margin fringed with hairs; mid-lobe 1.4–1.6 × 1.5 cm, ovate-cordate, mucronate, recurved, with silky hairs in the centre and towards the base, and a densely short-haired region corresponding to the V-shaped colour patch; callus ridges 2, tapering to the base of the lip, densely hairy, with a glabrous patch in front of the callus ridges over the mid-vein on the mid-lobe. *Column* 2.8 cm long, winged, narrowing towards the base, with inflated papillae ventrally towards the base. Pl. 28; Figs. 130: 3a–f, 133–135.

DISTRIBUTION. South-central Vietnam only (Map 34); 1350–1700 m (4430–5580 ft).

HABITAT. Averyanov and Christenson (1998) described its habitat in the Kontum area of southern Vietnam as open evergreen primary forest on north-west facing mountain slopes where 'it grows as a humus epiphyte on old trees along mountain streams at 1350 to 1700 m elevation. The canopy of these forests is dominated by cool-loving evergreen broad-leaved and coniferous trees. The climate is cool and moist and plants grow in large pockets of mossy humus where there is more than 4 inches (10 cm) of accumulated substrate on top of the tree bark. Under these conditions, the plants readily form massive clumps and each plant typically becomes a large specimen that bears 10 to 15 inflorescences. Somewhat surprisingly *Cymbidium schroederi* flowers well in either full sun or partial shade, but the flowers are noticeably more pigmented in the shade... *Cymbidium schroederi* exhibits good fruit set in nature, and so pollination is apparently quite effective. The large fruits resemble small lemons.' The local women make beautiful wreaths from the flowers.

It flowers from (December) February until June.

CONSERVATION STATUS. VU A1cd.

Fig. 135. *Cymbidium schroederi*. Cult. Munich B.G.

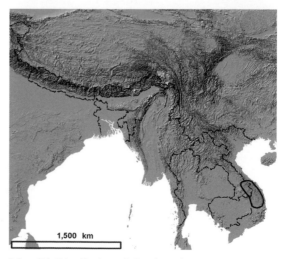

Map 34. Distribution of *C. schroederi*.

233

33. CYMBIDIUM SIGMOIDEUM

Wilhelm Micholitz discovered *C. sigmoideum* in central Sumatra, and sent specimens to Messrs Sander, who in 1905 sent one to Kew for identification, but Rolfe failed to recognise it as a new species. However, J.J. Smith described it shortly afterwards based on a collection by Connell from Java.

In 1993, Wong and Chan described *C. kinabaluense* based upon a collection made by Chan and Nais from 1700 m on Mt. Kinabalu in Sabah, Borneo. They distinguished it from *C. sigmoideum* because of its longer, more floriferous inflorescences and larger flowers with a longer pedicel and shorter column. Comber (2001) examined the evidence and considered it conspecific with *C. sigmoideum*, albeit with a longer, more floriferous inflorescence than typical plants from Java and Sumatra. The same phenomenon can be seen in *C. atropurpureum* where some Sumatran plants have much shorter, fewer-flowered inflorescences than the plants found elsewhere. Neither is the disjunction between Sabah and the islands farther west unique in the orchids: for example, *Paphiopedilum javanicum* is found in Sabah and also in Java, Sumatra, Bali and Flores.

It is similar in growth habit to *C. roseum*, which is its closest ally on these islands. The flower does bear some resemblance to *C. whiteae* from Sikkim, to which it appears to be most closely related. It has many distinguishing characters, some of them unique in the genus. The most distinctive of these are the shiny brown-

Fig. 136 (above). *Cymbidium sigmoideum*, W Java. **Fig. 137** (right). W Java.

spotted, green flowers; the narrow, curved petals; the short lip with a relatively long fusion of the margins of the lip and column-base; broad, falcate side-lobes; a narrow, ligulate, waxy, recurved mid-lobe; the bilobed, rounded callus apex; the broad, S-shaped column; and the short, backwards-projecting rostellum.

The fusion at the base of the lip and column, the two cleft pollinia, the annually produced pseudobulbs, the scape from the base of the pseudobulb, the pendulous rhachis, the narrow petals, the porrect dorsal sepal covering the column, the rostellum, the rectangular viscidium without hair-like processes from the lower corners and the quadrangular-pyriform pollinia support its inclusion in section *Cyperorchis*, close to *C. elegans*, *C. cochleare* and *C. whiteae*.

33. Cymbidium sigmoideum J.J. Sm. in *Bull. Agric. Indes Neerl.* 13, 52–53 (1907). Type: Java, Loemadjang, *Connell s.n.* (holotype BO!).

Cyperorchis sigmoidea (J.J. Sm.) J.J. Sm. in *Bull. Jard. Bot. Buitenzorg,* sér. 3, 9: 57, t. 8 (1927).
Cymbidium kinabaluense K.M. Wong & C.L. Chan in *Sandakania* 2: 86 (1993). Type: Sabah, Mt. Kinabalu, *C.L. Chan & Jamili Nais s.n.* (holotype SAN, isotype SRN).

A perennial, epiphytic *herb. Pseudobulbs* 4–6 cm long, 2–2.5 cm in diameter, not inflated, stem-like and inconspicuous within the leaf bases, produced annually, with about 9–12 leaves and 3 cataphylls that eventually become fibrous. *Leaves* up to 102 × 1.3–2.5 cm, linear-obovate, acute, articulated 5–8 cm from the pseudobulb. *Scape* 25–83 cm long, from the base of the pseudobulb; peduncle horizontal to pendulous, covered in about 9–10 overlapping, cymbiform, acute sheaths up to 17 cm long; rhachis 10–38 cm long, with 4–23 flowers; bracts 2–4 mm long, ovate or triangular, acute. *Flower* 3.5–4.5 cm across, waxy, inconspicuous, pendulous to horizontal; rhachis, pedicel and ovary purple; sepals and petals deep apple-green or olive-green, spotted and stained with dark or purple-brown; callus light green; column pale green, cream below; anther-cap cream. *Pedicel and ovary* 1.9–3.2 cm long. *Dorsal sepal* about 2.7–3.3 × 0.8–1.1 cm, narrowly obovate or lanceolate, acute, concave, porrect, closely covering the column; lateral sepals similar, falcate, spreading or reflexed. *Petals* about 2.4–3 × 0.4–0.5 cm, ligulate, acute, curved, sometimes spreading in the apical half. *Lip* 3-lobed, with the lip and the base of the column fused for 5–6 mm; side-lobes about 6 mm broad, broadly triangular-falcate, erect and clasping the column, minutely papillose, waxy, front margin S-shaped, subacute; mid-lobe small, about 6.5–9 × 2–2.5 mm, ligulate, glabrous, waxy, recurved, acute; callus a glabrous disc with 2–3 slightly raised ridges behind and a swollen, bilobed, rounded apex. *Column* broad, 1.8–2 cm long, S-shaped, glabrous, the basal quarter fused to the base of the lip; anther-cap and viscidium elongated into a projecting rostellum; pollinia quadrangular-pyriform, about 2 mm long, cleft, on a narrow rectangular viscidium without hair-like processes from the corners. Figs. 95: 4a–f, 136–137.

DISTRIBUTION. Java, Sumatra, and Borneo (Sabah) (Map 35); 800–1700 m (2625–5580 ft).

HABITAT. *Cymbidium sigmoideum* is an epiphyte growing on trunks and larger branches of trees in cooler montane rainforests, often in deep shade. It flowers sporadically almost throughout the year.

CONSERVATION STATUS. VU A1cd; B1ab.

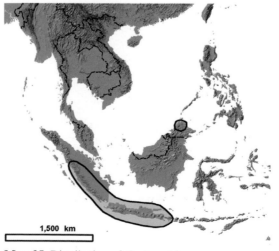

1,500 km

Map 35. Distribution of *C. sigmoideum*.

34. CYMBIDIUM TRACYANUM

Cymbidium tracyanum was first noticed in December 1890, when it flowered in the collection of Mr. Tracy in Twickenham, having been imported three years previously with *C. lowianum*, which it closely resembles vegetatively. It was exhibited at the Royal Horticultural Society, and described by Castle in the same year. A few days later, O'Brien described the plant again in the *Gardeners' Chronicle* where it was illustrated early in the following year. The plant was sold by auction, and no type specimen appears to have been preserved.

In 1895 a second plant was introduced, again in a collection of *C. lowianum*, and a year later a third that was said to have been collected in Upper Burma, giving the first clue to the origin of this species. Soon afterwards, several more plants appeared, imported with *C. lowianum* or occasionally *C. hookerianum*. In about 1900, several specimens, collected by Kerr near Chiang Mai in northern Thailand, flowered in cultivation. In 1902, Lucien Linden described and illustrated *C. zaleskianum* based upon a plant said to have been imported from Assam in a consignment of *C. iridioides* by the nursery of Moortebeek. The flowers are typical of *C. tracyanum* in shape and overall colouring, but the lip has bolder marginal blotches than is typical. Linden thought it was either a hybrid of *C. iridioides* and *C. hookerianum* or a variety of *C. tracyanum*. We are inclined to the latter view but do not exclude the possibility of introgression from *C. iridioides* accounting for the lip markings. A detailed account of the use and misuse of the name *C. zaleskianum* has been given by G. Russell (www.geocities.com/pennypoint9/zaleski/main.html).

Cymbidium tracyanum is similar to *C. erythraeum* in flower colour and shape, but can be readily distinguished by its much larger flower (up to 15 cm (6 in) in diameter), broader leaves and the long hairs on the callus, mid-lobe and margins of the side-lobes. It is similar in colour to *C. iridioides*, which is the only other species in the section that has long hairs in the centre and base of the mid-lobe (see discussion of *C. iridioides*).

Cymbidium hookerianum has flowers of a similar size and with hairy side-lobe margins, so that herbarium specimens of the two species are often confused. *C. hookerianum*, however, has clear apple-green sepals and petals, red-brown spotted side-lobes, and submarginal spots and blotches on the mid-lobe. *Cymbidium tracyanum* can also be distinguished by its strongly falcate petals and longer, denser indumentum. It has hairy callus ridges, lines of hairs in the centre and base of the mid-lobe, and lines of short hairs confined to the veins on the side-lobes (Du Puy *et al.*, 1984).

DAVID DU PUY

Fig. 138. *Cymbidium tracyanum, Seth* 82. Cult. Kew.

Plate 29. *Cymbidium tracyanum*. Cult. Kew, del. Mary Grierson, Kew collection.

34. Cymbidium tracyanum L. Castle in *J. Hort.*, ser. 3, 21: 513 (1890). Type: cult. Tracy (not located).

Cyperorchis traceyana (L. Castle) Schltr. in *Fedde, Repert. Sp. Nov. Regni Veg.* 20: 108 (1924) [*sphalm. pro tracyana*].

Cymbidium zaleskianum L. Linden, *Lindenia* 17: t. 728 (1902). Type: Assam, imp. Moortebeek, cult. Linden (illustration cited above).

A perennial, epiphytic or lithophytic *herb*. *Roots* fleshy, often with erect aerial roots present around the base of the plant. *Pseudobulbs* 5–15 cm long, 2–6 cm in diameter, elongate-ovate, bilaterally compressed, 5–11-leaved. *Leaves* up to 95 × 4 cm, linear-elliptic, acute, mid-green, articulated 6–13 cm from the pseudobulb to a yellow-green, broadly sheathing base. *Scape* up to 130 cm long, suberect, arching; peduncle covered in sheaths up to 20 cm long; rhachis robust, yellow-green, bearing 10–20 flowers; bracts triangular, 3 mm long. *Flowers* up to 15 cm across; scent strong, sweet; rhachis, pedicel and ovary green; tepals yellowish-green to olive-green, strongly stained with irregular, dull, red-brown veins and spots; lip pale yellow to cream, side-lobes veined dark red-brown, mid-lobe marked with scattered spots and vertical dashes of red-brown, with a central reddish stripe almost to the front of the callus; callus cream, spotted with red along the crests; column yellowish with many red-brown dashes on the ventral surface, and red-brown shading around the base and apex on the dorsal surface; anthercap cream. *Pedicel and ovary* 2.5–7.5 cm long. *Dorsal sepal* 5.5–8 × 1.4–1.9 cm, narrowly obovate, acute, concave, porrect; lateral sepals similar, asymmetric and twisted. *Petals* 5.4–7.2 × 0.8–1.4 cm, ligulate, usually falcate, spreading or reflexed, giving the flower a drooping appearance. *Lip* 3-lobed,

Plate 30. *Cymbidium tracyanum.* Cult. Kew, del. Stella Ross-Craig, Kew collection.

fused to the column for the basal 4–5.5 mm; side-lobes 1.1–1.6 cm broad, triangular, acute and porrect, margins hairy (hairs over 2 mm long), indumentum otherwise shorter with the hairs mainly pigmented red-brown and confined to the veins; mid-lobe 1.8–2.1 × 1.8–3 cm, elliptic, mucronate, recurved, with scattered hairs on the upper surface, except in the centre where there are 3 rows of long hairs continuing from the callus ridges to the centre of the mid-lobe, margin finely erose, undulate and partially reflexed; callus ridges 2, tapering towards the base and dilated towards the apex, densely covered in long hairs, with a third line of hairs between them. *Column* 3.4–4.4 cm long, winged to the base, with hairs present ventrally towards the base, and tufts of hairs at the apex on either side of the anther; pollinia 2, 3.2–4 mm long, triangular. *Capsule* up to 15 × 6 cm, fusiform-ellipsoid, stalked, with a 2 cm apical beak. Pls. 29 & 30; Figs. 106: 1a–d, 138 & 139.

Fig. 139. *Cymbidium tracyanum.* Thailand, Doi Inthanond.

DISTRIBUTION. China (southern and western Yunnan), eastern and northern Myanmar, northern Thailand (Map 36); 1200–1900 m (3940–6235 ft).

HABITAT. On moist rocks or on trees in damp, shaded evergreen forest, often near or overhanging streams; flowering in September–January. Kingdon Ward (1940) described its habitat in northern Myanmar [Burma] as 'in wet evergreen hill forests', and 'growing in the fork of a tree overhanging a stream in a deep gulley'. It grows in broken shade, in a constantly humid atmosphere, well protected from the occasional frosts. We have seen plants growing in similar situations in northern Thailand at elevations of 1500–1600 m. They can grow to considerable size in such habitats. *Cymbidium tracyanum* has characteristic erect, aerial roots that may be an adaptation to a moist environment.

CONSERVATION STATUS. VU A1cd.

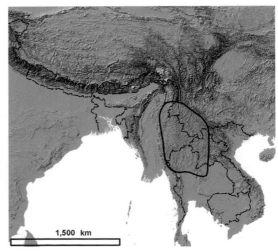

Map 36. Distribution of *C. tracyanum.*

239

35. CYMBIDIUM WENSHANENSE

This handsome orchid was described as recently as 1990 by Wu and Liu based upon a collection by the latter from near Wenshan in south-east Yunnan. The type plant persisted for some years in the living collection of the Kunming Botanic Garden, where one of the authors (PC) saw it in flower in the mid-1990s. Wu and Li compared it with *C. erythrostylum,* which has a similar flower shape, distinguishing it by its lip, which has a broad, emarginate, glabrous mid-lobe and two, separate, hairy callus keels that are fleshy at the apex, and by its column that is white marked with red and bears white hairs on its dorsal surface. Its flowers are also larger and are said to smell of bananas (Holger Perner, pers. comm.).

It is undoubtedly rare and threatened in the wild. One of the authors (PC) visited the type locality in submontane evergreen monsoon forest just south of Maguan in south-east Yunnan. Although the trees survived, no plants remained. Nurseries in Wenshan that he visited had many large plants in cultivation.

35. Cymbidium wenshanense Y.S. Wu & F.Y. Liu in *Acta Bot. Yunnanica* 12 (3): 291 (1990); Chen, Tsi & Luo, *Native Orchids of China in Colour*: 122 (1999). Type: China, SE Yunnan, Wenshan, March 1989, *Liu 8801* (holotype KUN!).

C. quinquelabrum Z.J. Liu & S.C. Chen in *Acta Bot. Yunnan.* 28(1): 13 (2006), **syn. nov.** Type: China, Yunnan, Maguan Co., *Z.J. Liu* 2848 (holotype SZWN).

A perennial, epiphytic herb, forming large clumps. *Pseudobulbs* ovoid, 3–4 cm long, 2–2.5 cm in diameter, covered by persistent leaf bases. *Leaves* arcuate, linear-lorate, acuminate, 60–90 × 1.3–1.7

Fig. 140. *Cymbidium wenshanense*, SE Yunnan.   **Fig. 141.** Type locality, SE Yunnan.

Fig. 142. *Cymbidium wenshanense*. **A.** Habit. **B.** Dorsal sepal. **C.** Lateral sepal. **D.** Petal. **E.** Lip, flattened. **F.** Column. **A.** × ²/₃. **B–F.** Natural size. Drawn from the type collection and photographs by Deborah Lambkin.

Fig. 143. *Cymbidium wenshanense*, SE Yunnan.

cm, dark green. *Inflorescences* suberect, arching, 32–40 cm long, 3- to 7-flowered. *Flowers* showy, scented, white with a lip having a yellow throat and callus and spotted and streaked with purple on the lobes, turning pink with senescence; column red-stained on the underside; pedicel and ovary 3–4 cm long. *Dorsal sepal* arching over column, 6.1–6.4 × 1.8–2 cm, oblong-elliptic, obtuse. *Lateral sepals* deflexed-incurved, 5.8–6.4 × 1.8–2.1 cm, elliptic-oblanceolate, obtuse. *Petals* incurved-porrect, arching forwards over column, 5.7–6.3 × 2–2.1 cm, narrowly obovate, obtuse. *Lip* fused at the base to the column for 2–3 mm, 3-lobed, arcuate, 5.6 × 5.8 cm; side-lobes erect, obovate, rounded in front, 2 cm wide; mid-lobe deflexed, incurved, 1.9 × 2.7 cm, broadly obovate to transversely oblong, emarginate; callus bilamellate, hairy. *Column* arcuate, 4–4.2 cm long, hairy on the dorsal side. Figs. 140–143.

DISTRIBUTION. Southern China (south-east Yunnan), possibly also in adjacent Vietnam (Map 37); 1000–1500 m (3280–4920 ft).

HABITAT. On trees in lower montane evergreen forest and in groves behind villages (Fig. 141). Flowering in March and April.

CONSERVATION STATUS. EN A1acd; B1ab.

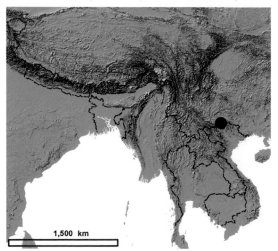

Map 37. Distribution of *C. wenshanense*.

242

36. CYMBIDIUM WHITEAE

Cymbidium whiteae was described and illustrated by George King and Robert Pantling in 1898 from a specimen collected in Sikkim, the name honouring Mrs. C. White who discovered it. Pradhan (1979) suggested that it is still found in the wild, albeit rarely. Trudel (1983) reported fresh collections being made in the vicinity of Gangtok in Sikkim and this remains its sole known locality.

In a letter from King that is attached to the type specimen, it was noted that *C. whiteae* was originally thought to be a variant of *C. mastersii*. It is, however, distinct from *C. mastersii*, lacking the strongly indeterminate growth habit and characteristic forked leaf apex of the latter, and the scape is produced from near the base of the pseudobulb. The flowers are bell-shaped and pendulous, and are held on a pendulous rhachis, whereas those of *C. mastersii* open more fully and are held more erect on a horizontal rhachis. The lip of *C. whiteae* is more strongly hairy, and its callus ridges are broadly

Fig. 144. *Cymbidium whiteae.* Sikkim. Cult. E. Young Orchid Foundation.

243

Plate 31. *Cymbidium whiteae*. Sikkim, del. C. Mukerjai from Annals of the Royal Botanic Garden, Calcutta 8: t. 258 (1898).

TABLE 14

THE CHARACTERS OF *C. WHITEAE* THAT COULD BE ATTRIBUTED TO
C. COCHLEARE AND *C. EBURNEUM*, IF A HYBRID ORIGIN WERE POSTULATED

C. cochleare	*C. eburneum*
pendulous rhachis	*long (to 8 mm) floral bracts
purple-coloured rhachis	*oblong-elliptic sepal shape
scape from near the	*broad sepals and petals
pseudobulb base	*white lip colour
bell-shaped flowers	long lip and-column fusion (5 mm)
pendulous flowers	callus ridges not well separated,
green-brown sepal and	strongly convergent at their apices
petal colour	callus ridges terminate well behind
numerous fine red spots	the mid-lobe
on the lip	callus with a broad, glabrous raised
mid-lobe with a deflexed tip	disc towards the base of the lip
anther-cap rostellate	*broad column
long, slender beak on the	*column white
capsule	*larger capsule, longer than the beak
leaf tips slightly unequal	

*Characters also found in *C. mastersii.*

separated on a glabrous, raised disc, converging suddenly at their apices; those of *C. mastersii*, lacking the broad, glabrous disc, are closer together and converge well behind their apices.

The green-brown perianth, the numerous small red spots on the whitish lip and the white callus ridges immediately distinguish *C. whiteae* in the living state. *Cymbidium sigmoideum*, from Java, Sumatra and Borneo (*q.v.*) is probably its closest relative. *Cymbidium cochleare*, which occurs sympatrically in Sikkim, has a similar scape with pendulous flowers and a similar sepal and petal colour, but the flowers are more tightly bell-shaped, as in *C. elegans* and the the lip and callus ridges are yellow, not white. It also has narrower leaves, shorter floral bracts, flower parts that are more slender, a lip with glabrous side-lobes, and callus ridges that are much shorter, closer together, and lack the broad, raised, glabrous disc.

The callus of *C. whiteae* is most strongly reminiscent of *C. eburneum*. It is possible that *C. whiteae* arose as a natural hybrid, the most likely parents being *C. cochleare* and either *C. eburneum* or *C. mastersii*. Table 14 gives the characters of *C. whiteae* that could be attributed to these species if a hybrid origin were postulated.

The pollinarium is intermediate in that the pollinia are quadrangular, but are elongated and tend towards the clavate shape of *C. cochleare*, and the viscidium has hair-like processes from its corners, but they are short.

However, several characters of *C. whiteae* could not be explained by this parentage (or by the substitution of *C. mastersii* for *C. eburneum*), including the long leaf, strong indumentum on the side-lobes of the lip, the lack of a dense patch of short hairs on the mid-lobe and the distinctive white callus ridges with hairy apices. These characters are significant enough to preclude the possibility that this parentage could have produced *C. whiteae*. The proposal that *C. erythraeum* could be one of the parents would explain some of these characters, but would leave even more inexplicable characters. Therefore the suggestion of a recent hybrid origin can be discounted.

Its close relationship to *C. elegans* and its allies is indicated by its ovoid pseudobulbs; acute, slightly unequal leaf tips; a scape produced from near the base of the pseudobulb; slender, pendulous, many-flowered rhachis; bell-shaped, pendulous flowers; an anther-cap with a beak; clavate pollinia; and the capsule with a beak almost as long as the capsule itself.

36. Cymbidium whiteae King & Pantl. in *Ann. Roy. Bot. Gard. Calcutta* 8: 193–194, t. 258 (1898). Type: Sikkim, Gangtok, *Pantling 425* (holotype K!).

Cyperorchis whiteae (King & Pantl.) Schltr. in *Fedde, Repert. Sp. Nov. Regni Veg.* 20: 107 (1924).

A perennial, epiphytic *herb. Pseudobulbs* up to 10 cm long, 2.5 cm in diameter, narrowly ovoid, lightly bilaterally flattened, with about 12–14 distichous leaves, growing in an indeterminate fashion for about two years before a new growth is produced, the base covered in persistent leaf bases. *Leaves* up to 90 × 1.3(1.5) cm, narrowly ligulate, arching, tapering to an acute, unequal apex; articulated 3–15 cm from the pseudobulb to a broad, sheathing base with a scarious margin up to 3 mm wide. *Scape* about 20–30 cm long, from near the base of the pseudobulb; peduncle suberect, arching, covered in inflated, cymbiform sheaths up to 14 cm long; rhachis pendulous, with 10–12 flowers; bracts 2–8 mm long, triangular. *Flower* about 3.5 cm in diameter, bell-shaped, suberect to pendulous; rhachis, pedicel and ovary purple; sepals and petals greenish, flushed and spotted with red-brown; lip white with numerous fine maroon or brownish spots; callus white, spotted behind with maroon; column cream, spotted ventrally with maroon; anther-cap white. *Pedicel and ovary* 1.5–2 cm long. *Dorsal sepal* 4.3–4.8 × 1.1–1.2 cm, narrowly oblong-elliptic, acute, concave, porrect, recurved at the tip; lateral sepals similar, often mucronate, somewhat spreading. *Petals* 4.1–4.5 × 0.7–0.9 cm, obovate-ligulate, acute, curved, porrect, spreading at the tip. *Lip* 3-lobed, elongated, fused to the base of the column for about 5 mm; side-lobes 1.1 cm broad, triangular, erect and lightly clasping the column, hairy, porrect, acute; mid-lobe 1.2–1.3 × 1.2 cm, cordate, acute, mucronate, deflexed near the apex, with sparse hairs, margin undulating; callus a raised glabrous ridge behind, swollen at the margins and forming 2 raised ridges that become inflated and confluent at the apex (although not forming the single cuneate apex found in *C. eburneum*), terminating well behind the junction of the mid- and side-lobes, hairy towards the apex. *Column* broad, 3.5 cm long, winged, decurved near the apex, minutely hairy below; anther-cap elongated with a rostellum; pollinia quadrangular-pyriform, about 2 mm long, on a rectangular viscidium with hair-like processes from the lower corners. *Capsule* 3.5 × 1.5 cm, oblong, pointed, with a beak that is shorter than the capsule. Pl. 31; Figs. 95: 2a–f & 144.

DISTRIBUTION. Sikkim (Map 38); about 1500–2000 m (4920–6560 ft).

HABITAT. Usually on *Schima wallichii* and occasionally on *Castanopsis* sp. in evergreen forest, often growing with other *Cymbidium* and *Bulbophyllum* species. The forest is wet during spring, summer and autumn, but in winter it is dry, and the temperature plummets (Trudel, 1983). It flowers in October and November.

CONSERVATION STATUS. EN A1cd; B1ab.

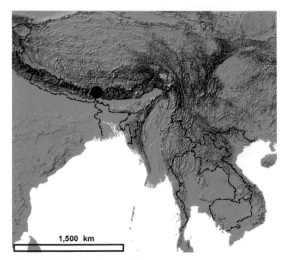

1,500 km

Map 38. Distribution of *C. whiteae*.

37. CYMBIDIUM WILSONII

Ernest Wilson, who discovered this species in 1901 near Mengzi in south-eastern Yunnan, sent plants to the nurseries of Messrs James Veitch, who flowered it in February 1904, and exhibited it at the Royal Horticultural Society as *C. giganteum* (= *C. iridioides*) var. *wilsonii* on the advice of Rolfe. Later the same month, Cook described it under that name in the February edition of *The Garden*. The Orchid Committee would not accept that name, but awarded it as *C. wilsonii*. Rolfe subsequently validated the name in the *Orchid Review* in March 1904.

Since then, there has been continuing debate about the status of *C. wilsonii*, and Taylor & Woods (1976) treated it as a cultivar of *C. iridioides*. There have been few herbarium collections of *C. wilsonii*. A single living specimen, probably from Wilson's original collection, has survived at the Royal Botanic Garden Edinburgh, and some pieces from this plant and selfed seedlings are being grown at Kew. A herbarium specimen collected from Hpimaw Fort on the border of western Yunnan and Myanmar, which closely resembles the type and is attributable to *C. wilsonii*, is preserved at Kew (R.A. 840). Its rarity and its superficial similarity to *C. iridioides* especially in lip shape and colour lends strength to the argument that *C. wilsonii* is a variant of *C. iridioides*. However, numerical taxonomic analyses (Du Puy *et al.*, 1984) indicated a more complex relationship. Analyses of the species in their section *Iridorchis* have shown *C. wilsonii* to be closely linked to *C. hookerianum* rather than to *C. iridioides* as might have been expected. Secondly, scatter diagrams of these three species showed *C. wilsonii* to be intermediate between *C. hookerianum* and *C. iridioides*. It therefore appears possible that hybridisation has been involved in the origin of *C. wilsonii*. The characters in Table 15 illustrate this intermediacy.

Cymbidium wilsonii occurs within the range of *C. iridioides*, but at higher elevations, whereas *C. hookerianum* occurs further to the north but at similar elevations to *C. wilsonii*. The last two flower in spring, *C. iridioides* in autumn. *Cymbidium hookerianum* and *C. iridioides* are therefore normally prevented from hybridising by both spatial and temporal isolation. Records indicate that the spatial isolation may be less distinct towards the southern extreme of the range of *C. hookerianum*, and occasional out-of-season flowering can occur in either species. A combination of these factors might allow occasional hybridisation between the two species.

Fig. 145. *Cymbidium wilsonii*, cult. Edinburgh B.G.   **Fig. 146**. *Cymbidium wilsonii*, cult. Edinburgh B.G.

TABLE 15

A COMPARISON OF *C. HOOKERIANUM, C. IRIDIOIDES*
AND *C. WILSONII*, SHOWING THE INTERMEDIACY OF *C. WILSONII*

Character	*C. hookerianum*	*C. wilsonii*	*C. iridioides*
Leaf size	up to 80 × 2.1 cm	up to 70 × 2.5 cm	up to 90 × 4.0 cm
Rhachis	slender	slender	robust
Number of flowers	6–15	5–15	7–20
Flower size	up to 12 cm	9–10 cm	up to 10 cm
Dorsal sepal size	5.6–6.0 × 1.7–1.9 cm	4.4–5.7 × 1.2–1.9 cm	4.5–4.7 × 1.2–1.8 cm
Petal and sepal colour	green, with some red-brown spots near the base	green, spotted red-brown over the veins to the middle	green to yellow, heavily veined and blotched red-brown
Side-lobe shape	apex acute	apex slightly rounded	apex slightly rounded
Side-lobe colour	spotted red-brown along the veins	veined red-brown, spotted towards the margins	veined red-brown
Side-lobe margin	hairy (hairs 1 mm + long)	fringed with short hairs	fringed with short hairs
Mid-lobe shape	broadly ovate-cordate	ovate	ovate
	(B = L × 1.5)	(B = L)	(B = L)
Mid-lobe indumentum	papillose with scattered hairs	papillose with scattered hairs, glabrous at the base and centre	hairy, with lines of long hairs extending from the callus ridges to the centre
Callus length	to the base of the lip	to the base of the lip	not to the base of the lip
Callus indumentum	densely short-hairy	densely short-hairy	densely long-hairy
Pollinium length	3.2–4.0 mm	2.4–2.7 mm	2.1–2.5 mm
Flowering period	January–March	February–April	August–December
Altitude	1500–2600 m	2300 m	1200–2000 m

Note: B = breadth; L = length.

It was originally collected in south-eastern Yunnan, distant from the known range of *C. hookerianum*. The original hybrid must have been able to disperse and establish itself as a reproductively viable population, occupying the high elevation niche of *C. hookerianum*, but farther south than that species. It is therefore suggested that *C. wilsonii* is a distinct species, probably originating as a hybrid between *C. hookerianum* and *C. iridioides*.

Cymbidium wilsonii can be distinguished from *C. iridioides* by its green sepals and petals, lack of long hairs in the centre and base of the mid-lobe, narrower leaves and spring flowering season. *Cymbidium hookerianum* differs in its narrower leaves and larger flowers with more strongly falcate sepals and petals, and a spotted lip with more acute side-lobes and hairy side-lobe margins.

37. Cymbidium wilsonii (Rolfe ex Cook) Rolfe in *Orchid Rev.* 12: 97 (1904). Type: China, Yunnan, cult. Veitch, *Wilson Cym. Sp. 2* (holotype K!).

C. giganteum Wall. ex Lindl. var. *wilsonii* Rolfe ex Cook in *The Garden* 65: 158, 189 + fig. (1904).
Cyperorchis wilsonii (Rolfe) Schltr. in *Fedde, Repert. Sp. Nov. Regni Veg.* 20: 108 (1924).
Cymbidium giganteum Wall. ex Lindl. cv. Wilsonii (Rolfe) P. Taylor & P. Woods in *Curtis's Bot. Mag.* n.s. 181: t. 704 (1976).

A perennial, epiphytic *herb*. *Pseudobulbs* up to 6 cm long, 3 cm in diameter, elongate-ovoid, bilaterally flattened, about 7-leaved. *Leaves* up to 90 × 2.5 cm, linear-elliptic, acute, articulated 6–11 cm above the pseudobulb, to a broadly sheathing base. *Scape* 25–70 cm long, suberect to horizontal, arching; peduncle loosely covered by sheaths up to 11 cm long; rhachis slender, pendulous at the tip, bearing 5–15 flowers; bracts triangular. *Flowers* 9–10 cm across; scented; rhachis, pedicel and ovary green; sepals and petals green with some pale brown shading over the veins, and distinct red-brown speckles on the veins in the basal half; lip cream, flushing purplish pink on pollination, side-lobes veined dark red-brown, becoming broken and spotted towards the margins and the tips, mid-lobe with a submarginal ring of confluent red-brown blotches, and a reddish median line to the front of the callus ridges; callus cream, spotted red-brown along the crests; column pale yellow-green with a conspicuous dark maroon apex above, cream with spots and dashes of red-brown below; anther-cap pale yellow. *Pedicel and ovary* 2.2–4.2 cm long. *Dorsal sepal* 4.4–5.7 × 1.2–1.9 cm, narrowly obovate, acute, porrect; lateral sepals similar, spreading. *Petals* 4–5.3 × 0.7–1.3 cm, ligulate or narrowly obovate, curved, acute; spreading. *Lip* 3-lobed, fused to the column base for 3.5–5 mm; side-lobes 1.2–1.4 cm broad, triangular, acute and porrect, hairy, margin fringed with hairs; mid-lobe 1.5–1.8 × 1.5–2.1 cm, ovate, tapering to a fine point, recurved, papillose with scattered hairs, margin erose, undulating; callus ridges 2, tapering to the base and dilated at the apex, densely hairy. *Column* 2.7–3.2 cm long, broadly winged at the tip, narrowly at the base, with sparse hairs and papillae present ventrally towards the base; pollinia 2, 2.40–2.65 mm long, triangular. Figs. 106: 5a–e, 145 & 146.

DISTRIBUTION. China (southern and western Yunnan) (Map 39); about 2400 m (7875 ft).

HABITAT. On tall trees in deeply shaded forest; flowering February to April.

CONSERVATION STATUS. VU A1cd; B1ab.

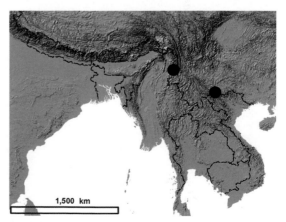

Map 39. Distribution of *C. wilsonii*.

SECTION PARISHIELLA

Section **Parishiella** (Schltr.) P.F. Hunt in *Kew Bull.* 24: 94 (1970). Type: *C. tigrinum* Parish ex Hook. *Cyperorchis* section *Parishiella* Schltr. in *Fedde, Repert. Sp. Nov. Regni Veg.* 20: 108 (1924).

This section contains a single, highly distinctive species and was established by Schlechter in 1924 in the genus *Cyperorchis*, later transferred to *Cymbidium* by P. Hunt in 1970. In the phylogenetic analyses of DNA sequences of Yukawa *et al.* (2002) it is sister to section *Cyperorchis*. It is distinct vegetatively, having lens-shaped pseudobulbs that are not covered by leaf bases and has only 2–4 apical leaves, usually less than 17 cm long. A new pseudobulb is produced annually, and the 2–5 flowered scape is basal. The flower shape is also unusual, with a slender, spidery appearance, porrect petals and a lip with a rectangular, cuspidate mid-lobe and highly unusual horizontal markings. The lip and column are basally fused. Its pollinarium shape closely resembles that of some of the large-flowered species of section *Cyperorchis*.

38. CYMBIDIUM TIGRINUM

Cymbidium tigrinum was described by W.J. Hooker in 1864 from specimens collected in Burma by the Rev. Charles Parish, who sent herbarium material and a watercolour sketch by his wife along with living plants to the nursery of Messrs Low. The plants were collected on the summit of Mulayit (about 2000 m, 6560 ft) near Moulmein on the Thai-Myanmar border. Major Robson Benson (1870) also reported it as growing on trees between 6500 and 7000 ft in the mountains near Moulmein. This species has since been collected several times on this and some surrounding mountains in the Dawna Range. Ghose (1972) and Hynniewata (1979) also reported it from Nagaland, on the border of north-east India and northern Myanmar.

Fig. 147. *Cymbidium tigrinum.* Cult. Kew.

Plate 32. *Cymbidium tigrinum*. NE India, Rittershausen ex. Kew, del. Claire Smith.

Parish (1883) described an interesting floral dimorphism in the flowers. He reported that the lower flowers on the spike were of a 'rich red colour', and that the column had an unusual structure. He wrote: 'The column is quite abnormal, being unusually thickened and less curved. There is no anther at all, and there are no pollen masses; but the edges of the column at the top are turned inwards so as to form a sort of hood, and underneath those edges is a small quantity of a waxy yellow substance (pollen) in an amorphous state. And, occasionally, the intermediate flowers are intermediate also in condition, having no anther, but perfect pollen masses, though without any triangular gland'. The type specimen contains a scape that is annotated by Parish as having these dimorphous flowers. An examination of these flowers, and of the observations quoted above both indicate that the lower flowers had simply been pollinated, rather than that they were of a different structure. It is normal in this genus for the flower to 'blush' red after pollination and for the column tip to curve inwards to close the stigmatic cavity once the pollinia have been deposited. The pollinia then start to disintegrate and lose their original structure. The 'intermediate' flowers are also interesting, as it seems as though the anther-cap was dislodged and the viscidium removed by the pollinator, but the pollinia were left intact on the end of the column. It may be that these flowers were visited in an immature state, resulting in only partial removal of the

F. HOECK

Fig. 148. *Cymbidium tigrinum*. Cult. Munich Botanical Garden.

pollinarium. As the upper flowers were intact, it appears that the lower flowers mature earliest (see also Seidenfaden, 1983).

It is different from all other *Cymbidium* species, especially in its habit, in which respect it is more reminiscent of a species of *Coelogyne*. Indeed, Hooker, when he described it, was not certain that it should be included in *Cymbidium*. However, the large flowers, the fusion of the base of the lip and the column and the two cleft pollinia place this species close to section *Cyperorchis*. The relatively broad leaves, basal flower spike, short lip, two distinct callus ridges and triangular pollinia on a broad triangular viscidium show an affinity with *C. iridioides* and its allies, but *C. tigrinum* has several unusual characters that place it in a separate monotypic section. These characters include the small size of the plant (less than 15 cm (6 in) tall), lithophytic habit, lens-shaped pseudobulbs exposed and not covered by leaf bases, few (2–4) leaves less than 17(22) cm (6.7–8.7 in) long that are articulated close to the pseudobulb, peduncle with distant, spreading sheaths, spreading sepals but porrect petals, lip with almost entirely purple-brown side-lobes, mid-lobe marked with horizontal rather than vertical markings and with a cuspidate apex, and well-defined, parallel, glabrous callus ridges.

38. Cymbidium tigrinum *Parish ex Hook.* in *Bot. Mag.* 90: t. 5457 (1864). Type: Burma [Myanmar], Moulmein, Mulayit, *Parish 144* (holotype K!).

Cyperorchis tigrina (Parish ex Hook.) Schltr. in *Fedde, Repert. Sp. Nov. Regni Veg.* 20: 108 (1924).

A perennial lithophytic *herb. Pseudobulb* up to 3(6) × 3(3.5) cm, sub-sphaerical to broadly ovoid, bilaterally compressed, lens-shaped, with (1)2–6 distichous leaves and 4–5 cataphylls; the lower leaves and cataphylls deciduous, the persistent leaf bases becoming scarious and eventually fibrous with age; mature pseudobulbs wrinkled, usually with 2–4 apical leaves, not covered by the leaf bases, with nodes towards the base and the apex only. *Leaves* up to 17(22) × 3.3 cm, narrowly elliptic, twisted, articulated 1–1.5 cm from the pseudobulb to a broadly sheathing base with a narrow (1 mm) membranous margin. *Scape* 12–23 cm long, from the base of the pseudobulb; peduncle suberect to horizontal, with 5–9 sheaths up to 3.7 cm long, the upper sheaths distant, spreading, cymbiform, with a cylindrical base; rhachis usually longer than the peduncle, with 2–5 large, distant flowers; bracts triangular, acute, (2)4–9(17) mm long. *Flower* 4–5 cm across, of a spidery appearance; honey-scented; rhachis, pedicel and ovary green; sepals and petals olive-green to mustard, shaded red-brown and spotted towards the base with purple-brown; lip white, turning pink on pollination, with almost entirely purple-brown side-lobes, and spots and transverse dashes of red-purple on the mid-lobe; callus ridges white, spotted purple; column olive-green to mustard and tipped with purple-brown above, white, streaked with purple below; anther-cap cream. *Pedicel and ovary* 1.7–4.3 cm long. *Dorsal sepal* 3.7–4.2 × 0.8–1.3 cm, narrowly obovate with recurved margins, obtuse or subacute, mucronate, suberect, curved over the column; lateral sepals similar, narrowly elliptic, downcurved, not fully spreading. *Petals* 3.4–4 × 0.5–1 cm, narrowly elliptic, acute, porrect and often covering the column. *Lip* 3-lobed, fused to the base of the column for about 3 mm; side-lobes 1–1.3 cm broad, rounded to subacute, tapering sharply towards the base, erect but not clasping the column, papillose; mid-lobe 1.2–1.5 × 1.1–1.5 cm, oblong, papillose, recurved, mucronate, margin weakly undulate; callus of 2 well-defined, almost parallel, glabrous, inflated ridges extending from the base of the lip to the base of the mid-lobe. *Column* 2.5–3 cm long, arcuate, papillose, broadly winged towards the apex; pollinia triangular, about 2 mm long; viscidium broadly triangular, tapering to 2 hair-like processes at the lower corners. Pl. 32; Figs. 95: 6a–d, 147 & 148.

DISTRIBUTION. North-east India (Nagaland), Myanmar, China (western Yunnan) (Map 40); (1000) 1500–2700 m ((3280) 4920–8860 ft). It has a similar disjunct distribution to that of several other *Cymbidium* species (see also *C. devonianum* and *C. mastersii*).

HABITAT. It grows on open rocks and in rock crevices in open situations and has also been recorded as an epiphyte; it is often subjected to frosts in winter (Ghose, 1972). It flowers from March until July.

CONSERVATION STATUS. VU A1cd; B1ab.

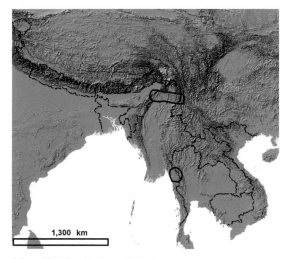

1,300 km

Map 40. Distribution of *C. tigrinum*.

SECTION JENSOA

Section **Jensoa** (Raf.) Schltr. in *Fedde, Repert. Sp. Nov. Regni Veg.* 20: 102 (1924). Type: *C. ensifolium* (L.) Sw.

Jensoa Raf., *Fl. Tellur.* 4: 38 (1836).

Cymbidium subgenus *Jensoa* (Raf.) Seth & P.J. Cribb in Arditti (ed.), *Orchid Biol.: Rev. Persp.* 3: 283–322 (1984). Type: *Jensoa ensata* (Thunb.) Raf. (= *Cymbidium ensifolium* (L.) Sw.)

C. section *Maxillarianthe* Schltr. in *Fedde, Repert. Sp. Nov. Regni Veg.* 20: 101 (1924). Type: *C. goeringii* (Rchb.f.) Rchb.f., lectotype chosen by P. Hunt (1970).

Jensoa was established as a genus in 1836 by C.S. Rafinesque-Schmaltz. Rudolf Schlechter (1924) reduced it to a section of *Cymbidium*, whereas Seth and Cribb (1984) gave it a broader definition and subgeneric status. Van den Berg *et al.* (2002) and Yukawa *et al.* (2002) analysed DNA sequences in *Cymbidium* from the chloroplast gene *matK* and the ITS region of nuclear ribosomal DNA (see Chapter 11). Both showed that subgenus *Jensoa* is monophyletic, providing *C. aliciae* is excluded. In a combined tree, the former placed it sister to *C. devonianum* in a clade itself sister to section *Cyperorchis*. The latter placed it sister to a clade that includes *C. chloranthum* and *C. hartinahianum* from the eastern Malay Archipelago and *C. floribundum* and *C. suavissimum* from mainland South-east Asia (section *Floribundum*).

The section includes terrestrial species characterised by flowers with four pollinia in two unequal pairs, and a lip with the two callus ridges converging towards the apex, forming a short tube at the base of the mid-lobe. Species of section *Borneense* also have four pollinia, whereas the leaf surface morphology closely resembles that of species in section *Cymbidium*, and they are included by Yukawa *et al.* (2002) as a sister clade to section *Cymbidium*.

In addition to the above, many other characters have been shown to be characteristic of this section. The micromorphology of the abaxial leaf surface is highly characteristic; the epidermal cells are papillose, the stomatal covers project beyond the surface of the epidermis, and the apertures in these covers are almost circular. The long, thread-like seed shape is also characteristic, and the outer walls of the testa cells have transverse rather than longitudinal secondary striations. The seed capsules are erect rather than pendulous and the protocorms are rhizomatous rather than spherical.

Within the larger clade of Yukawa *et al.* (2002), two sister clades are recognisable, one that comprises sections *Jensoa* and *Maxillarianthe*, the other sections *Geocymbidium* and *Pachyrhizanthe*. Bootstrap support for each clade is strong. Species in each pair are also similar morphologically. It seems reasonable to treat each of these well-defined clades at sectional level to reduce the infrageneric categories in the genus. Thus, section *Jensoa*, as understood here, includes section *Maxillarianthe* (sensu Du Puy & Cribb 1988). When Schlechter (1924) established these sections he distinguished the latter by its single-flowered scape. However, some varieties of *C. goeringii* have two or three-flowered inflorescences. More detailed analyses that sample more of the species may further resolve the infrageneric classification.

Two separate groups are apparent from the vegetative morphology; the first, containing *C. cyperifolium, C. faberi, C. goeringii* and *C. tortisepalum,* has growths with 5–10 leaves, the outer ones shorter and transitional with the cataphylls; the other, comprising the remaining species, has 2–4 leaves per growth that are distinct from the cataphylls.

39. CYMBIDIUM CYPERIFOLIUM

Cymbidium cyperifolium was first collected in the Khasia Hills and named by Nathaniel Wallich. However, John Lindley formally described it in 1833 using Wallich's epithet. William Griffith subsequently collected this species in Bhutan and described it in 1835 under the name *C. viridiflorum*. His type specimen agrees well with that of *C. cyperifolium*, and the flower colour is described as green with a reddish-spotted lip. A Griffith specimen (5264) in the Edinburgh herbarium is annotated with the name '*C. tesserte*', but this name does not appear to have been validly published, and the specimen at Kew with the same number is annotated *C. viridiflorum*.

In 1851, Griffith also described *C. carnosum* based upon a specimen that he had collected in the Khasia Hills. This is also attributable to *C. cyperifolium*. The description includes several characters that are diagnostic of *C. cyperifolium* subsp. *cyperifolium*, notably its long bracts that exceed the ovary in length, its connivent petals, its green sepals and petals, its connivent callus ridges and its four sessile pollinia.

Cymbidium cyperifolium is most closely allied to *C. faberi*, both having numerous leaves per shoot and characteristic inflated papillae on the mid-lobe of the lip. The latter may be reduced or absent in *C. cyperifolium* and the mid-lobe of its lip is shorter and usually does not have a fimbriate-undulate margin. *Cymbidium ensifolium*, *C. kanran*, *C. sinense* and their close allies can be readily distinguished by only having 2-4 leaves per shoot.

39. Cymbidium cyperifolium Wall. ex Lindl., *Gen. Sp. Orchid. Pl.*: 163 (1833). Type: India, Khasia Hills, Sylhet, *Wallich* 7353 (holotype K- LINDL!).

C. viridiflorum Griff., *Itin. Not. Bhotan*: 53 (1835). Type: Bhutan, *Griffith* s.n. (holotype K!).
C. carnosum Griff., *Notul.* 3: 339–40 (1851). Type: India, Khasia Hills, *Griffith* 185 (holotype ?CAL).
Cyperorchis wallichii Blume, *Orchid. Archip. Ind. Jap.*: 92 (1858). Type: as for *C. cyperifolium*.

A perennial, terrestrial *herb*. *Pseudobulbs* inconspicuous, covered by several cataphylls and sheathing leaf bases. Cataphylls leaf-like, with a membranous margin 2 mm broad, becoming scarious and fibrous with age. *Leaves* 5–10, distichous, up to (30)50–90 × 0.9–1.5 cm, linear-elliptic, acute, erect, margin entire, conduplicate at the base but not petiolate, obscurely articulated 4–6 cm from the base, the lowest leaves making a gradual transition between the cataphylls and the true leaves. *Scape* 23–43 cm tall, erect, arising basally from within the cataphylls, with (2)4–7 flowers produced in the apical third; peduncle covered by slender sheaths up to 8 cm long, which are distant except towards the base; bracts subulate or narrowly ovate, (0.3)1.5–4 cm long, often exceeding the pedicel and ovary in length, becoming scarious. *Flower* about 4–5 cm across; lemon-scented; rhachis, pedicel and ovary greenish, often stained dull purple; sepals and petals apple-green, fading to yellow-green, occasionally pale yellow or straw-coloured with 5–7 longitudinal red-brown lines (in Indo-China); lip pale green or whitish, sometimes pale yellow, with red-purple streaks on the side-lobes that become confluent at the margin, and red-purple spots and blotches on the mid-lobe; column green or yellow, spotted purple below; anther-cap cream. *Pedicel and ovary* 1.3–3.5(4.3) cm long. *Dorsal sepal* (20)25–35 × (4)5–8(11) mm, narrowly oblong-elliptic, acute, erect; lateral sepals similar, spreading. *Petals* 19–29 × (5)6–9(10) mm, ovate to elliptic, acute, usually broader than the sepals, porrect, usually closely shading the column. *Lip* (14)17–22(24) mm long when flattened; side-lobes erect, rounded, usually angled at the apex, pubescent or papillose; mid-lobe (7)9–13 × (6)8–11 mm, oblong to broadly ovate, obtuse or subacute, often as broad as the side-lobes when the lip is flattened, recurved, with some small papillae that are occasionally sparse and almost confined to the apical region, margin entire or minutely fibriate; callus ridges converging in the apical half to form a short tube at the base of the mid-lobe. *Column* 1–1.5 cm long, arching, winged; pollinia 4, in 2 pairs, broadly ovate, on a broadly crescent-shaped viscidium. *Capsule* 5–7 cm long, fusiform, held erect and parallel to the rhachis, retaining the column as an apical beak.

DISTRIBUTION. Nepal, north-east India (Nagaland, Manipur, Mizoram, Meghalaya, Sikkim), Bhutan, southern China, Myanmar (Burma), Thailand, Cambodia, Philippines (Luzon) (Map 41);

1500–2750 m (4920–9020 ft) but 300–900 m (985–2950 ft) in Indo-China. Wu & Chen (1980) stated that the distribution of *C. cyperifolium* in China includes the provinces of Yunnan, Hainan, Guangdong, Guangxi and Guizhou, but some of these collections may be attributable to *C. faberi*.

It is best known from specimens collected in the Himalaya and Khasia Hills of northern India. However, the distribution of this species continues south into south-west China. In Indo-China there is a distinctive variant, which is recognised here at subspecies rank.

HABITAT. Temperate rainforest or bamboo forest, on steep banks of boulders and loam, in shade (Himalaya, Meghalaya); flowering November–January (Himalaya, Meghalaya), May–July (Thailand, Cambodia).

CONSERVATION STATUS. NT.

Key to the subspecies of *C. cyperifolium*

Leaves 32–51 cm long; floral bracts 3–22(25) mm long; sepals and petals pale yellow with 5–7 red-brown longitudinal lines; petals equal to or narrower than the sepals...............subsp. **indochinense**
Leaves 50–90 cm long; floral bracts 18–35(43) mm long; sepal and petals green with red-brown at the base over the mid-vein; petals broader than the sepals.......................................subsp. **cyperifolium**

subsp. **cyperifolium**

Leaves numerous (usually 7–10 per pseudobulb) and strongly distichous giving a fan-like appearance, grass-like, stiffly arching and with an entire margin. Scape with up to 6 flowers; bracts usually equalling or exceeding the pedicel and ovary in length. Flowers with apple-green sepals and petals, with a red-brown streak over the mid-vein at the base. Petals usually pointing forward, covering the column. Lip with an ovate mid-lobe with an entire margin, pale green or whitish, with a few red-brown blotches, and papillose. Figs. 149–151 & 154: 3a-g.

DISTRIBUTION. Nepal, north-east India, Bhutan, southern China.

The relatively long, slender floral bracts are usually characteristic of subsp. *cyperifolium*. *Cymbidium kanran* also has long floral bracts and may have green flowers and a papillose mid-lobe, but its sepals are more slender and acuminate, and it has fewer leaves per pseudobulb.

Cymbidium faberi, which has a similar number of leaves to *C. cyperifolium*, is distinguished by its often finely serrated leaf margins, usually much shorter floral bracts, usually more robust scape with more numerous (up to 20) flowers, and longer, narrowly ligulate or tapering mid-lobe of the lip, which has an undulating and fimbriate margin. The mid-lobe is covered by inflated papillae in both *C. faberi* and *C. cyperifolium* subsp. *cyperifolium*, but those in *C. faberi* are more numerous, larger and much more conspicuous. Specimens of *C. faberi* from Nepal and the western Himalaya of northern India often have bracts that are similar in length to those of *C. cyperifolium* subsp. *cyperifolium*, and they closely resemble each other in habit and flower colour. This has led to the confusion of these two species by several authors including Duthie (1906), Banerjee & Thapa (1978) and Raizada *et al.* (1981). The distinguishing characters are discussed more fully under *C. faberi* subsp. *szechuanicum*.

Fig. 149. *Cymbidium cyperifolium* subsp. *cyperifolium,* SW Guizhou.

TABLE 16

A COMPARISON OF THREE SIMILAR TAXA THAT OCCUR SYMPATRICALLY IN NORTHERN THAILAND

Character	*C. cyperifolium* subsp. *indochinense*	*C. ensifolium* subsp. *haematodes*	*C. sinense*
Leaf number	*(5)6–8	2–4(5)	2–4(5)
Leaf breadth (mm)	*9–14	14–17	*(15)20–35
Leaf length (cm)	32–51	52–94	40–100
Flower number	5–7	3–6	*usually 8–15
Dorsal sepal length (mm)	(20)26–33	*16–26(31)	26–39
Petal/dorsal sepal breadth comparison	*petals almost equal to or narrower than the sepals	petals broader than the sepals	petals broader than the sepals
Lateral sepal orientation	slightly drooping	*almost horizontal,	slightly drooping
Petal orientation	porrect, forming a hood over the column	*weakly porrect, not forming a hood over the column	porrect, forming a hood over the column
Mid-lobe breadth (mm)	(6)7–9	8–10	9–13
Mid-lobe margin	entire or slightly kinked	*tightly undulating	entire or slightly kinked
Mid-lobe indumentum	*some inflated papillae especially near the apex	minute papillae only	minute papillae only
Flowering period	May–July	(January) February–March	October–March

*indicates key characters for each taxon.

258

subsp. **indochinense** Du Puy & P.J. Cribb, *The Genus Cymbidium*: 176 (1988). Type: Thailand, Chiengrai, *Put* 3972 (holotype K!).

This subspecies resembles *C. ensifolium* in its habit and flower colour (sepals and petals pale yellow with 5–7 longitudinal red-brown lines). Its habit is similar to that of subsp. *cyperifolium*, although it is usually a smaller plant. Leaves 32–51 cm long. Lip mid-lobe similar in shape to that of subsp. *cyperifolium*, with angled side-lobe apices and similar papillae, although they are more sparsely scattered and are often only evident towards the apex. Otherwise, it differs from subsp. *cyperifolium* as indicated in the key. Fig. 154: 5a–f.

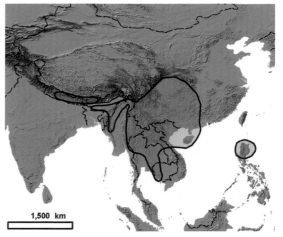

Map 41. Distribution of *C. cyperifolium*.

DISTRIBUTION. Myanmar, Thailand, Cambodia, Philippines (Map 41).

Subspecies *indochinense* differs from *C. faberi*, which has a similar leaf number, in its few-flowered scape and several characters of the mid-lobe of the lip, which is ovate rather than ligulate or tapering. The indumentum of inflated papillae is less pronounced than in *C. faberi*. The margin of the lip of subsp. *indochinense* is entire, whereas the lip of *C. faberi* is characteristically minutely fimbriate and undulate.

The floral bracts, which are characteristically long in subsp. *cyperifolium*, are much shorter in subsp. *indochinense*. Flower colour may also cause confusion with related species. *Cymbidium cyperifolium*, *C. sinense* and *C. ensifolium*, all found in Thailand, are similar in habit, size and flower colour. The differences between them are summarised in Table 16, the key characters of each being indicated by an asterisk.

Seidenfaden (1983) highlighted the confusion surrounding this group of related species in Thailand. His *Cymbidium sinense* (his Fig. 39) appears to be correctly identified, whereas his *C. ensifolium* (his Fig. 38) shows many of the features of *C. ensifolium* subsp. *haematodes*, except that the flowers drawn on the spike differ from the flower drawn in close-up in having broader petals. The plant with six leaves and the petals the same breadth as the sepals that he referred to *C. siamense* (his Fig. 40) is probably *C. cyperifolium* var. *indochinense*. *Cymbidium siamense* usually has broad petals and is considered to be conspecific with *C. ensifolium* subsp. *haematodes*.

DAVID DU PUY

Fig. 150. *Cymbidium cyperifolium* subsp. *cyperifolium,* Nepal, *Bailes*.

Subsp. *indochinense* is found in Myanmar, Thailand and Cambodia, between 300 and 900 m (985–2950 ft) elevation. It grows in open, probably deciduous forest, in grassy undergrowth, and flowers between May and July. In common with some specimens of *C. ensifolium* subsp. *haematodes* in Thailand, the pseudobulbs grow beneath the soil surface and are further protected by their fibrous cataphylls and sheathing leaf bases from the numerous fires in their habitat. All of the specimens of this taxon that have been examined have lost their old leaves and show signs of scorching on the older pseudobulbs, with only the new growths bearing leaves.

Fig. 151. *Cymbidium cyperifolium* subsp. *cyperifolium*, Nepal, *Bailes*.

40. CYMBIDIUM DEFOLIATUM

This little-known species was described by Y.S. Wu and S.C. Chen in 1991 and is based on a cultivated plant collected in Guizhou province in southern China. Nothing is known of its exact provenance or ecology. The authors compared this nondescript plant with *C. ensifolium*, distinguishing it by its deciduous habit, rhizome-like pseudobulbs, deciduous leaves, shorter, few-flowered inflorescence and smaller flowers with an obscurely 3-lobed lip with callus ridges reduced to the apical half of the disc of the lip. We have examined the type collection, but its only surviving flower has lost its lip. Other collections from Guizhou (*Bodinier 1503*; *Tsi et al. ASBK 106*) agree well with the description of *C. defoliatum* and probably belong here. The latter was collected in south-west Guizhou in degraded evergreen forest on karst limestone at an altitude of 1025 m. Its flowers resemble those of *C. ensifolium*, but the leaves are narrower, the inflorescence shorter and with fewer flowers and perianth shorter, two-thirds the size of those of *C. ensifolium*. The lip is papillose, suggesting an affinity with *C. faberi* and the mid-lobe almost as broad as the basal part.

It is obviously closely related to *C. ensifolium* subp. *ensifolium* and may prove on further study to be no more than a variety of that widespread taxon. The lack of preserved material in collections means that little can be deduced about its relationships. Living material is in cultivation in China and, possibly, elsewhere. Chen (1999) gave its range in China as northern Fujian, Guizhou, Sichuan and Yunnan. We have only seen herbarium material from Guizhou.

40. Cymbidium defoliatum Y.S. Wu & S.C. Chen in *Acta Phytotax. Sinica* 29 (6): 549 (1991); Chen, *Flora of China Orchidaceae* 2: 216 (1999). Type: China, Guizhou, cult. Beijing BG (holotype PE!).

A terrestrial herb with small indistinct pseudobulbs arranged in a row and stout fleshy roots, 0.5–0.8 cm in diameter. *Leaves* 2–4, deciduous in winter, erect-spreading or suberect, linear, subobtuse, 10–40 × 0.5–1 cm, articulated at the base, 5- to 7-veined; sheathing bases brownish yellow. *Inflorescence* erect, laxly 3- or 4-flowered, 10–20 cm long; peduncle 10–15 cm long, sheathed below; bracts lanceolate or linear-lanceolate, acuminate, 0.5–1 cm long. *Flowers* 2.5–3 cm in diameter, fragrant, whitish, greenish or yellowish, marked with red or purple spots on the mid-lobe, more or less fragrant; pedicel and ovary 1.3–1.7 cm long. *Sepals* narrowly oblong, acute, 1.2–2 × 0.3–0.6 cm. *Petals* suberect near the column, 1–1.6 × 0.25–0.5 cm, shorter and broader than the sepals. *Lip* obscurely 3-lobed, 1–1.2 × 0.5–0.6 cm; side-lobes semi-elliptic; mid-lobe ovate, revolute, acute, 0.3–0.4 mm long and wide; callus of two lamellae convergent towards the apex, 3–4 mm long. *Column* 0.7–0.8 cm long. Fig. 152.

DISTRIBUTION. China only (northern Fujian, Guizhou, Yunnan and Sichuan) (Map 42); 1000–1100 m (3280–3610 ft).

HABITAT. In evergreen forest on limestone hills; flowering from May to August.

CONSERVATION STATUS. DD.

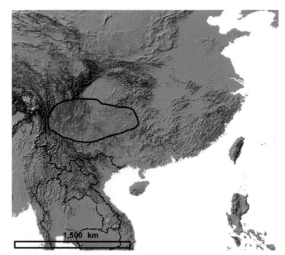

Map 42. Distribution of *C. defoliatum*.

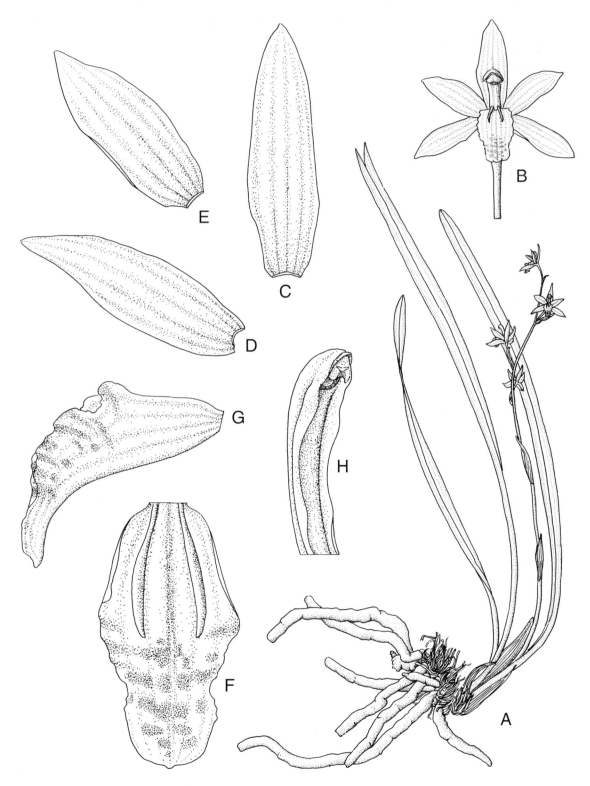

Fig. 152. *Cymbidium defoliatum.* **A.** Habit. **B.** Flower. **C.** Dorsal sepal. **D.** Petal. **E.** Lateral sepal. **F.** Lip, flattened. **G.** Lip, side view. **H.** Column. **A.** ×1/3. **B.** ×1. **C–H.** ×3. Drawn from the type collection by Deborah Lambkin.

41. CYMBIDIUM ENSIFOLIUM

This well-known and widespread orchid was originally described by Linnaeus in 1753 as *Epidendrum ensifolium*, based on a specimen bought in Canton [Guangzhou] by Pehr Osbeck, who noted that it was cultivated in China for its scent, as it is to the present day. Olof Swartz transferred it in 1799 to the genus *Cymbidium*. Thunberg named this species *Limodorum ensatum* in his *Flora Japonica* in 1784 but later transferred it to *Epidendrum* in his accompanying book of illustrations (Thunberg, 1794). It is probably not native to Japan, but it is widely cultivated there and has been for centuries. Rafinesque transferred Thunberg's species to the genus *Jensoa* in 1838, and his publication is the source of the sectional name *Jensoa*.

Cymbidium ensifolium has only 3 to 4, occasionally 5, leaves on each pseudobulb, and the leaves are distinct from the cataphylls that surround the base of the plant, a habit shared by *C. sinense*, *C. munronianum* and *C. kanran*. *Cymbidium faberi* and *C. cyperifolium* are easily distinguished by their more numerous leaves. *Cymbidium kanran* differs from *C. ensifolium* in having longer leaves and longer, more slender, acuminate sepals, and petals that are about twice as broad as the sepals. *Cymbidium sinense* has broader leaves, dark green in colour and glossy on the upper surface, more numerous flowers on the scape, porrect petals forming a hood over the column and a larger mid-lobe to the lip. *Cymbidium munronianum*, which replaces *C. ensifolium* in the Himalaya of northern India, has a more robust habit and broader leaves, similar to *C. sinense*, but has smaller flowers with a smaller lip mid-lobe than *C. ensifolium*. In addition, *C. defoliatum, C. nanulum* and *C. omeiense* (q.v.) are tentatively recognised as distinct but closely related species, following Chen (1999) and Chen & Liu (2003).

In this study, three subspecies of *C. ensifolium* are recognised, but further research, especially of living material, will be necessary to clarify the taxonomy of this variable species. The characters which that distinguish these subspecies are listed in Table 17.

41. Cymbidium ensifolium (L.) Sw. in *Nov. Act. Soc. Sci. Upsal.* 6: 77 (1799). Type: China, Canton, *Osbeck* (holotype LINN!).

Epidendrum ensifolium L., *Sp. Pl.* 2: 954 (1753).

Limodorum ensatum Thunb., *Fl. Japan*: 29 (1784) & *Icon. Fl. Japan* 1: 28 (1974). Type: as for *C. ensifolium* (L.) Sw.

Epidendrum sinense Redouté, *Lilac.* 2: t. 113 (1805). Type: illustration cited (lectotype selected here).

C. xiphiifolium Lindl., *Gen. Sp. Orchid. Pl.* 7: t. 529 (1821). Type: China, *Hume* (holotype K!).

C. ensifolium (L.) Sw. var. *estriatum* Lindl. in *Bot. Reg.* 13: t. 1976 (1837). Type: cult. Hort. Soc. London (holotype K!).

C. ensifolium (L.) Sw. var. *striatum* Lindl. in *Bot. Reg.* 23: t. 1976 (1837). Type: China or Japan, *Fothergill s.n.* (holotype K!).

Jensoa ensata (Thunb.) Raf., *Fl. Tellur.* 4: 38 (1838). Type: as for *C. ensifolium* (L.) Sw.

C. ecristatum Steud., *Nomencl.*, ed. 2, 1: 460 (1840), *nom nud.*

C. micans Schauer in *Nov. Act. Nat. Cur.* 19 (suppl. 1): 433 (1843). Type: China, Macao (holotype K!).

C. albo-marginatum Makino, *Iinuma, Somoku-Dzusetsu*, ed. 3, 4(18): 1183 (1912). Type: not cited, but probably in MAK.

C. gyokuchin Makino, *loc. cit.* 1181, t. 7 (1912). Type: not cited, probably in MAK, but see the above reference for an illustration of the type.

C. gyokuchin var. *soshin* Makino, *loc. cit.* 1182, t. 8 (1912). Type: not cited, probably in MAK, but see the above reference for an illustration of the type.

C. koran Makino, *loc. cit.* 1179, t. 3 (1912). Type: not cited, probably in MAK, but see the above reference for an illustration.

C. niveo-marginatum Makino., *loc. cit.* 1183 (1912). Type: not cited, probably in MAK.

C. shimaran Makino, *loc. cit.* 1183, t. 10 (1912). Type: not cited, probably in MAK, but see the above reference for an illustration of the type.

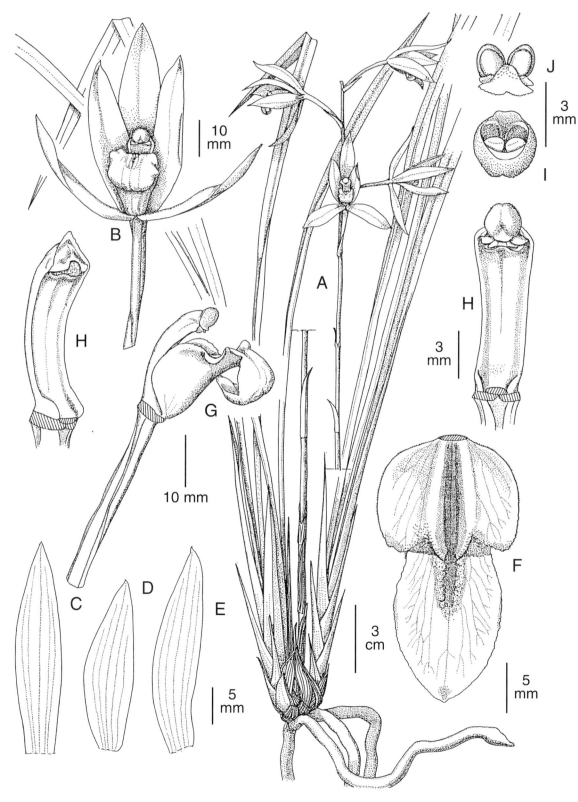

Fig. 153. *Cymbidium ensifolium* subsp. *ensifolium* (as *C. koran*). **A.** Habit. **B.** Flower. **C.** Dorsal scpal. **D.** Petal. **E.** Lateral sepal. **F.** Lip, flattened. **G.** Column, ovary and lip, side view. **H.** Column, two views. **I.** Anther-cap. **J.** Pollinarium. Drawn from a living collection from Nagasaki Pref., Japan, by Mutsuko Nakajima.

Plate 33. *Cymbidium ensifolium* subsp. *ensifolium*. China, del. H. Weddell, Curtis's Botanical Magazine: t. 1751 (1815).

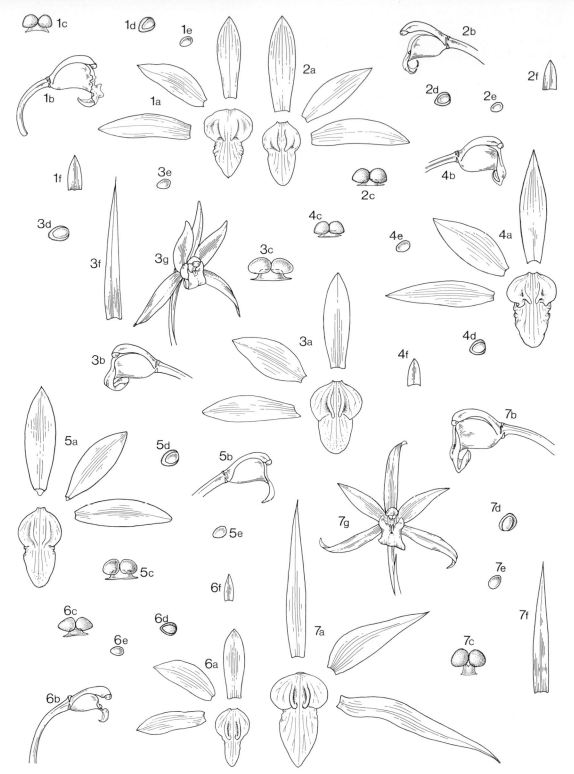

Fig. 154. *Cymbidium ensifolium* subsp. *haematodes* (Kew spirit no. 47263/47265/48347). **1a.** Perianth, × 1. **1b.** Lip and column, × 1. **1c.** Pollinarium, × 3. **1d.** Pollinia (one pair, reverse), × 3. **1e.** Pollinium, × 3. **1f.** Bract, × 1. *C. ensifolium* subsp. *ensifolium* (Kew spirit no. 47259). **2a.** Perianth, × 1. **2b.** Lip and column, × 1. **2c.** Pollinarium, × 3. **2d.** Pollinia (one pair, reverse), × 3. **2e.** Pollinium, × 3. **2f.** Bract, × 1. *C. cyperifolium* subsp. *cyperifolium* (Kew spirit no. 6192). **3a.** Perianth, × 1. **3b.** Lip and column, × 1. **3c.** Pollinarium, × 3. **3d.** Pollinia (one pair, reverse), × 3. **3e.** Pollinium, × 3. **3f.** Bract, × 1. **3g.** Flower, × 0.8. *C. sinense* (Kew spirit no. 47253/47931). **4a.** Perianth, × 1. **4b.** Lip and column, × 1. **4c.** Pollinarium, × 3. **4d.** Pollinia (one pair, reverse), × 3. **4e.** Pollinium, × 3. **4f.** Bract, × 1. *C. cyperifolium* subsp. *indochinense* (Thailand, *Put* 3972). **5a.** Perianth, × 1. **5b.** Lip and column, × 1. **5c.** Pollinarium, × 3. **5d.** Pollinia (one pair, reverse), × 3. **5e.** Pollinium, × 3. **5f.** Bract, × 1. *C. munronianum* (Kew spirit no. 36360). **6a.** Perianth, × 1. **6b.** Lip and column, × 1. **6c.** Pollinarium, × 3. **6d.** Pollinia (one pair, reverse), × 3. **6e.** Pollinium, × 3. **6f.** Bract, × 1. *C. kanran* (Kew spirit no. 40631). **7a.** Perianth, × 1. **7b.** Lip and column, × 1. **7c.** Pollinarium, × 3. **7d.** Pollinia (one pair, reverse), × 3. **7e.** Pollinium, × 3. **7f.** Bract, × 1. **7g.** Flower, × 0.8. All drawn by Claire Smith.

C. yakibaran Makino, *loc. cit.* 1182, t. 9 (1912). Type: not cited, probably in MAK, but see the above reference for an illustration of the type.

C. arrogans Hayata, *Icon. Pl. Formos.* 4: 76 (1914) & in *op. cit.* 6: 79, t. 12 (1916); Mark, Ho & Fowlie in *Orchid Dig.* 50: 31 (1986). Type: Taiwan, Kusukusu, *Hayata & Sasaki s.n.* (holotype TI).

C. misericors Hayata, *loc. cit.* 4: 79, t. 386 (1914); Mark, Ho & Fowlie in *Orchid Dig.* 50: 32 (1986). Type: Taiwan, Mt. Kwannonzan, near Tamsui, cult. in seminario Taihoku, *Hayata & Soma s.n.* (holotype TI).

C. rubrigemmum Hayata, *op. cit.* 6: 81, t. 15 (1916); Mark, Ho & Fowlie in *Orchid Dig.* 50: 31 (1986). Type: Taiwan, cult. in seminario Taikhoku, *Soma s.n.* (holotype TI).

C. gonzalesii Quisumb. in *Philippine J. Sci.* 72: 485 (1940). Type: Philippines, Luzon, *Quisumbing 5783E* (holotype PNH).

C. ensifolium (L.) Sw. var. *misericors* (Hayata) T.P. Lin, *Nat. Orchids Taiwan* 2: 105, t. 36–39 (1977); Liu & Su in *Fl. Taiwan* 5: 942 (1978).

C. gyokuchin Makino var. *arrogans* (Hayata) S.S. Ying in *Coll. Ill. Indig. Orchids Taiwan* 1: 126, t. 43 (1977).

C. kanran Makino var. *misericors* (Hayata) S.S. Ying, *loc. cit.* 1: 440 (1977).

C. ensifolium (L.) Sw. var. *rubrigemmum* (Hayata) T.S. Liu & H.J. Su in *Fl. Taiwan* 5: 940–941 (1978).

C. ensifolium (L.) Sw. var. *yakibaran* (Makino) Y.S. Wu & S.C. Chen in *Acta Phytotax. Sin.* 18(3): 296 (1980).

C. ensifolium (L.) Sw. var. *susin* T.C. Yen, *Icon. Cymbid. Amoyens.* D.b. 1 (1964). Type: China, Fukien, An-shee Hsien, *Yen 2039* (holotype not located).

A perennial, terrestrial *herb*. *Pseudobulbs* to 3 × 1.5 cm, ovoid, often inconspicuous, occasionally subterranean, covered in leaf bases and scarious cataphylls that become fibrous with age, both with membranous margins about 1 mm broad. *Leaves* 2–4(5), up to 29–94 × 0.8–2.5 cm, distichous, not merging with the cataphylls, erect, arching, linear-elliptic, acute, the margin sometimes serrulate towards the apex, articulated 1.5–5 cm from the pseudobulb. *Scape* 15–67 cm long, produced from the base of the pseudobulb inside the cataphylls, with 3–9 flowers in the apical third; peduncle covered in cymbiform sheaths up to 5–6.5 cm long, the uppermost sheath often distant from the others; bracts 0.4–2.2(2.9) cm long, triangular or linear-ovate. *Flowers* 3–5 cm across; often strongly scented; rhachis, bracts, pedicel and ovary greenish, often stained red-brown; petals and sepals straw-yellow to green with 5–7 longitudinal red or red-brown veins, the petals often with a stronger central stripe and red-brown spots and blotches towards the base; lip pale yellow or green, occasionally white, side-lobes streaked red, with a red margin, mid-lobe with red blotches or transverse spots; column pale yellow, with red dashes beneath; anther-cap cream: the red-brown pigment occasionally absent from the flowers, which are then pale green and white in colour. *Pedicel and ovary* 1–3.7 cm long. *Dorsal sepal* (16)19–31 × 4.6–8.8(9.8) mm, narrowly elliptic, acute to oblong, obtuse or mucronate, erect; lateral sepals similar, spreading, horizontal or drooping. *Petals* 14–26 × 5.5–9 mm, ovate-elliptic, subacute, shorter and either almost equal in breadth, or broader than the sepals, porrect, but not closely covering the column. *Lip* 1.4–2.2 cm long when flattened; side-lobes rounded, papillose; mid-lobe 6–12 × (5)6–10 mm, ovate to triangular, mucronate or acute, minutely papillose, margin kinked to undulate, entire; callus of 2 ridges, converging in the apical half to form a short tube at the base of the mid-lobe. *Column* 1–1.5(1.8) cm long, arching, winged; pollinia 4, broadly ovate, in two unequal pairs, on a semi-circular viscidium. *Capsule* about 6 × 2 cm, fusiform, strongly ridged, tapering at each end, with a beak 1–5 cm long, held erect and parallel to the rhachis.

DISTRIBUTION. Sri Lanka, southern India, central and southern China, Taiwan, Korea, Ryukyus (southern Japan), Indo-China, Thailand, Vietnam, western Malaysia, Sumatra, Java, Borneo, Philippines (Luzon), New Guinea (Map 44); 300–1800 m (985–5905 ft).

HABITAT. Terrestrial in lightly shaded, broad-leaved forest, thickets and grassy places, often in damp situations; it flowers January–April, but in equatorial regions flowering is sporadic throughout the year.

CONSERVATION STATUS. NT (but vulnerable in some parts of its range).

Key to the subspecies of *C. ensifolium*

1. Petals acuminate; lip acuminate, 14–15 mm long .... subsp. **acuminatum**
   Petals obtuse to rounded; lip obtuse to rounded, 12 mm or less long **2**

2. Leaves arching, usually less than 50 cm long and 0.8–1.5 cm broad, usually with an entire margin; petals and sepals almost equal in breadth; mid-lobe of the lip ovate, with a lightly kinked margin, blotched with red .... subsp. **ensifolium**
   Leaves almost erect, usually longer than 50 cm, and 1.5–1.9(2.5) cm broad, usually with a serrated margin; petals broader than the sepals; mid-lobe of the lip triangular to narrowly elliptic, with an undulating margin, finely spotted with red .... subsp. **haematodes**

subsp. **ensifolium**

Leaves arching, up to 55 cm long, 0.8–1.6 cm broad, usually lacking serrations on the margin. *Inflorescence* usually held clear of the foliage, with up to 9 flowers. *Flowers* 3.5–4.0 cm across, with drooping lateral sepals. *Sepals* and *petals* straw-yellow to light brown in colour, with a strong central stripe of red-brown, and several weaker stripes often only distinct towards the base; petals as broad as

Fig. 155 (left). *Cymbidium ensifolium* subsp. *ensifolium* 'Soshin', cult. Kew. **Fig. 156** (right). *Cymbidium ensifolium* subsp. *ensifolium*. Nanning, Guangxi.

the sepals, or slightly narrower. *Lip* with an ovate or elongate-ovate mid-lobe that is obtuse or mucronate at the apex, with a few kinks in the margin, but not strong undulations. Plate 33, Figs. 153, 154: 2a–f, 155 & 156.

DISTRIBUTION. China, Taiwan, northern Vietnam, Korea, Ryukyu Islands, Philippines (Luzon). Subspecies *ensifolium* is widely distributed throughout southern China and adjacent northern Vietnam. Wu & Chen (1980) indicated that it is found as far west as Yunnan and Sichuan, and as far north as Henan and possibly South Korea (Map 44). It is also found in Hong Kong and Taiwan, but has not been collected on Hainan. Despite the large number of names and descriptions published from Japanese material, it seems unlikely that this species is native to Japan. Garay & Sweet (1974) gave the most northern locality as Okinawa, where it is uncommon. Nagano (1955) noted that specimens are found in the wild in Kyushu, the southernmost of the Japanese islands, but these may be naturalised seedlings that have escaped from cultivation. In China, it is commonly found between 500 and 1000 m (1640 and 3280 ft) elevation, but it probably also occurs at higher elevations. In Hong Kong it is found from 250 m (820 ft) to the highest peaks in the territory at about 1000 m (3280 ft).

HABITAT. In southern China, it is most common in ravines, in broad-leaved or bamboo forest, and in grasslands or steep, exposed mountain slopes where the forest cover has retreated or has been burnt, although these are usually weak plants (G. Barretto, pers. comm.). In Hong Kong it grows in partial shade near streams or water seepage on sloping land, in humus-rich soil. It flowers in autumn and early winter, and is most commonly found in flower during August. In Taiwan it is uncommon, growing in hardwood forest in dry conditions, between 300 and 3000 m (985–9840 ft), flowering mainly between July and October, but also sporadically throughout the rest of the year (Lin, 1977; Liu & Su, 1978).

This subspecies has been in cultivation in China and Japan for over 2000 years, and is prized for its shape and the perfume of its flowers. The variants that lack red pigment in the flowers are particularly valued, and have been described several times under different names, including *susin*, *soshin*, *gyokuchin* and *misericors*, and popular names such as 'Goddess of Mercy' and 'Iron bone/Iron stick' and the Japanese name 'Kwannon'. Of these only var. *susin* (Yen, 1964; Wu & Chen, 1980) has been validly published as a variety of *C. ensifolium*, whereas the others have been published as distinct species or varieties of species now synonymised under subsp. *ensifolium*. *Cymbidium xiphiifolium* is a white-and-green-flowered form, which Lindley described in 1821 from a specimen collected in Hong Kong. The name refers to the leaves that appear stiff, reminiscent of *Iris xiphium*. Several specimens from Hong Kong have been examined, and they are similar in flower to those from mainland China, all being referable to subsp. *ensifolium*. These variants occur rarely in the wild, without forming distinct populations, and have been selectively maintained in cultivation. This type of mutation also occurs sporadically in many other *Cymbidium* species.

Apart from this type of colour variation, there is also some variation in the amount of red-brown striation of the petals and sepals, in the ground colour of the petals and sepals, and in the amount of red pigment on the lip. Lindley recognised var. *estriatum* in 1837 based on a specimen with weak striations on the sepals and petals, the striations being distinct only at the base of the perianth instead of extending to near the apex. Variants with all degrees of striation occur, and it seems undesirable to recognise these as distinct taxa.

Several Makino (1912) names are used in Japan to distinguish cultivated variants, but are not applied to wild plants. The form of the plant is important horticulturally in Japan, and although many of the names are accompanied by a line drawing, the flowers and scape are often omitted, and the form of the plant is highly stylised. Reference to other works that use these names illustrates how some of them have been applied. Following the treatment of other authors (Garay & Sweet, 1974; Wu & Chen, 1980; Seidenfaden, 1983; Seth & Cribb, 1984), the names published by Makino (1912) have been placed under the synonymy of *C. ensifolium*. Wu & Chen (1980) recognised var. *yakibaran* as distinct, but so little information is available about this taxon that it is at present included in the synonymy.

Hayata (1914, 1916) distinguished three species from Taiwan. *Cymbidium arrogans* was reported to differ from the specimens of *C. ensifolium* from mainland China in its falcate, semi-oblong lateral sepals. The sepal shape is variable, and the lateral sepals tend to droop in *C. ensifolium*. Consequently, the lateral sepals are usually oblique and tend to curve, so neither of these characters exclude this variant from *C. ensifolium*. *Cymbidium misericors* was described by Hayata in 1914 from a green-and-white-flowered specimen collected in Taiwan. This is another of the horticulturally desirable variants, lacking the red and brown pigmentation normally present in the flower, which occur sporadically in wild populations. Finally, Hayata described *C. rubrigemmum* in 1916. Lin (1977) considered it to be conspecific with *C. misericors*, but with red pigmentation in the flowers, and he distinguished these Taiwanese plants from *C. ensifolium* in mainland China by their slender, arcuate leaves and the reddish cataphylls of the former. These all fall within the range of variation of *C. ensifolium* in China, and the Taiwanese taxa are therefore included in it, although further study and comparison of specimens from China and Taiwan may indicate a more suitable treatment at the infraspecific level. The photographs of *C. ensifolium* and these three taxa by Mark *et al.* (1986) illustrated the variation in the strength of red and brown colouring in the flowers of *C. ensifolium* in Taiwan, but otherwise the flowers and plants are similar.

Cymbidium gonzalesii was described by Quisumbing in 1940 from Luzon in the Philippines. He suggested that this was allied to *C. faberi* but did not compare it with *C. ensifolium*, which the description and plate closely resemble. The petals are narrower than the sepals, and the margin of the lip flat, allowing its inclusion in the type subspecies of *C. ensifolium*.

Two further names, *C. micans* described from Macao by Schauer in 1843 and *C. ecristatum*, a name lacking any description and listed by Steudel in 1814, both agree well with subsp. *ensifolium*.

Liu & Chen described *C. micranthum* in 2004 based on a collection from Maguan in SE Yunnan. They distinguished it from *C. nanulum* by its small pseudobulbs, lack of an elongated rhizome, articulated leaves and the incurved tip to its lip. The small flowers and an incurved lip are common aberrations in *Cymbidium*, especially of plants from the wild flowering in cultivation for the first time. This taxon is close to *C. ensifolium* and may belong here. Good photographs of the habit and flowers have been published by Liu *et al.* (2006)

Cultivated specimens from Japan and China include variants with variegated leaves. The names 'setsugetsu', 'hohrai' and 'gyokuryu' do not appear to have been validly published, but they are commonly used with reference to these variants in Japan (Nagano, 1955).

A specimen from Hong Kong, known there as 'Golden Line' and collected from the wild (*Fowlie s.n.*), has erect leaves and pure white flowers with a single maroon stripe up the centre of each sepal and petal and a white lip spotted with red. It is a striking departure from the drab colours normally found in *C. ensifolium*. This population may possibly result from hybridisation, perhaps with *C. lancifolium* (or *C. dayanum*) as the other parent, passing on clear colours to the flower.

subsp. **haematodes** (Lindl.) Du Puy & Cribb, *The Genus Cymbidium*: 161 (1988). Type: Sri Lanka, *Macrae 12* (holotype K!).

C. ensifolium (L.) Sw. var. *haematodes* (Lindl.) Trimen, *Cat.* 89 (1885) & *Handb. Fl. Ceylon* 4: 180, t. 90 (1898).
C. ensifolium sensu J.J. Sm., *Orch. Java*: 478 (1905).
C. sundaicum Schltr. in *Fedde, Repert. Sp. Nov. Regni Veg., Beih.* 4: 266 (1919). Type: Java, *J.J. Smith s.n.* (cited in J.J. Smith, *Orch. Java*: 478, 1905).
C. sundaicum Schltr. var. *estriata* Schltr. *loc. cit.* Type: Java?, cult. Bot. Gard. Buitenzorg (not located).
C. munronianum sensu Ridl., *Fl. Malay Peninsula* 4: 46 (1924), *non* King & Pantling.
C. siamense Rolfe ex Downie in *Kew Bull.*: 382 (1925). Type: Thailand, Doi Suthep, *Kerr 242* (holotype K!).

C. munronianum auct. non King & Pantling.; Holttum, *Fl. Malaya* 1: 515 (1953); J.J. Wood in *Orchid Rev*. 85: 94–6 (1977); Comber in *Orchid Dig*. 164–8 (1980); Seidenfaden in *Opera Bot*. 72: 71–3, f. 38, t. 4A (1983); Du Puy & Lamb in *Orchid Rev*. 92: 352, f. 291 (1984).

Usually a stronger, larger plant than subsp. *ensifolium*. *Leaves* stiff and erect, longer and broader than subsp. *ensifolium*, up to about 2 cm broad, arching near the tip, and usually with a finely serrulate margin. *Inflorescence* weak in comparison, with up to 8 but usually fewer flowers. *Flowers* 3–3.5 cm across, with almost horizontal, spreading lateral sepals. *Sepals* and *petals* pale straw-yellow or light

Fig. 157. *Cymbidium ensifolium* subsp. *haematodes*, Sumatra.

271

brown to green in colour, usually with about 5 red lines over the veins; petals broader than the sepals. *Lip* often with a triangular, recurved mid-lobe terminating in an acute apex, with an undulating margin. Figs. 154: 1a–f, 157 & 159.

DISTRIBUTION. Sri Lanka, southern India, Thailand, western Malaysia, Sumatra, Java, Borneo, Sulawesi (Map 44).

HABITAT. Jayaweera (1981) and Trimen (1885) give the habitat in Sri Lanka as the submontane or mid-country tropical, wet, evergreen forests, from 300 to 1800 m (985 to 5905 ft) elevation. Comber (1980) reported that it is more common in western Java, where the dry season is less severe, than in the east. It is found in semi-open submontane forest in good light, from 500–1300 m (1640–4265 ft). In Sabah it is found from 600–1800 m (1970–5905 ft). The plants found at the lower elevations grow in light to heavy shade in deep leaf litter, often under *Casuarina sumatrana*, whereas those from the higher elevations prefer peaty or mossy ground in cool, damp moss forest, often near a river, and have occasionally been noted growing in thick moss on the bases of tree trunks (Du Puy & Lamb, 1984).

In Thailand, subsp. *haematodes* grows in tall grass in semi-shaded positions in open oak and dipterocarp forest, at elevations of 300–1500 m (985–4920 ft). The plants were only visible among the dry and brown grass because the leaves of the orchid remained green. The pseudobulbs grew about 3 cm (1.2 in) below the surface of the heavy clay soil and were covered in fibrous leaf sheaths and cataphylls, giving protection from forest fires that were frequently started to clear the undergrowth or for hunting. The distribution of subsp. *haematodes* spans the Equator and reduces the effect of seasonality upon flowering time, so that specimens may flower at any time of year in various parts of the range. However, Comber (1980) reported that in Java the most common flowering time is at the end of the wet season, in February–May. In the more seasonal climate of Sri Lanka, Jayaweera (1981) recorded December–April as the flowering time and Trimen (1885, 1898) gave January–March, although in southern India it can flower as late as July. The flowering period in Thailand is also during February and March, at the start of the warm, dry season.

Subspecies *haematodes* is widely distributed and variable over its range, particularly in the presence and strength of the leaf serrations, the length and breadth of the leaf, the number and size of the flowers, the breadth of the petals and the shape of the mid-lobe of the lip.

Its distribution comprises several geographically isolated islands and peninsulas. Some of these have populations that contain a dominant variant, which appears to be distinct in one or more characters. The plants with the broadest leaves with light serrations on the margins are found on Java and Sumatra, and have been named as *C. sundaicum*. Plants from Thailand are comparatively small, with narrow leaves, but the petals are sometimes broad in comparison to the sepals, and the leaves are long and erect. This variant was named *C. siamense* by Rolfe in 1925.

The type of *C. haematodes* was collected on Sri Lanka by Macrae. This is one of the more isolated islands in the range. It has 4–8 flowers (up to 4 cm across), and decurved lateral sepals. The sepals are larger than is normally found in subsp. *haematodes*, and they are narrowly elliptic rather than oblong. However, in all other respects it agrees well with south-east Asian material of subsp. *haematodes*.

Subspecies *haematodes* has broader leaves than the typical subspecies, and the petals are broader than the sepals (Fig. 158). The scattered cluster of subsp. *haematodes* reflects the large variation found, but there is no break in variation between the specimens from Thailand (1), and Sri Lanka (2), which are most similar to subsp. *ensifolium*, and the more extreme variants from Sumatra (3). It can be seen in Map 43 that subsp. *ensifolium* from China is distinctly smaller in all three parameters used. Table 17 lists the differences between these two subspecies.

The two subspecies cannot be reliably distinguished by any single character. The characters vary in their degree of expression in both subspecies, and there is some overlap in some characters. Further study is necessary to clarify the patterns of variation in both subspecies.

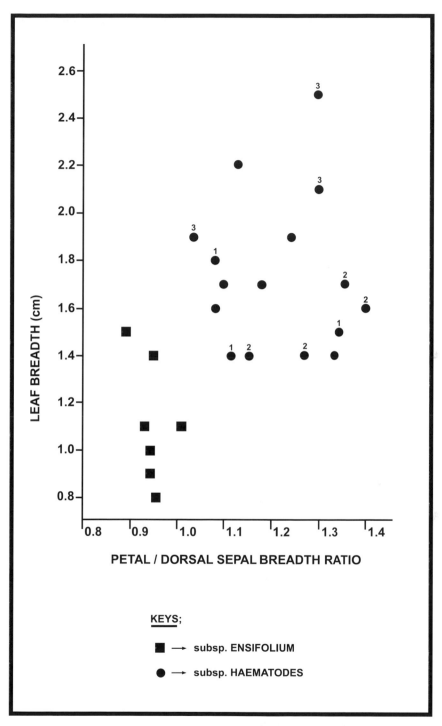

Fig. 158. Scatter diagram displaying the range of variation in leaf breadth and comparative breadth of the petals in *C. ensifolium*. Two groups are apparent, those indicated as subsp. *ensifolium* originating in China and Taiwan, and those indicated as subsp. *haematodes* are from southern India, Indo-China and Malesia. **1.** Sri Lanka. **2.** Thailand. **3.** Sumatra. Subspecies *haematodes* usually has broader leaves than subsp. *ensifolium*, and its petals are 1.0–1.4 times as broad as its sepals.

TABLE 17

A COMPARISON OF THE CHARACTERS THAT CAN BE USED TO DISTINGUISH THE SUBSPECIES OF *C. ENSIFOLIUM*

	subsp. *ensifolium*	**subsp. *haematodes***	**subsp. *acuminatum***
leaves	29–56 cm long, up to 0.8–1.1 (1.6) cm broad, arching	44–94 cm long, up to 1.4–1.9 (2.5) cm broad, erect, arching at the tip	30–120 cm long, 1.5–2.3 cm broad, arching
leaf margin	usually entire	usually serrate	serrate
leaf base	with membranous margin about 1 mm broad	with membranous margin about 2 mm broad	with membranous margin about 2 mm broad
dorsal sepal	27–31 × (5.8) 7–8.8 (9.8) mm	16–26 (31) × 4.6–8.7 mm	23–28 × 3–4 mm
sepals	narrowly-elliptic	oblong	linear-lanceolate
lateral sepals	curved downwards	horizontal	horizontal
petals	21–26 × 7.1–7.7 (8.8) mm; petals and sepals almost equal in breadth	14–26 × 5.5–9 mm; petals broader than sepals	18–22 × 3–4 mm; sepals and petals equal in breadth
lip mid-lobe	usually ovate, obtuse or mucronate 9–12 × (5) 6.2–10 mm,	rounded triangular to ligulate-elliptic, 6–11 × (5.1) 6.2–8.2 (9.1) mm, acute, obtuse or apiculate	ovate-lanceolate, 14–15 × 9 mm, acuminate
mid-lobe margin	kinked	undulate	undulate
mid-lobe coloration	with few large red blotches	with more numerous red spots	with irregular red blotches
Distribution	China, Taiwan, Hong Kong, Ryukyu Isl. (Japan), Philippines (Luzon)	S India, Sri Lanka, Thailand, Malaya, Sumatra, Java, Borneo.	?Sulawesi, New Guinea.

Specimens from southern China have broad, erect leaves, and the flowers appear to be typical of subsp. *haematodes*, with an undulating lip margin, but the petals are similar in breadth to the sepals, indicating subsp. *ensifolium*. Examination of further specimens from southern China and Taiwan is necessary to determine the variation present in the regions where the distributions of the two subspecies meet.

Colour variation is encountered in subsp. *haematodes*, equivalent to that found in subsp. *ensifolium*. In Sabah, two colour variants have been found. One, from lower elevations (600 m, 1970

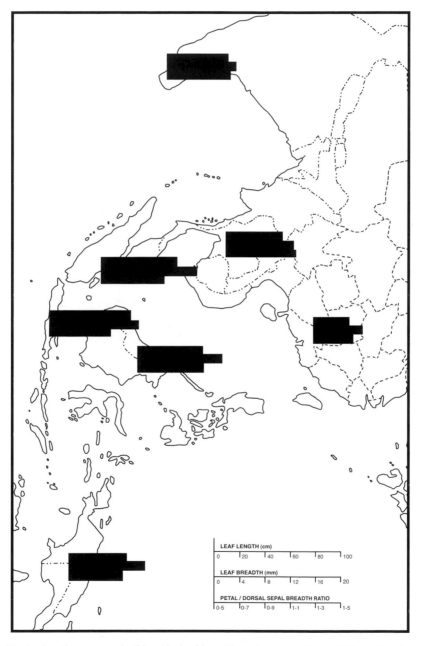

Map 43. Map illustrating the average leaf length, leaf breadth and comparative petal breadth of specimens from various parts of the distribution of *C. ensifolium*. Subsp. *ensifolium* (from China) is distinctly smaller in all three parameters than subsp. *haematodes* (SE Asia and Malay Archipelago) and subsp. *acuminatum* from New Guinea.

ft), has the striped petals and sepals typical of the species. The other, from higher elevations (1800 m, 5905 ft), has green sepals, and petals with a single central stripe. This latter does not appear to differ from the other specimens of subsp. *haematodes*, except in colour, and cannot be treated as a separate taxon (Du Puy & Lamb, 1984).

subsp. **acuminatum** (M.A. Clem. & D.L. Jones) P.J. Cribb & Du Puy **comb. nov.**

C. acuminatum M.A. Clem. & D.L. Jones in *Lasianthera* 1, 1: 28 (1996). Type: Papua New Guinea, Morobe Prov., Waria River Distr., Garassa, 6 April 1990, *M.A. Clements 6387* (holotype CANB).

C. ensifolium sensu T.M. Reeve in *Orchadian* 8, 2: 33 (1984).

Similar in habit to subsp. *haematodes*. *Pseudobulbs* 2.5–3.5 × 1.5–2.5 cm. *Leaves* 3–4, linear-ensiform, 30–120 × 1–2.3 cm. *Inflorescence* erect, 30–80 cm tall, laxly 2- to 7-flowered; bracts subulate 1–1.4 cm long. *Flowers c.* 5 cm across, pale green to yellow or brownish with 5 to 7 prominent, dark red, longitudinal lines on the sepals and petals; lip spotted and suffused dark red. *Sepals* linear to ensiform, acuminate, 2.3–2.8 × 0.3–0.4 cm. *Petals* narrowly linear-lanceolate, acuminate, 1.8–2.2 × 0.3–0.4 cm. *Lip* 3-lobed, 1.8–2 × 1–1.2 cm; midlobe recurved, acuminate. *Column* 1–1.2 cm long. See *Lasianthera* 1, 1: Plate 7B, Fig. 7 (1996).

DISTRIBUTION. New Guinea and possibly Sulawesi (Map 44),

HABITAT. Terrestrial in evergreen forests.

Clements and Jones described *C. acuminatum* in 1996 based upon New Guinea plants previously referred by Reeve (1984) to *C. ensifolium* and by Du Puy & Cribb (1988) to its subsp. *haematodes*. It differs only, as far as we can see, in having unusually acuminate sepals and petals and an acuminate lip. The flower colour and posture of the lateral sepals and lip side-lobes, indicated by Clements and Jones as distinctive, are seen to a lesser degree in *C. ensifolium* elsewhere in its distribution. Specimens from New Guinea typically have stiff, erect leaves with a serrated margin, and the mid-lobe of the lip is triangular with an undulating margin.

Fig. 159. *Cymbidium ensifolium* subsp. *haematodes*, Sabah, *Bailes & Cribb* 746.

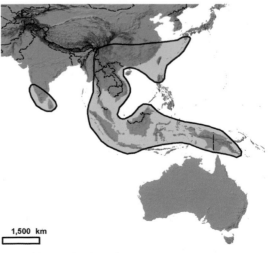

Map 44. Distribution of *C. ensifolium*.

276

42. CYMBIDIUM FABERI

The habit of *C. faberi*, its 6–8 slender leaves without subepidermal strands of lignified sclerenchyma, the drooping flowers and porrect petals (and often dorsal sepal also) forming a close covering over the column place it along with *C. cyperifolium* and *C. goeringii* in section *Jensoa*. Flower colour, shorter bract length, larger flower number, the longer lip mid-lobe covered with numerous inflated papillae and the undulating, fimbriate lip margin distinguish it from *C. cyperifolium*.

Cymbidium faberi was described based upon two specimens, *Faber 94* and *Henry 5515*, both from southern China. The former is a mixed collection from Zhejiang, composed of *C. goeringii* (right-hand specimen) and *C. faberi* (left-hand specimen), which was selected by Du Puy & Cribb (1988) as the lectotype. Both it and *Henry 5515* agree with the type description, but the latter specimen has broader leaves, longer floral bracts and flowers with more acuminate perianth parts and a broader lip mid-lobe.

Cymbidium scabroserrulatum, considered here to be conspecific with *C. faberi*, was described by Makino in 1902 from a plant cultivated in Japan but originally imported from China. In its leaf number (up to 9), minutely serrulate leaf margins, flower number (5–10), flower colour, and lip features it agrees well with *C. faberi*. The habit of the plant is likened, by Makino, to *C. virescens* (= *C. goeringii*), which is also in this group of species, and the description again agrees well with *C. faberi*.

In 1916, Hayata described *C. oiwakense* from a plant collected in Taiwan, differentiating it from all other Taiwanese species by its 'manifestly denticulate lip' described more fully in the type description as a crisped, undulating and minutely erose lip margin, a set of characters diagnostic of *C. faberi*. The rest of the description also agrees well with *C. faberi*, and Su (1975), Lin (1977), Liu & Su (1978) and Wu & Chen (1980) included *C. oiwakense* in its synonymy.

In 1922, Schlechter described *C. cerinum* from a specimen collected by Limpricht in east Tibet (Xizang or Sichuan Province), where it was in cultivation. Schlechter noted that it was without doubt closely related to *C. faberi* but that it differed in its waxy yellow flower colour, narrower sepals, wide, parallel (rather than S-shaped) keels and in the mid-lobe of the lip. The colour is not unusual in *C. faberi,* and the other features all fall within the range of variation of *C. faberi*. This variant is therefore placed in the synonymy of *C. faberi*, following the treatment of several previous authors, notably Wu & Chen (1980).

Its wide distribution and variability have undoubtedly led to it being described several times under different names. Wu & Chen (1980) recognised three varieties, var. *faberi*, var. *szechuanicum* and var. *omeiense*. Variety *omeiense* has fewer leaves (4–5) than *C. faberi*; a shorter scape not exceeding the leaves in length; smaller and fewer flowers; and an ovate (not ligulate), acute mid-lobe with a margin that is not fimbriate or undulating. These characters prevent its inclusion in *C. faberi* but place it closer to *C. ensifolium*. Following Chen and Liu (2003), it is treated here as a distinct species.

This is typically a small plant resembling *C. goeringii*, with 6–9 grey-green, arching leaves, often with serrulate margins. It has a robust scape with up to 20 flowers that are olive-green to yellow-green in colour, and are held above the foliage. The sepals are usually narrowly obovate, and the petals are porrect and cover the column. The mid-lobe of the lip is yellow

Fig. 160. *Cymbidium faberi* var. *faberi,* SW Guizhou.

or green, usually with red markings, elongated, covered in many glossy, inflated papillae and with a tightly undulating and erose margin.

It is similar to *C. cyperifolium* in that it has narrow leaves, the lowest short and merging with the cataphylls, and a short mid-lobe with some inflated papillae. It is easily distinguished from *C. cyperifolium*, however, by its long mid-lobe in comparison with the total length of the lip, its undulating and fimbriate mid-lobe margin, much more numerous and dense, inflated, shiny papillae on the mid-lobe, larger number of flowers on a more robust spike and much shorter floral bracts.

Cymbidium ensifolium is found in China in the same regions as *C. faberi* but usually at lower elevations. It is easily distinguished by its fewer (usually broader) leaves (about four), which are all long and distinct from the cataphylls, its weaker, fewer-flowered spike and the lip without either inflated papillae or an undulating and fimbriate mid-lobe margin.

The distribution, and the high-elevation preference, indicates that it is cold-tolerant, probably as hardy as *C. goeringii*, which will survive outdoors in a sheltered position in warmer parts of the British Isles.

Fig. 161. *Cymbidium faberi* var. *szechuanicum*, Yunnan. **Fig. 162** (opposite). *Cymbidium faberi* var. *szechuanicum*. Vietnam, Cao Bang.

L. AVERYANOV

42. Cymbidium faberi Rolfe in *Kew Bull.*: 198 (1896). Type: China, Chekiang (Zhejiang), Mt. Tientai, *Faber* 94, in part (lectotype K!, chosen by Du Puy & Cribb, 1988).

C. scabroserrulatum Makino in *Jap. Bot. Mag.* 16: 154 (1902). Type: China, Musashi, cult. Tokyo, *Makino* s.n. (holotype MAK).

C. oiwakensis Hayata, *Icon. Pl. Formos.* 6: 80, t.14 (1916), Mark, Ho & Fowlie in *Orchid Dig.* 50: 22 (1986). Type: Taiwan, Gokwanzan, Oiwaka, *Hayata* s.n. (holotype TI).

C. cerinum Schltr. in *Fedde, Repert., Beih.* 12: 350–1 (1922). Type: E Tibet, Xizang Province, *Limpricht* 1392 (holotype B).

C. fukiense T.C. Yen, *Icon. Cymid. Amoyens* AI (1964). Type: China, Fukien, Changchow, *Yen* 3001 (holotype not located).

A perennial, terrestrial *herb. Pseudobulbs* inconspicuous, covered in leaf-like cataphylls that become fibrous with age and sheathing leaf bases, both with a narrow (less than 1 mm) membranous margin, with 5–9(13) distichous leaves. *Leaves* up to (30)40–100 × 0.4–1.1 cm, linear-elliptic, acute, arching, often grey-green in colour, often with a serrulate margin, conduplicate at the base but not tapering into a petiole, obscurely articulated 2–6 cm from the pseudobulb, the shortest leaves merging with the cataphylls. *Scape* 26–62 cm tall, erect, arising basally from within the outermost cataphylls, with 4–20 flowers produced in the apical half to third of the scape; peduncle covered in sheathing sterile bracts up to 6.5 cm long, usually overlapping towards the base of the spike; bracts triangular, (0.8)1–3(4) cm long. *Flowers* about 6 cm across, often drooping and not opening fully; lightly scented; rhachis dull green; pedicel and ovary greenish, stained red-brown; petals and sepals green to yellowish, sometimes stained reddish, especially over the mid-vein of the petals; lip yellowish or green, often with a narrow white margin, side-lobes lined red, usually with a red margin, mid-lobe with many reddish spots and blotches; column yellowish; anther-cap pale yellow. *Pedicel and ovary* 1.4–2.9(3.9) cm long. *Dorsal sepal* (22)26–36(44) × 5.8–10.4 mm, narrowly obovate to elliptic or oblong-elliptic, acute to acuminate, suberect; lateral sepals similar, spreading. *Petals* 20–33 × 6.4–10.9 mm, slightly shorter than the sepals but almost equal in breadth, similar in shape or more ovate, usually porrect and covering the column. *Lip* 1.9–3.3 cm long when flattened; side-lobes erect, rounded, often reduced, papillose or pubescent; mid-lobe long, (9)11–17 × 5.1–12.2 mm, ligulate, often tapering to a mucronate apex, occasionally oblong with a broad, mucronate apex, recurved, often as broad as the side-lobes when the lip is flattened, covered in glossy, inflated papillae, margins fimbriate or erose and undulating and crisped; callus in 2 ridges, converging in the apical half to form a small tube at the base of the mid-lobe. *Column* 1.2–1.9 cm long, arching, winged; pollinia 4, broadly ovoid. *Capsule* 5 cm long, fusiform or oblong-fusiform, erect with an apical beak.

DISTRIBUTION. Nepal, northern India (northern Uttar Pradesh), China (southern Shaanxi, southern Gansu, Anhui, Zhejiang, Jiangxi, Fujian, Henan, Hubei, Hunan, Guangdong, Guangxi, Guizhou, Yunnan, Sichuan), Taiwan, northern Vietnam (Map 45); (700–3000 m (2295–9515 ft)).

HABITAT. It is often found growing in moist and well-drained places in open, sunny situations on steep land or cliffs in soil-filled crevices and ledges among stands of *Miscanthus* that has leaves similar in appearance; flowering (January) March–June.

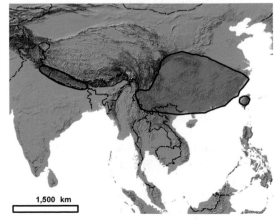

1,500 km

Map 45. Distribution of *C. faberi.*

280

CONSERVATION STATUS. NT (but vulnerable in some parts of its range).

Key to the varieties of *C. faberi*

Plants small to medium in size; leaves 5–8, usually up to 8 mm broad, grey-green; scape robust, usually with 9–20 flowers; floral bracts usually much shorter than the pedicel and ovary; sepals somewhat oblanceolate, obtuse to acute, mucronate; mid-lobe of lip narrowly ovate, tapering to an acute apex ....var. **faberi**

Plants medium to robust in size; leaves 8–9 (15), 8–11 mm broad, green; scape slender, usually with 2–8 flowers; floral bracts exceeding the pedicel and ovary in lowermost flowers; sepals lanceolate, acuminate; mid-lobe of lip oblong, with a broad mucronate apex ....var. **szechuanicum**

var. **faberi**

This is typically a small plant resembling *C. goeringii*, with 6–9 short, narrow, grey-green, arching leaves, often with serrulate margins. It has a surprisingly robust scape with numerous (up to 20) flowers which are olive-green to yellow-green in colour, and are held above the foliage. The sepals are usually narrowly oblanceolate, and the petals are porrect and cover the column. The mid-lobe of the lip is yellow or green, usually with red markings, elongated, covered in many small, glossy, inflated papillae and has a tightly undulating and erose margin. Figs. 160 & 178: 2a-e.

DISTRIBUTION. Southern and central China, Taiwan.

HABITAT. Var. *faberi* is a high elevation plant, growing at up to 3000 m (9840 ft) in mountainous regions in the southern and central provinces of China and in Taiwan. It is often found growing in open, sunny situations amongst stands of *Miscanthus* which has leaves similar in appearance. It often grows on steep land or on cliffs, often near streams in soil-filled crevices and ledges, and it has often been collected in more shaded, moist forests. The distribution, and the high altitude preference, suggests that it is cold-tolerant, probably as hardy as *C. goeringii*, which will survive outdoors in a sheltered position in warmer parts of the British Isles.

This variety is somewhat similar to *C. cyperifolium* in that they both usually have 5–9 or so leaves, the lowest leaves short and merging with the cataphylls, and the mid-lobe has some inflated papillae. It is easily distinguished by its long mid-lobe in comparison with the total length of the lip, its strongly undulating and minutely fimbriate mid-lobe margin, its much more numerous and dense, inflated, shiny papillae on the mid-lobe, its larger number of flowers on a more robust spike and its much shorter floral bracts.

Cymbidium ensifolium is found in China in the same regions as var. *faberi*, but usually at lower altitudes, and is easily distinguished by its fewer (usually broader) leaves (about four), which are all long and distinct from the cataphylls, its weaker, fewer flowered spike and the lip without either inflated papillae or an undulating and fimbriate mid-lobe margin.

var. **szechuanicum** (Y.S. Wu & S.C. Chen) Y.S. Wu & S.C. Chen in *Acta Phytotax. Sinica* 18: 299 (1980). Types: China, Sichuan, Chion-lai-shan, *Wu 2040*, *Wu & Fee 2055*, *Fee 2061* (syntypes PE).

C. cyperifolium sensu Duthie in *Ann. Roy. Bot. Gard. Calcutta* 9: 135 (1906); Banerjee & Thapa, *Orchids of Nepal*: 89 (1978) & Raizada, Naithani & Saxena, *Orchids of Mussoorie*: 39–40 (1981), *non* Wall. ex Lindl. (1833).

C. szechuanicum Y.S. Wu & S.C. Chen in *Acta Phytotax. Sinica* 11: 33 (1966).

C. cyperifolium var. *szechuanicum* (Y.S. Wu & S.C. Chen) S.C.Chen & Z.J.Liu in *Acta Phytotax. Sinica*
   41, 1: 83 (2003).

Var. *szechuanicum* differs from var. *faberi* in that it often has fewer flowers, on a slender scape, its floral bracts are longer and often exceed the ovary in the lower flowers, it has longer, more elliptic sepals tapering to an acuminate apex and the lip is broader and more oblong in shape, with a broad, mucronate apex. The plant is larger and more luxuriant in its growth habit, closely resembling *C. cyperifolium*. Figs. 161, 162, 178: 1a-e & 3a-f.

DISTRIBUTION. China (Anhui, Gansu, Guangdong, Guizhou, Henan, Hubei, Hunan, Shaaanxi, Sichuan, Yunnan, Zhejiang), Nepal, India (northern Uttar Pradesh), northern Myanmar, northern Vietnam.

Wu & Chen (1966) originally described this as a distinct species, but later (1980) reduced it to varietal status within *C. faberi*. The type specimen was collected in Sichuan, and the second specimen cited in the description of *C. faberi*, *Henry* 5515, was also collected there, and it is included in var. *szechuanicum*.

   Chen & Liu (2003) argued for this taxon to be treated as a variety of *C. cyperifolium* and provided a table to show that in many characters it is closer to *C. cyperifolium* than it is to *C. faberi*. However, its flowers are much closer morphologically to *C. faberi* and distinct from those of *C. cyperifolium* in the features mentioned in the key to the species. We do not follow Chen and Liu's treatment here. Further work is undoubtedly needed to elucidate the relationships of these variable and widespread taxa.

   Several collections, from W China and the Himalaya of N India and Nepal, which had previously been identified as *C. cyperifolium*, are included here in *C. faberi* var. *szechuanicum* (Duthie, 1906; Banerjee & Thapa, 1978; Raithada, Naithani & Saxena, 1981). Vegetatively they closely resemble the often sympatric *C. cyperifolium*, but their leaves are slightly narrower and have a minutely serrulate margin. The scapes of specimens from N India and Nepal are slender and usually carry 2–5 flowers, while those from China are usually robust, and more similar to those of var. *faberi*. The few-flowered scapes, the long floral bracts and the similar-shaped green flowers of var. *szechuanicum* are very suggestive of *C. cyperifolium*, but the mid-lobe of the lip has a tightly undulating and minutely erose margin, and is ornamented with numerous, inflated, shiny papillae in *C. faberi*. These characters of the mid-lobe are the most constant characters which distinguish these two species. They further differ in their flowering period; from January until April for *C. faberi* and during November and December for *C. cyperifolium* in this region.

   Collections from the Kathmandu Valley in E Nepal (*Bailes 1040*), at about 200 m (656 ft), in temperate oak/rhododendron forest, and also in the NW Himalaya of N India, at about 2000 m (6560 ft) near Mussoorie (*Du Puy 558*), in similar vegetation, have proved to be identical with those collected by Duthie (1906). The Mussoorie collection was made in early January at which time there was frost at night, and patchy snow lay on the ground, suggesting that they may be hardy in sheltered areas of Britain. In this region, *C. faberi* occurs on steep north or north-west facing slopes in deep shade under dense scrubby woodland cover. The major constituent species in the woodland were the evergreen oak *Quercus leucotricophora*, *Rhododendron arboreum*, *Pieris ovalifolia*, *Viburnum coriaceum* and *Berberis* species, with a wild *Rosa* species and a white-leaved bramble, *Rubus paniculatus* scrambling through these low trees. *C. faberi* was often found on rocky outcrops, or at the base of *Mahonia nepaulensis* shrubs which provided protection from grazing animals. Ground cover was sparse, including ferns and sedges amongst the leaf litter. Plants flowered at Kew during February, at the same time as the specimens collected in E Nepal.

43. CYMBIDIUM GOERINGII

John Lindley first described this species as *C. virescens*, based on a specimen collected in Japan in 1838. However, Willdenow (1805) had previously used this name for a South American orchid. The next available name for this species was *C. goeringii*, originally described as *Maxillaria goeringii* by H.G. Reichenbach in 1845 and based on a specimen collected in Japan by Goering. The name *C. virens* Rchb.f., which is based on the same type specimen as *C. virescens* Lindl., appears to be a mis-spelling of the latter name introduced by Reichenbach into circulation, and unfortunately taken up by later authors.

It is readily recognised by its grass-like leaves and usually one-flowered inflorescence. The flowers do not usually open widely and the sepals are incurved and usually spatulate. Flower colour varies from green to orange-brown with a white lip marked with purple spots.

Cymbidium goeringii varies considerably over its entire range. The variation in Japanese plants was well illustrated by *Garden Life Magazine* (1979). Some of the most distinctive variants are found in Taiwan, including that described by Fukuyama as *C. formosanum*. This name was based on a variant with slender, elliptic, acute sepals and is often used for large-flowered variants from Taiwan. Wu & Chen (1980) regarded this as synonymous with *C. goeringii*, with no discontinuities in variation to

Fig. 163. *Cymbidium goeringii*, SW Guizhou.

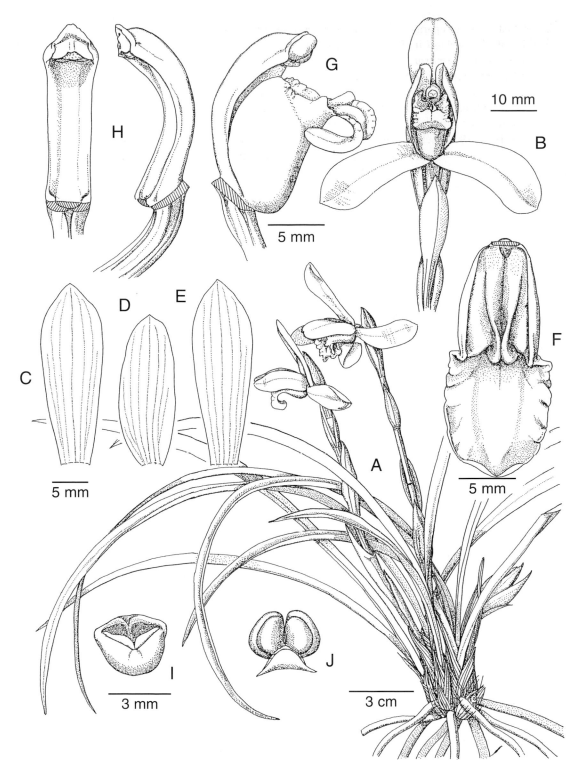

Fig. 164. *Cymbidium goeringii.* **A.** Habit. **B.** Flower. **C.** Dorsal sepal. **D.** Petal. **E.** Lateral sepal. **F.** Lip, flattened. **G.** Column, ovary and lip, side view. **H.** Column, two views. **I.** Anther-cap. **J.** Pollinarium. Drawn from a living collection from Gunma Pref., Japan by Mutsuko Nakajima.

indicate that it is a separate taxon. Similar variants are found in southern China and Japan. The variability of *C. goeringii* in Taiwan is well demonstrated by Mark *et al.* (1986), who published four photographs (as *C. formosanum*).

Cymbidium yunnanense Schltr. from south-west China is a variant with small flowers, a glabrous lip and a supposedly distinct callus structure. The first two characters are highly variable in *C. goeringii*, and the description of the callus agrees well with that of *C. goeringii*. Wu & Chen (1980) also considered this to be synonymous with *C. goeringii*.

Similarly, in 1922, Schlechter distinguished *C. pseudovirens* from *C. goeringii* by the presence of two outgrowths on the outer side of the keels, and a distinctive mid-lobe shape. The pads of callus on the side-lobes of the lip are typical of *C. goeringii*, although they may vary in prominence. The mid-lobe, from its description, falls within the range of variation normally found in *C. goeringii*.

There has been some discussion of the differences between Chinese and Japanese specimens, and the possibility of recognising two distinct taxa. Nagano (1955) suggested that they differ in that the sepals and petals of the Chinese plants are more slender and red-brown in colour, and the flowers have a stronger fragrance. Summerhayes (1963) reiterated these differences, adding that the Chinese specimens 'may be distinguished by the sepals which are broadest in the middle instead of the upper part ...'. In general, such patterns of variation are encountered, but specimens with obovate or elliptic petals have been collected in both countries. Variation in colour and scent also occurs, but these characters are not constant and cannot be used to separate two taxa. Thirty-six photographs of different Chinese forms of *C. goeringii* (sensu Du Puy and Cribb, 1988) were illustrated by Perner (2002). Of these, nine are probably referable to *C. tortisepalum* var. *longibracteatum*. Flower colour, sepal shape

Plate 34. *Cymbidium goeringii.* Cult. Kew, del. Margaret Stones.

Fig. 165. Variation in cultivated clones of *Cymbidium goeringii*. Cult. Sichuan, China.

and posture and lip markings are particularly variable in those referable to *C. goeringii*. For example, sepal and petal colour ranges from green and yellow to orange and purple. The sepals and petals can be purple-striped or not and the lip streaked or spotted with purple or unspotted.

An apparently widely disjunct portion of the distribution is in the Himalaya of north-west India. The plants collected from around Mussoorie, and described by Duthie in 1902 as *C. mackinnoni*, are similar to the type of *C. goeringii*, except that the leaf margins are not serrated. Various degrees of serration can be found in *C. goeringii* elsewhere in its range and, therefore, this cannot be used reliably to differentiate these taxa.

In 1913, Rolfe described *C. forrestii* from a specimen collected in Yunnan, and differentiated it from the Japanese *C. goeringii* by its prominent lip side-lobes, and the prominent mounds of tissue on the side-lobes adjacent to the usual two callus ridges. In the type specimen of *C. virescens* the lip is also three-lobed, and the mounds on the side-lobes are present but not as prominent. Therefore, these characters are considered insufficient to distinguish *C. forrestii* as a distinct species.

Schlechter described *C. serratum* in 1919, based upon a specimen collected in southern China. Wu & Chen (1980) recognised it as a distinct variety of *C. goeringii*, and their treatment was followed by Du Puy and Cribb (1988). It also occurs in Taiwan, where it has variously been named *C. gracillimum*, *C. formosanum* var. *gracillimum*, and *C. goeringii* var. *angustatum*. The variability in flower shape and colour in Taiwan is illustrated by Mark *et al.* (1986). They also noted that it has an earlier flowering season than var. *goeringii*. It has also been reported from Japan (Maekawa, 1971). It occurs mainly at high elevations (up to 3000 m, 9840 ft) in Taiwan, in steep rocky localities, often in broad-leaved forest (Mark *et al.*, 1986).

Chen and Liu (2003) distinguished var. *serratum* from the typical variety in having narrower leaves, 2–4 mm broad, and olive-green, thick-textured sepals and petals and scented flowers. However, we have examined many specimens of *C. goeringii* and have not found clear-cut distinctions between it and the typical variety. The variation in leaf width is continuous, similar flower colour variation is found in specimens assigned to each variety and floral texture and scent are indeterminable in herbarium material. Furthermore, flower scent is difficult to assess, because its production can be temperature- and time-dependent.

Cymbidium goeringii has been cultivated in Japan and China for several centuries, and many variants have been selected there and maintained in cultivation. Some of these, such as those with variegated leaves, or albino or peloric flowers, are very distinct and highly prized.

Wu & Chen (1980), in their revision of *Cymbidium* in China, recognised four distinct varieties of *C. goeringii*. Du Puy and Cribb (1988) accepted three of them. The fourth, var. *longibracteatum*, was considered insufficiently different from var. *tortisepalum* to warrant taxonomic recognition. Chen and Liu (2003) reconsidered the species and its varieties and recognised *C. tortisepalum* with two varieties, var. *tortisepalum* and var. *longibracteatum*, as distinct from *C. goeringii*. They concluded that *C. tortisepalum* differs in having longer leaves that are articulated near the base and in having 3- to 7-flowered inflorescences (Table 19). We follow their treatment here.

43. Cymbidium goeringii (Rchb.f.) Rchb.f. in *Walp. Ann. Bot.* 3: 547 (1852). Type: Japan, *Goering 592* (holotype W!).

Maxillaria goeringii Rchb.f. in *Bot. Zeit.* 3: 334 (1845).
Cymbidium virescens Lindl. in *Bot. Reg.* 24: misc. 37 (1838), *non* Willd. (1805). Type: Japan, cult. Rollinsons, *Siebold* (holotype K!).
C. virens Rchb.f. in *Walp. Ann. Bot.* 6: 626 (1861), *sphalm. pro C. virescens* Lindl.
C. mackinnoni Duthie in *J. Asiat. Soc. Bengal* 71(2): 41 (1902). Type: NW India, nr. Mussoorie, *Mackinnon s.n.* (isotype K!).
C. formosanum Hayata in *J. Coll. Sci. Tokyo* 30: 335 (1911) & in *Mater. Fl. Formos.*: 335 (1911). Type: Taiwan, *Nakahara s.n.* (holotype TI!).

C. forrestii Rolfe in *Notes Roy. Bot. Gard. Edinburgh* 8: 23, t. 11 (1913); Schltr. in *Fedde, Repert. Beih.*
    4: 267 (1919). Type: China, Yunnan, *Forrest 415* (holotype E!).

C. yunnanense Schltr. in *Fedde, Repert. Sp. Nov. Regni Veg., Beih.* 4: 74 (1919). Type: China, Yunnan,
    E of Teng-Tchouan [Tengyueh], *Maire 6425* (holotype B†).

C. serratum Schltr. in *loc. cit.* 73 (1919). Type: China, Guizhou, *Esquirol* (holotype B†; photo. K!).

C. pseudovirens Schltr., *loc. cit.* 12: 351 (1922). Type: China, Zhejiang, Ningpo, *Limpricht 304*
    (holotype B†).

C. gracillimum Fukuy. in *Trans. Nat. Hist. Soc. Formosa* 22: 413–415, t. 1 & 2 (1932); Mark, Ho &
    Fowlie in *Orchid Dig.* 50: 24, + figs. (1986). Type: Taiwan, prov. Sintiku, Mt. Tyotui-zan
    Fukuyama 3220 (holotype KANA).

C. tentyozanense Masam. in *Trans. Nat. Hist. Soc. Formosa* 25: 14 (1935). Type not located.

C. uniflorum T.C. Yen, *Icon. Cymbid. Amoy.*, A2 (1964). Type: China, Fukien, *Lin 7001* (not located).

C. chuen-lan C. Chow, *Formosan Orchids* ed. 1: 21 (1968); *nom. inval.*, holotype not indicated.

C. goeringii Rchb.f. var. *angustatum* F. Maek., *The Wild Orchids of Japan in Colour*: 416, t. 171 (1971),
    nom. nud.

C. formosanum Hayata var. *gracillimum* (Fukuy.) T.S. Liu & H.J. Su in *Flora of Taiwan* 5: 943 (1978).

C. goeringii var. *serratum* (Schltr.) Y.S. Wu & S.C. Chen in *Acta Phytotax. Sinica* 18: 300 (1980).

A terrestrial *herb* with fleshy roots. *Pseudobulbs* ovoid, enclosed in about 6 scarious cataphylls. *Leaves* 5–7(8), up to 80 × 0.2–1 cm, but usually less than 40 cm long, the shortest merging with the cataphylls, linear-elliptic, acute, usually V-shaped in section, arching, usually with a serrated margin, narrowed towards the base. *Scape* erect 1-(occasionally 2–4-) flowered; peduncle up to 15(21) cm long, covered in 4–8 sheaths; sheaths up to 6(9) cm long, becoming scarious, cylindrical in the basal half, expanded and cymbiform in the upper half, acute; bract 2.5–6 cm, usually exceeding the ovary, scarious, cymbiform, acute, often red-tinted. *Flower* about 4–5 cm across, porrect or nodding; sometimes scented; pedicel and ovary green to purplish; sepals and petals apple-green to red-brown, stained red towards the base, especially over the mid-vein of the petals; lip cream with crimson spots and margins on the side-lobes and sparse red blotches on the mid-lobe; callus cream to pale yellow; column usually pale green, cream towards the base, spotted lightly above and densely below with maroon; anther-cap cream, sometimes purple or yellow below. *Pedicel and ovary* 2.5–6 cm, curved behind the flower. *Sepals* usually obovate to elliptic, obtuse or apiculate, occasionally elliptic, acute, margins often incurved; dorsal sepal often porrect, 2.5–3.9 × 0.85–1.25 cm; lateral sepals similar, spreading. *Petals* 1.85–3 × 0.85–1.1 cm, oblong-ovate to oblong-elliptic, obtuse, oblique, usually closely covering the column. *Lip* 1.7–2.6 × 1–1.6 cm when flattened, subentire to 3-lobed; side-lobes sometimes reduced, erect, rounded, sometimes angled at the front, papillose; mid-lobe 0.7–1 × 0.75–1.1 cm, broadly ovate to oblong, obtuse or rounded, recurved, papillose, the margin sometimes undulate; callus 2-ridged, convergent towards the apex and forming a short tube that extends to the base of the mid-lobe, with two large pads of callus on the side-lobes adjacent to the callus ridges. *Column* 1.3–1.9 cm long, broadening into two wings towards the apex; pollinia 4, in two unequal pairs. *Capsule* up to 8 cm long, fusiform, erect, and parallel to the rhachis, pedicellate, beaked. Pl. 34, Figs. 163–166, 178: 5a–e, 6a–e.

DISTRIBUTION. Japan, Korea, Ryukyu Islands, widespread in southern China (Anhui, Fujian, southern Gansu, Guangdong, Guangxi, Guizhou, southern Henan, Hubei, Hunan, Jiangsu, southern Shaanxi, Sichuan, Taiwan, Yunnan and Zhejiang) and rare in north-west India (northern Uttar Pradesh). (Map 46); 300–3000 m (985–9840 ft). Chen (1999) gave its distribution in China as including all of the southern provinces as far north as Sichuan, Henan and Jiangsu, and including Taiwan but not Hainan Island or Hong Kong. Y.N. Lee (1976) noted that it was abundant in forests in the southern part of Korea, and it had also been collected on Tsushima Island in the Straits of Korea.

DAVID DU PUY

Fig. 166. *Cymbidium goeringii*, cult. Tokyo, Japan.

HABITAT. Terrestrial in open forest, usually on lightly shaded cliffs or slopes, often in coniferous forests near the sea in Japan. Maekawa (1971) noted that *C. goeringii* occurs in southern Japan in the warm, temperate vegetation zone, in broadleaved, evergreen forest dominated by *Castanopsis cuspidata* and *Machilus thunbergii*, often with *Podocarpus* and *Cephalotaxus* and *Camellia japonica*. Chen (1999) reported it from rocky slopes, forest margins and open places in forests from 300 to 3000 m (985–9850 ft) elevation. It is usually found in open forest, on lightly shaded cliffs or rocky slopes, often amongst grasses or bamboos. It flowers from January until March.

CONSERVATION STATUS. NT (but vulnerable in some parts of range).

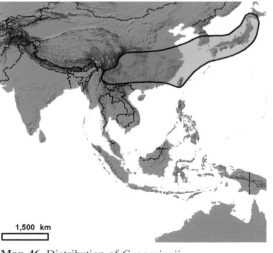

1,500 km

Map 46. Distribution of *C. goeringii*.

289

44. CYMBIDIUM KANRAN

Cymbidium kanran was described in 1902 by Tomitaro Makino from a wild-collected Japanese specimen that had flowered in cultivation. The specific epithet means 'orchid which flowers in winter', referring to its flowering time between November and February. Its leaves are slender, glossy and deep green in colour. The flowers are elegant and scented and are known in a wide range of colours and, combined with the graceful form of the plant, make it a desirable plant for pot cultivation. For many centuries it has been cultivated in Japan and Taiwan, and several selected variants are highly prized.

It is closely allied to *C. sinense*, both having 3–4 glossy, dark green leaves and scapes with several comparatively large flowers, with petals that are strongly porrect, forming a hood over the column. It can be distinguished from the latter and other related species by its long, narrow, acuminate leaves, relatively long floral bracts that are almost equal in length to the pedicel and ovary in the lowest flower in the spike, but are shorter in the upper flowers, and its long, narrow, finely tapering sepals that give the flower a spidery appearance. The sepals are usually seven or more times longer than broad, whereas in *C. sinense* and *C. ensifolium* the ratio is four times.

Four taxa, *C. oreophilum* Hayata (1914), *C. purpureo-hiemale* Hayata (1914), *C. linearisepalum* Yamamoto (1930) and *C. sinokanran* Yen (1964), are considered by most authors to be synonyms of *C. kanran* (Su, 1975; Lin, 1977; Ying, 1977; Liu & Su, 1978; Wu & Chen, 1980).

Cymbidium oreophilum Hayata was described as having a 'botryoideo-tuberculate' mid-lobe. Specimens of *C. kanran* usually have some thickening of the veins in front of the callus ridges, and the indumentum varies from minutely papillose to shortly pubescent, or with a few swollen papillae. The indumentum type seems to vary independently of flower colour, and it is impossible to recognise distinct taxa on the basis of this character.

Cymbidium tosyaense (as '*tosyaenus*') was described by Masamune from a specimen cultivated in Japan. This name has possibly been corrupted to *C. tentyozanense* by Lin (1977). The description of this taxon does not distinguish it from *C. kanran*, and it is now placed in the synonymy of the latter.

Flower colour in *C. kanran* varies greatly, usually olive-green with some red-brown on the central vein of the petals and sepals, and a pale green or yellow lip lightly spotted with red. Nagano (1955) listed six colour variants recognised in cultivation in Japan, and these are well illustrated by Maekawa (1971). These can be grouped into the following categories on the basis of the colour of the petals and sepals: 1. green with weak to strong red-brown veins, and red-brown markings on the lip—the most widespread colour; 2. purple-brown; 3. pink; 4. pale yellow; 5. green, without any red-brown markings on the sepals, petals or lip; and 6. multi-coloured, known as the 'sasara' colour variant. This wide colour range has led to the description of several taxa. Makino named four forms in Japan to cover this colour variation; f. *purpurascens* (1902), f. *viridescens* (1912), f. *rubescens* (1912) and f. *purpureo-viridescens* (1912). When Yamamoto described *C. linearisepalum* in 1930 he also described two colour variants, naming them f. *atropurpureum* and f. *atrovirens*. In 1933, Masamune raised their taxonomic rank from form to variety. In 1964, Yen also noted two varieties of his *C. sinokanran* from Taiwan: the typical green one and var. *atropurpureum*.

In 1914 Hayata described a purple variant from Taiwan as *C. purpureo-hiemale*, meaning the 'winter-flowering purple' *Cymbidium*. He further distinguished it by the lines of short hairs on the mid-lobe of the lip. Ying (1976) reduced it to forma *purpureo-heimale* of *C. kanran*, but later (1977) raised it to varietal status. Mark *et al.* (1986) illustrated the range of flower colour of *C. kanran* in Taiwan. Distinguishing these colour variants may be horticulturally useful, but they do not warrant recognition at specific rank.

Cymbidium kanran has leaves less than 1.5 cm (0.6 in) broad, but there are variants with leaves up to 2 cm (0.8 in) broad. In 1912, Makino published var. *latifolium*, a variant with broader leaves. However, the variation in leaf width is continuous, does not appear to have any geographic pattern, and is probably not taxonomically significant.

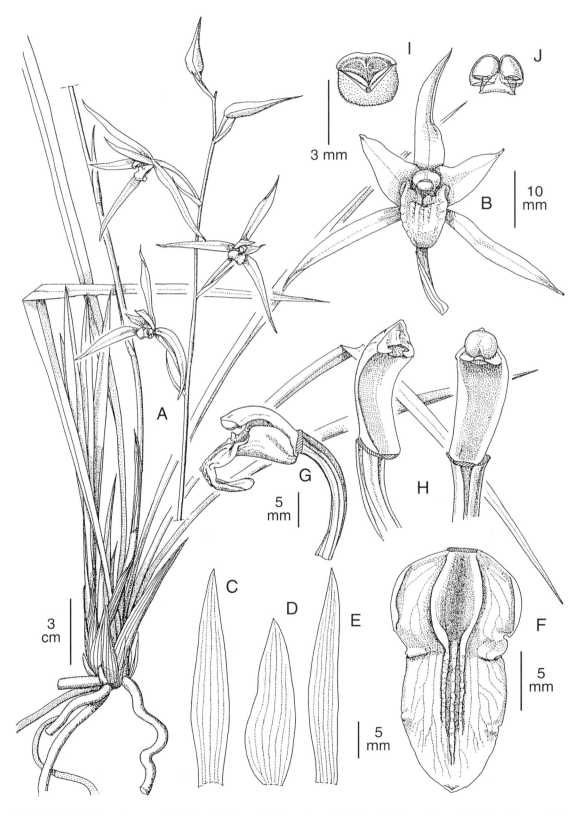

Fig. 167. *Cymbidium kanran.* **A.** Habit. **B.** Flower. **C.** Dorsal sepal. **D.** Petal. **E.** Lateral sepal. **F.** Lip, flattened. **G.** Column, ovary and lip, side view. **H.** Column, two views. **I.** Anther-cap. **J.** Pollinarium. Drawn A: Okinawa Island (TI); B–J: Kochi Pref., Japan (spirit coll. TI) by Mutsuko Nakajima.

DAVID DU PUY

Fig. 168. *Cymbidium kanran*. Cult. E. Young Orchid Foundation.

Further variation, in cultivated plants, includes those with variegated leaves with narrow white margins that are known as 'Takachiko' or 'Nangoku' in Japan (Nagano & Nagano, 1957). A detailed photographic account of the variation in Japanese *C. kanran* is provided by Anon. (1980, 1984).

Two variants are known in Hong Kong: one has the normal, long sepals, but the other has shorter, less acuminate petals and sepals but is unmistakably close to *C. kanran* in its other vegetative characters of the plant and the flower spike (G. Barretto, pers. comm.).

Maekawa (1971, p. 411, 479, t. 168) illustrated two hybrids between *C. kanran* and *C. goeringii*. The name *C. × nishiuchianum* was reputedly given to these by Makino and validated by Shaw (2002). The hybrid has been reported from the central part of the province of Tosa in Japan. Cheng (1981) also illustrated several natural hybrids of *C. kanran* with *C. ensifolium*, *C. goeringii* and *C. sinense* in Taiwan. Hybridisation between *C. kanran* and *C sinense* may account for the variation in *C. kanran* in Hong Kong, as they are often found growing together there (G. Barretto, pers. comm.).

Cymbidium qiubeiense from southern China is a closely related species.

44. Cymbidium kanran Makino in *Bot. Mag. Tokyo* 16: 10 (1902) & in *Iinuma, Somoku-Dzusetsu*, ed. 3, 4(18): t. 4, 5 & 6 (1912). Type: Japan, Musashi Prov., cult. Koishikawa Bot. Gard., *Makino s.n.* (holotype MAK).

C. kanran var. *latifolium* Makino, in *Iinuma, Somoku-Dzusetsu* 4: 5, t. 5 (1912). Type: not indicated, but probably in MAK.

C. oreophyllum Hayata, *Icon. Pl. Formos*. 4: 80, t. 38c (1914). Type: Formosa, [Taiwan], cult. Tiahoku, *Hayata s.n.* (holotype TI).

C. misericors Hayata var. *oreophyllum* (Hayata) Hayata, *loc. cit.*: 81 (1914).

C. purpureo-hiemale Hayata, *loc. cit.*: 81 (1914). Type: Formosa [Taiwan], cult. Taihoku, *Hayata s.n.* (holotype TI).

C. linearisepalum Yamam. in *Trans. Nat. Hist. Soc. Formosa* 20: 40 (1930). Type: not located.

C. tosyaense Masam., *loc. cit.* 25: 14 (1935). Type: Japan, Takao-Syu Heitogum Tosga, cult. Watanabe (holotype MAK).

C. sinokanran T.C. Yen, *Icon. Cymid. Amoy.*: G–1 (1964). Type: Fukien [Fujian], Nan-gien Hsien, *Cheng 4001* (holotype not located).

C. sinokanran T.C. Yen var. *atropurpureum* T.C. Yen, *loc. cit.*: G–2 (1964). Type: China, Fukien [Fujian], Nan-gien Hsien, *Cheng 4002* (holotype not located).

C. tentyozanense T.P. Lin, *Nat. Orch. Taiwan* 2: 112 (1977); *sphalm.* for *C. tosyaense* Masam.

A perennial, terrestrial *herb*. *Pseudobulbs* 4–6 × 1–1.8 cm, narrowly ovoid, often conspicuous, covered in scarious cataphylls and broad, sheathing leaf bases, with 3–4(5) almost distichous leaves. *Leaves* up to (20)30–90(120) × 0.5–1.5(2) cm, linear to linear-elliptic, acuminate, arching, dark green, shining, margins often minutely serrulate towards the apex, articulated about 5 cm from the pseudobulb, the lowest leaves long and distinct from the cataphylls. *Scape* (15)25–60(80) cm tall, erect, arising basally from inside the cataphylls, with (3)5–12(20) distant flowers produced in the apical half to one-third of the scape and held above the leaves; peduncle covered in about 3–6 distant sheaths up to 4.5 cm long; bracts (0.8)1.5–3.5(5) cm long, narrowly ovate, acuminate. *Flowers* 5–7 cm across, spidery in appearance; scented; rhachis, pedicel and ovary green to purple; flower colour variable, but usually sepals and petals olive to clear green, with a short maroon stripe over the mid-vein at the base; lip pale yellow or pale green, side-lobes streaked with red with a solid red margin, mid-lobe spotted and blotched with red, with a narrow cream margin; column usually pale green with purple spots towards the apex, and cream streaked with red below; anther-cap cream. *Pedicel and ovary* 2–3.5 cm, usually longer than the bracts. *Dorsal sepal* (25)30–45(50) × (3)3.5–5(7) mm, linear-elliptic, acuminate, erect; lateral sepals similar, spreading. *Petals* 22–30(35) × (3)5–8(10) mm, ovate, acuminate, porrect, almost forming a hood over the column. *Lip* (1.6)2–2.7 cm long when flattened; side-lobes erect, rounded, minutely papillose or minutely pubescent; mid-lobe (10)12–16 × 7–2 mm, oblong-ovate to triangular-ovate, obtuse to subacute, recurved, papillose, hairy or with some papillae, margin entire or erose, kinked but not undulating; callus of 2 ridges, converging in the apical half to form a small tube at the base of the mid-lobe. *Column* 10–15(17) mm long, arching, winged; pollinia 4, broadly ovate in two unequal pairs, on a small crescent-shaped viscidium. *Capsule* about 5 cm long, fusiform, ridged, beaked, held erect and parallel to the rhachis. Figs. 154: 7a–g, 167 & 168.

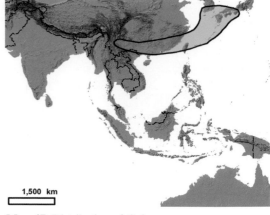

1,500 km

Map 47. Distribution of *C. kanran.*

DISTRIBUTION. Southern China, Hong Kong, Taiwan, Ryukyus, southern Japan, South Korea (Map 47); 800–1800 m (2625–5905 ft). It is found in the south-east tip of Honshu, and in Shikoku and Kyushu in southern Japan. It is also found in the Ryukyu Islands, including Yakusima, Amami-Osima and Okinawa (Garay & Sweet, 1974; Ohwi, 1965). It has been reported from one locality in Korea (Lee, 1976). This northerly distribution limit reflects its relatively high tolerance of cold. Wu & Chen (1980) gave its distribution in China as the southern provinces from Sichuan and Yunnan to Hong Kong and Taiwan.

HABITAT. In open hardwood forest, in shade. Makino (1902) reported that *C. kanran* grows in shady forests in the warmer parts of southern Japan. In Taiwan it is found in ravines or on mountain ridges on south-east facing slopes, usually in hardwood forests. In Hong Kong it is also found at high elevations, usually in association with *C. sinense*. It is probably a high-elevation plant in southern China also. It flowers between (October) November and February.

CONSERVATION STATUS. VU A1cd; B1ab.

45. CYMBIDIUM MUNRONIANUM

Cymbidium munronianum is closely related to *C. ensifolium*, particularly to subsp. *haematodes* which has similarly broad leaves and petals broader than the sepals. Their flowers are similar, 3.0–3.5 cm across and usually pale yellow in colour with five red stripes on the sepals and petals, with similar red markings on the lip. The mid-lobe of the lip of *C. munronianum* is narrower than that of *C. ensifolium* subsp. *haematodes*, but both have an undulating margin. *Cymbidium munronianum* has more numerous flowers in its inflorescence, and is similar to *C. sinense* in this respect. Vegetatively it is also closer to *C. sinense,* the range of which range extends from China into north-east India (Meghalaya), but is not recorded from the Himalayas. All three taxa have comparatively broad leaves, and although *C. sinense* has the broadest leaves they can be difficult to distinguish on this character alone. The flower of *C. munronianum* is distinct from that of *C. sinense* in size, colour and the manner in which the petals are held (Table 18). It can be further distinguished from both of these related taxa by the characters of the inflorescence. Its peduncular sheaths are short and distant from each other, clasp the peduncle closely, and do not spread out even at their tips. Its floral bracts are short, and its numerous flowers small. Table 18 compares these three taxa.

The type specimen was collected by Robert Pantling in the Teesta Valley in Sikkim. It was described and illustrated by Pantling and George King in 1895, who dedicated it to Mr. Munro, resident in Sikkim at that time. Hooker (1891) did not differentiate between the Himalayan specimens of this species and *C. ensifolium*, but it now appears that *C. munronianum* replaces *C. ensifolium* in northern India. Further collections from Nepal, Bhutan, Meghalaya (Khasia Hills) and eastern Assam (Nagaland) are needed to determine the complete distribution of this species.

Cymbidium cyperifolium is also found in the same region, but it is readily distinguished by its more numerous, narrower leaves, much longer floral bracts, its larger flowers, its green petals and sepals lacking red stripes and mid-lobe with inflated papillae.

TABLE 18

A COMPARISON OF THE CHARACTERS WHICH SEPARATE THE THREE CLOSELY RELATED TAXA *C. MUNRONIANUM, C. ENSIFOLIUM* SUBSP. *HAEMATODES* AND *C. SINENSE*

	C. munronianum	*C. ensifolium* subsp. *haematodes*	*C. sinense*
Leaf breadth	1.8–2.7 cm	1.4–2.5 cm	1.5–3.5 cm
Spike length	up to 60 cm	17–67 cm	41–72 cm
Number of flowers	8–13	3–8	(6)8–26
Sheath length	up to 5.0 cm	up to 6.5 cm	up to 10.0 cm
Sheath arrangement	sheaths distant, amplexicaul	sheaths overlapping, cymbiform at the apex	sheaths overlapping, cymbiform at the apex
Floral bract length	0.2–0.8	0.2–2 cm	0.4–2.5 cm
Flower size	2.5–3.5 cm across	3–3.5 cm across	about 5 cm across
Tepal colour	pale straw-yellow, with longitudinal red-brown stripes	pale straw-yellow to pale green, with longitudinal red-brown stripes	dark, chocolate-brown
Dorsal sepal length	16–26 mm	16–26(31) mm	26–39 mm
Lip length	12–20 mm	14–22 mm	21–29 mm
Mid-lobe size	7.1–8(10) × 4.2–7.1 mm	6–11 × (5.1)6.2–8.2(9.1) mm	12–16 × 9–13 mm
Column length	7–11 mm	10–14(18) mm	12–16 mm
Petal arrangement	petals spreading, not covering the column	petals spreading, not covering the column	petals usually closely covering the column
Flowering period	December–May	January–March, but also sporadic throughout the year	October–November

45. Cymbidium munronianum King & Pantling in *J. Asiatic Soc. Bengal* 64: 338 (1895) & in *Ann. Roy. Bot. Gard. Calcutta* 8: 187, t. 249 (1898). Type: Sikkim, Teesta Valley, *Pantling s.n.* (holotype CAL).

C. ensifolium var. *munronianum* (King & Pantling) Tang & Wang in *Acta Phytotax. Sin.* 1: 91 (1951).

A perennial, terrestrial *herb. Pseudobulbs* small, up to 3 × 2 cm, ovoid, prominent, covered in scarious cataphylls up to 11 cm long that become fibrous with age, and broad sheathing leaf bases, both with a 2 mm broad membranous margin, with 3–4(5) distichous leaves. *Leaves* 60–80 × 1.8–2.7 cm, linear-elliptic, erect, articulated 3–5 cm from the pseudobulb, the lowest leaves distinct from the cataphylls. *Scape* up to 60 cm tall, erect, arising basally, with 8–13 flowers produced in the apical third of the spike; peduncle with about 5 short, amplexicaul sheaths which are closely adpressed to the peduncle, up to 5 cm long, distant, only the lowest sheaths overlapping each other; bracts 0.2–0.8 cm long, triangular. *Flowers* 2.5–3.5 cm across; scented; rhachis, pedicel and ovary green; sepals and petals pale green or yellow to cream, with about 5 pale purple-brown, broken, longitudinal lines; lip pale yellow, side-lobes pink, streaked with red, with a solid red margin, mid-lobe blotched with red; column pale green, streaked maroon below; anther-cap cream. *Pedicel and ovary* 1.2–2.5 cm long. *Dorsal sepal* 16–26 × 5–6.5 cm, oblong-elliptic, rounded and mucronate at the apex, erect; lateral sepals similar, spreading, almost

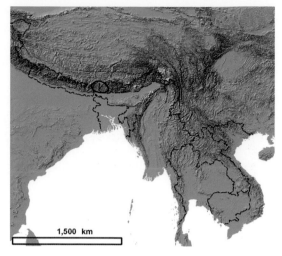

Map 48. Distribution of *C. munronianum*.

horizontal. *Petals* 16–22 × 6–9.8 mm, ovate, subacute, porrect but not closely shading the column, broader than the sepals. *Lip* 1.2–2 cm long when flattened; side-lobes erect, prominent, minutely pubescent; mid-lobe 7.1–8(10) × 4.2–7.1 mm, ligulate or oblong, narrower than the side-lobes when the lip is flattened, not recurved, papillose, margin entire and undulating or kinked; callus of 2 ridges, converging in the apical half to form a small tube at the base of the mid-lobe. *Column* 7–11 mm long, arching, winged; pollinia 4, broadly ovate, in two unequal pairs. *Capsule* about 5 cm long, fusiform, ridged, held erect and parallel to the rhachis. Fig. 154: 6a–f.

DISTRIBUTION. Northeast India (Sikkim), Bhutan (Map 48); about 500 m (1640 ft).

HABITAT. Few habitat notes are available for this species. It flowers between December and May. The type description notes its habitat as 'in the Teesta Valley on dry knolls; at an elevation of 1500 ft (*c.* 500 m)'. It flowers from December until May.

CONSERVATION STATUS. VU A1cd; B1ab.

46. CYMBIDIUM NANULUM

In 1991 Y.S. Wu and S.C. Chen described the diminutive terrestrial, *C. nanulum*, based upon a collection from near Liuku in western Yunnan. Liuku lies on the Nu Jiang [Salween River] in the shadow of the Nu Shan to the east and Gaoligong Mountains to the west, which adjoin the Burmese border. They compared it with the widespread *C. ensifolium*, to which it is closely allied: it has a subterranean rhizome rather than a pseudobulb, 2 or 3 leaves per shoot, a 2- to 9-flowered inflorescence and a lip with a shorter mid-lobe. Unfortunately, captions for the illustration that accompanied the type descriptions of *C. nanulum* and *C. defoliatum* were inadvertently switched (Chen, pers. comm.).

We have seen the type and a few cultivated plants, and further study is needed to establish its exact relationship to *C. ensifolium*. Chen (1999) gave its range as south-east and south-west Yunnan, south-west Guizhou and Hainan. The latter may refer to small plants of *C. ensifolium* rather than to *C. nanulum*.

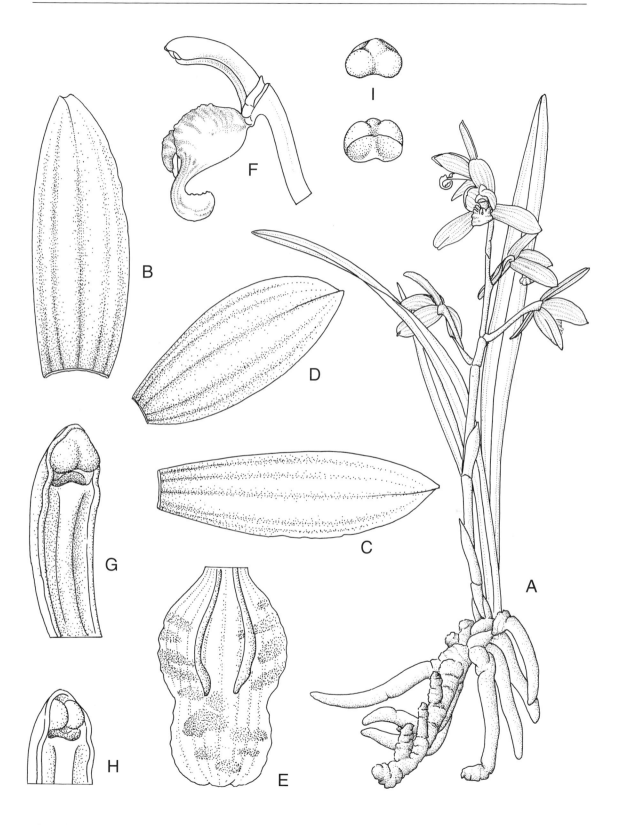

Fig. 169. *Cymbidium nanulum*. **A.** Habit. **B.** Dorsal sepal. **C.** Lateral sepal. **D.** Petal. **E.** Lip, flattened. **F.** Lip and column, side view. **G.** Column, ventral view. **H.** Column apex. **I.** Anther cap, two views. A, ×1; B–E, H–I, ×6; F, × 3. Drawn from the type collection by Deborah Lambkin.

46. Cymbidium nanulum Y.S. Wu & S.C. Chen in *Acta Phytotax. Sinica* 29 (6): 551 (1991); Chen *et al.*, *Native Orchids of China in Colour*: 118 (1999); Chen, *Flora of China Orchidaceae* 2: 216 (1999). Type: China, Yunnan, near Liuku, 800–1600 m, 15 June 1989, cult. Beijing B.G., *D.P. Yu 66* (holotype PE!).

A terrestrial herb with a fleshy, flattened-terete, several-noded, sub-terranean rhizome over 1 cm in diameter; roots fleshy, 7–8 mm in diameter. *Pseudobulbs* reduced. *Leaves* suberect, 2–3, linear-lorate, acute, 25–30 × 0.6–1.2 cm. *Inflorescence* erect, laxly 2- to 9-flowered, 10–15 cm long; bracts linear-lanceolate, acuminate, 4–9 mm long, green. *Flowers* 2.5–3.2 cm across, fragrant; sepals and petals pale yellow or pale green with 5 red-purple stripes on the sepals and petals; lip side-lobes purple-striped and mid-lobe boldly purple-spotted; pedicel and ovary 1.6–2 cm long. *Dorsal sepal* suberect-arching forwards, 1.3–1.6 × 0.6–0.7 cm, narrowly oblong, rounded or obtuse at apex. *Lateral sepals* spreading, falcate, 1.3–2 × 0.5–0.6 cm, oblong-lanceolate, acute. *Petals* porrect, enclosing column, 1.1–1.4 × 0.6–0.7 cm, narrowly oblong-elliptic, acute. *Lip* obscurely 3-lobed, 0.8–1 cm long, 0.5–0.6 cm wide, oblong-ovate, obtuse or rounded at apex; side-lobes obscurely semi-elliptic; mid-lobe recurved, 3 × 2 mm, subcircular; callus of 2 connivent ridges forming a tube between them at the base of the mid-lobe. *Column* semi-terete, arcuate, 0.6–0.7 cm long. Figs. 169 & 170.

DISTRIBUTION. South-west China (south-west Guizhou, southern Yunnan, ?Hainan) (Map 49).

HABITAT. Terrestrial, in rocky places in open forest; flowering in June.

CONSERVATION STATUS. VU A1cd; B1ab.

Fig. 170. *Cymbidium nanulum*. SE Yunnan.

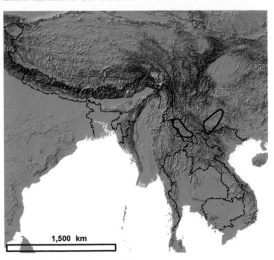

1,500 km

Map 49. Distribution of *C. nanulum*.

298

47. CYMBIDIUM OMEIENSE

Wu & Chen (1966) described *Cymbidium omeiense* based on a specimen from Emei Shan (Mt. Omei) in Sichuan. The type was destroyed or not preserved. On the basis of the original description, Du Puy & Cribb (1988) treated it as a synonym of *C. kanran*. The description included many characters that are found in *C. kanran*. Wu & Chen (1966, 1980) described it as more 'feathery' than *C. faberi*, a description that seemed to fit the spidery flowers of *C. kanran*. It is a smaller plant with shorter leaves, smaller flowers and less acuminate sepals, but it has the narrow sepals, long floral bracts and porrect petals characteristic of *C. kanran*.

In contrast, Wu and Chen (1980) considered it a variety of *C. faberi* (q.v.) and Chen *et al.* (1999) followed that treatment for the account of *Cymbidium* in the *Flora of China Orchidaceae*. However, it has fewer shorter leaves and a differently shaped lip with an entire margin, lacking a strongly papillose mid-lobe. Chen and Liu (2003) recently resurrected *C. omeiense* as a distinct species on the basis of its leaf venation not being translucent, the fewer, smaller leaves, shorter inflorescence, semi-annual flowering and less papillose lip mid-lobe.

A specimen, *Farges 944* (K, P), from Emei Shan fits well the type description of *C. omeiense* and indicates that it is closer to *C. ensifolium* than to *C. faberi* and *C. kanran*. It is a smaller plant and has a shorter, fewer-flowered inflorescence than the other species. It has narrower leaves than *C. ensifolium* and longer bracts, as long as or slightly longer than the the pedicel and ovary. Hybridisation in the origin of this taxon cannot be discounted because several allied species have been found on this mountain.

47. Cymbidium omeiense S.Y. Wu & S.C. Chen in *Acta Phytotax. Sin.* 11, 1: 32, pl. 5, t. 4–8 (1966). Type: China, Szechuan [Sichuan], Omeishan, *Fee 2009* (holotype PE†); Sichuan, Omeishan, *Z.J. Liu 22319* (neotype PE).

C. faberi Rolfe var. *omeiense* (Y.S. Wu & S.C. Chen) Y.S. Wu & S.C. Chen in *Acta Phytotax. Sin.* 18: 299 (1980).

A perennial, terrestrial *herb. Pseudobulbs* obscure, covered in scarious cataphylls and broad, sheathing leaf bases, with 4 or 5 almost distichous leaves. *Leaves* arching, up to 15–30(35) × 0.6–1.0 cm, linear to linear-elliptic, acuminate, dark green, margins often serrulate towards the apex, the lowest leaves distinct from the cataphylls. *Scape* 15–17 cm tall, erect, arising basally from inside the cataphylls, laxly 3–5-flowered in the apical third of the scape and held above the leaves; peduncle covered in about 3–6 sheaths up to 4.5 cm long; bracts (0.8)1.5–3.5(5) cm long, linear-lanceolate, acuminate. *Flowers* 5–7 cm across, spidery in appearance, fragrant; sepals and petals yellow-green, with a median purple-red stripe over the mid-vein at the base; lip yellow green, side-lobes streaked with red with a solid red margin, mid-lobe marked with purple-red, column yellow marked with red. *Pedicel and ovary* 2–3.5 cm, usually as long as the bracts. *Dorsal sepal* 25–30 × 3–5 mm, linear-elliptic, acuminate, erect; lateral sepals similar, spreading. *Petals* 16–18 × 3–4 mm, obliquely rhombic, acuminate, porrect, almost forming a hood over the column. *Lip* 2 cm long, ovate when flattened, 18–20 mm long; side-lobes erect, semi-elliptic, papillose or pubescent; mid-lobe 11 × 8 mm, oblong-ovate to triangular-ovate, obtuse to subacute, recurved, papillose,

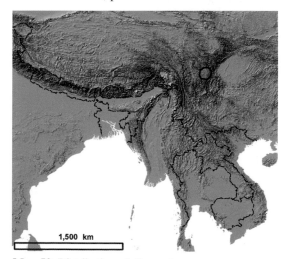

Map 50. Distribution of *C. omeiense*.

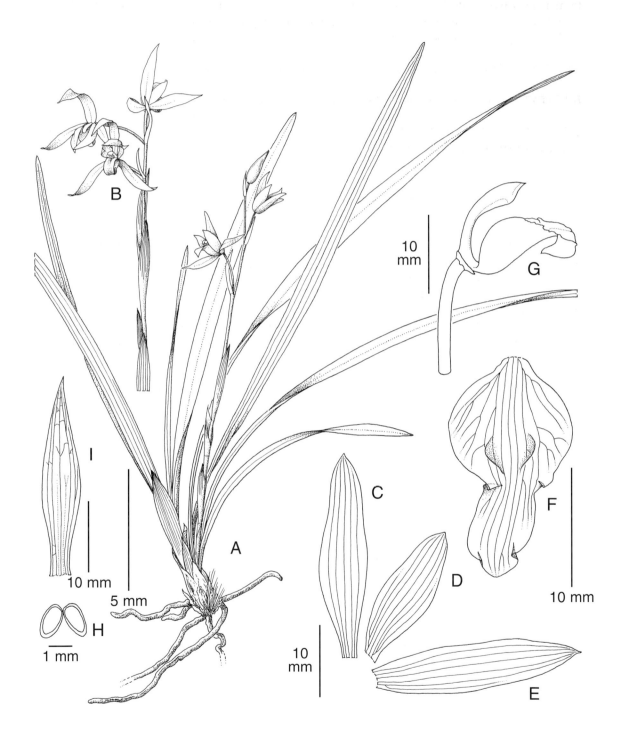

Fig. 171. *Cymbidium omeiense.* **A.** Habit. **B.** Flower. **C.** Dorsal sepal. **D.** Petal. **E.** Lateral sepal. **F.** Lip, flattened. **G.** Column, ovary and lip, side view. **H.** Pollinarium. **I.** Bract. Drawn from *Farges* 944 by Andrew Brown.

hairy or with some papillae, margin entire; callus of 2 parallel ridges, converging in the apical half to form a small tube at the base of the mid-lobe. *Column* 10–11 mm long, arching, winged; pollinia 4, broadly ovate in two unequal pairs, on a small crescent-shaped viscidium. *Capsule* about 5 cm long, fusiform, ridged, beaked, held erect and parallel to the rhachis. Fig. 171.

DISTRIBUTION. Western China: Sichuan (Map 50); 800–1800 m (2625–5905 ft).

HABITAT. In open forest, in shade; flowering in March and April.

CONSERVATION STATUS. DD, insufficient information to make an assessment, although apparently restricted to a single mountain.

48. CYMBIDIUM QIUBEIENSE

Feng & Li described this species in 1980 based upon a plant from Qiubei in south-east Yunnan. This taxon was originally stated to have some affinity with *C. faberi*. However, the leaf number, its short, ovate mid-lobe with a margin that is entire and not undulating, and its indumentum of minute papillae (not strongly inflated), all prevent its inclusion in *C. faberi*.

Du Puy & Cribb (1988) treated *C. qiubeiense* as a synonym of *C. kanran*. It is close to *C. kanran*, but it has 2 or 3 leaves with a wiry petiole and shorter sepals than are normal in that species, but they are still slender and acuminate. It also differs in its vegetative habit and the dark green leaves that are spotted with purple below on the sheaths. The variation between this and *C. kanran* needs to be

Fig. 172. *Cymbidum qiubeiense*, Guangxi.   **Fig. 173.** *Cymbidum qiubeiense*, Cult. Kunming B.G.

explored, but we have no evidence for their conspecificity. Consequently, this species is recognised as distinct in this study, following Chen (1999). Further investigation may clarify its status and distribution.

48. Cymbidium qiubeiense K.M. Feng & H. Li in *Acta Bot. Yunnanica* 2 (3): 334 (1980). Type: China, Yunnan, Qiubei Xian, 1800m, *Qiu Pin-yun 59140* (holotype KUN!).

A terrestrial *herb* with stout roots. *Pseudobulbs* ovoid, 1–1.5 cm long, 0.6–0.9 cm in diameter, covered by 3 or 4, purplish brown to violet, caducous sheaths, 3–5 cm long. *Leaves* 2 or 3, linear, acuminate, 30–80 cm long, 0.5–1 cm wide, serrate, dark green tinged with dull purple; petiole wiry, filiform, 10–20 cm long, violet. *Inflorescence* 5- or 6-flowered; peduncle purple or violet, 25–30 cm long, bearing 2 violet sheaths along length; bracts lanceolate, acuminate, the lowermost 1.5 × 0.5 cm, violet. *Flowers* fragrant, greenish, the petals purple-spotted at base within, the lip white with rose-coloured side-lobes and a purple-spotted mid-lobe; pedicel and ovary 2.5–3.5 cm long. *Sepals* 2.5 × 0.6 cm, linear-lanceolate, acuminate or acute. *Petals* 2.1–2.2 × 0.7 cm, obliquely

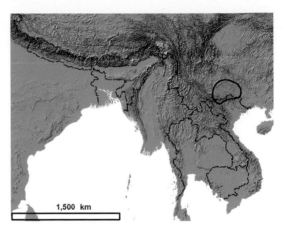

Map 51. Distribution of *C. qiubeiense*.

lanceolate, acute. *Lip* obscurely 3-lobed, 1.9–2 × 1 cm, pandurate when flattened; side-lobes semi-circular; mid-lobe ovate-oblong, obtuse; callus of two fleshy, sub-parallel ridges. *Column* arcuate, 1.2–1.3 cm long, violet-spotted. Figs. 172 & 173.

DISTRIBUTION. China: South-east Yunnan, western Guangxi and south-west Guizhou (Map 51); 700–1800 m (2295–5905 ft).

HABITAT. In woods, especially on limestone; flowering from October to December.

CONSERVATION STATUS. VU A1cd; B1ab.

49. CYMBIDIUM SINENSE

George Jackson described this fine orchid in 1802, as *Epidendrum sinense*, based on a specimen grown in England by G. Hibbert of Clapham Common, which flowered in September 1801. He recorded that the species had been introduced into cultivation into Britain by J. Slater of Leytonstone in 1793, but had not flowered previously. No type specimen has been preserved, and the illustration in the *Botanical Repository* must therefore serve as the type. In 1812, Salisbury had concluded that the *Epidendrum sinense* of Jackson, and also the specimen figured in the *Botanical Magazine* (1805), should be transferred to the genus *Cymbidium*. He named it *C. fragrans*, not realising that the transfer had already been made by Willdenow earlier that same year.

It is a robust plant, usually with 3 or 4 arching, dark green shiny leaves that are comparatively broad (usually 2.5–3.5 cm, 1.0–1.4 in), and large cataphylls (to 13 cm (5 in) long). The flower spike is robust, commonly with 8–15 flowers held well clear of the foliage.

Cymbidium hoosai, described by Makino in 1902 and based on a Chinese specimen cultivated in Japan, is conspecific with *C. sinense* (Wu & Chen, 1980), the original description agreeing well with *C. sinense*.

Plate 35. *Cymbidium sinense*. Cult. Kew, del. Claire Smith.

DAVID DU PUY

Fig. 174. *Cymbidium sinense.* Hong Kong, ex G. Barretto, cult. R.B.G. Kew no. 429-83-05476.

Cymbidium sinense is a widespread orchid, variable in flower colour, which ranges from deep purple-brown through pale and striped to green and lacking any red pigmentation. Several of these variants have been described as distinct taxa. The most common variant in China and Taiwan has dark, purple-brown petals and sepals, with a yellowish, red-spotted lip, but lighter brown variants are known that have dark veins visible, especially in the petals. This latter colour is typical of the specimens from northern India. Specimens from Thailand have light straw-coloured petals and sepals, all with red veins.

Hayata (1914) described *C. albo-jucundissimum* from a specimen cultivated in Taiwan that had green-and-white flowers with a more hirsute lip mid-lobe. This is an alba variant, uncommon in the

Plate 36. *Cymbidium sinense*. Hong Kong, del. J. Eyre.

305

wild, but of horticultural value and common in cultivation. This type of colour variation is found in many species in section *Jensoa*. Although Wu & Chen (1980) recognise this as a distinct variety, it is treated here as a synonym of *C. sinense*. Variety *margicoloratum* Hayata (1916) differs in the colour of the mid-lobe of the lip: it is not spotted, but has a purple-red margin instead. This is a minor colour variant that was again described from a cultivated specimen.

Cymbidium sinense has been cultivated for several centuries in China and Japan, and is particularly popular for its attractive foliage, tall, elegant flower spikes with many strongly scented flowers and its ease of cultivation. Colour variants are highly prized, and are consequently maintained in cultivation, but they may be rare or absent in the wild. Several of these variants have been named as species or varieties. Although they may merit horticultural names, they are not taxonomically significant. Lin (1977) and Su (1975) reported a variant with variegated leaves of horticultural value in Taiwan, where it is known as the 'Golden thread orchid'.

A specimen from Hong Kong, ascribed to *C. ensifolium* var. *munronianum* by Tang & Wang in 1951, has a spike similar to *C. munronianum*, with short, clasping, peduncular sheaths and short, triangular floral bracts. However, the leaves are broader than in *C. munronianum*, and the large size of the flower, and especially of the lip, indicates that the specimen should be included in *C. sinense*. In fact, the length of the floral bracts is a variable character in *C. sinense*. Specimens collected in Meghalaya of northern India and in Yunnan have acuminate bracts up to 3.5 cm (1.4 in) long, whereas those from eastern China and Taiwan are usually shorter, up to 2.0 cm (0.8 in), and those already noted from Hong Kong are only 1 cm (0.4 in) long or less.

It is most readily distinguished from the other species in section *Jensoa* by its vegetative habit and characters of the scape. *Cymbidium munronianum* also has broad leaves, but they are usually narrower than in *C. sinense*, and the scape is weaker, and the individual flowers smaller. *Cymbidium sinense* flowers are usually larger than those of either *C. ensifolium* or *C. munronianum*, especially in the size of the lip. Its sepals are long, narrow and elliptic, and its petals are broader than the sepals, strongly porrect and closely shading the column. Its lip is large, with a broad mid-lobe, and an entire margin that may be kinked, but is never strongly undulating.

Cymbidium kanran has similar narrow sepals and broader petals, but the sepals are much longer, more slender and more acuminate than those of *C. sinense*. Its leaves are longer and narrower, and the scape has fewer flowers.

Cymbidium cyperifolium and *C. faberi* can both be distinguished by their more numerous, narrower leaves, and by the presence of shiny, inflated papillae on the mid-lobe of the lip. The former can usually be further distinguished by its clear green sepals and petals and long floral bracts.

It can be difficult to distinguish the species in section *Jensoa* from the flowers alone. Those of *C. ensifolium* and *C. sinense* are both variable, and are often similar to each other, so that the broad, dark green, shiny leaf of *C. sinense* is usually the easiest character by which these species can be separated. The large, numerous flowers on a long and robust scape, the long, narrow sepals, the petals that are broader than the sepals (see also *C. ensifolium* subsp. *haematodes*) and porrect, covering the column, and the large, broad mid-lobe of the lip of *C. sinense* usually serve to separate the flowers of the two species.

It flowers during January–March in Taiwan and China, and in October–November in northern India. Therefore, the flowering time in northern India does not appear to coincide with the flowering of *C. munronianum* to which it is clearly related. These two species are not recorded as being sympatric, but further study may show that their distributions coincide (see also the discussion under *C. munronianum*, and Table 18 for a comparison between these two species and also *C. ensifolium* subsp. *haematodes*).

49. Cymbidium sinense (Jacks.) Willd., *Sp. Pl.*, ed. 4: 111 (1805). Type: ex China, cult. G. Hibbert, Icon. in Andr., *Bot. Rep.* 3: t. 216 (1802).

Epidendrum sinense Jacks. in Andr., *Bot. Rep.* 3: t. 216 (1802).

C. fragrans Salisb. in *Trans. Hort. Soc.* 1: 298 (1812), *nom. superfl.* Type: as for *C. sinense*.

C. chinense Heynh., *Nomencl.* 2: 179 (1846), *sphalm.* for *C. sinense*.

C. ensifolium sensu Hook.f., *Fl. Brit. India* 6: 13 (1891), *pro parte, non* (L.) Sw.

C. hoosai Makino in *Bot. Mag. Tokyo* 16: 27 (1902). Type: Japan, Musashi Prov., cult. Tokyo Bot. Garden, *Makino s.n.* (holotype TI or MAK).

C. albo-jucundissimum Hayata, *Icon. Pl. Formos.* 4: 74 (1914) & op. cit. 6: 80, t. 13 (1916). Type: Taiwan (Formosa), cult. Taihoku, *Hayata s.n.* (holotype TI).

C. sinense (Jacks.) Willd. var. *margicoloratum* Hayata, *loc. cit.* 6: 82, t. 16B & 17 (1916). Type: Taiwan, cult. Taikhoku, *Soma s.n.* (holotype TI).

C. sinense (Jacks.) Willd. var. *albo-jucundissimum* (Hayata) Masam., in *Trop. Hort.* 3: 31 (1933).

C. ensifolium var. *munronianum sensu* T. Tang & F.T. Wang in *Acta Phytotax. Sin.* 1: 91 (1951), *non* King & Pantling.

C. sinense (Jackson in Andr.) Willd. var. *album* T.C. Yen, *Icon. Cymbid. Amoy.* (1964). Type: China, Kwangtung [Guangdong], *Yen & Hong 1006* (holotype not located).

C. sinense (Jackson in Andr.) Willd. var. *bellum* T.C. Yen, *loc. cit.* (1964). Type: China, Fukien [Fujian], *Cheng & Cheng 1019* (holotype not located).

C. yakusimense Masam., *Native orchids of Nippon* 1: 79 (1984), *nom. nud.*

C. yamagawaense Masam., *Native orchids of Nippon* 6: 19 (1984), *nom. nud.*

A perennial, terrestrial *herb. Pseudobulbs* up to 3 × 2 cm, ovoid, prominent, covered in scarious cataphylls up to 13 cm long that become fibrous with age, and broad, sheathing leaf bases, both with a membranous margin 2 mm broad, with 2–4(5) distichous leaves. *Leaves* broad, (25)40–103 × (1.5)2–3.2 cm, linear-elliptic, arching, dark green, glossy, articulated 4–9 cm from the pseudobulb, the lowest leaves distinct from the cataphylls. *Scape* 40–80 cm tall, erect, robust, arising basally from inside the cataphylls, usually with (6)8–26 flowers produced in the apical half or third of the scape and held above the leaves; peduncle covered in 3–5 or more sheaths up to 10 cm long, usually overlapping each other; bracts 0.4–2.5(4.1) cm long, triangular, acuminate. *Flowers* about 5 cm across; scented; peduncle, rhachis, floral bracts, pedicel and ovary greenish, often stained purple-brown; sepals usually purple-brown, sometimes dark, petals paler, often showing darker veins, occasionally sepals and petals straw-yellow with dark red veins; lip cream or pale yellow, side-lobes streaked red with a solid red margin, mid-lobe heavily spotted and blotched with dark red; column reddish above, cream with red spots below; anther-cap cream. *Pedicel and ovary* 1.9–4 cm long. *Dorsal sepal* 26–39 × 5.7–8.7 mm, elliptic, acute, erect to porrect, margins revolute; lateral sepals similar, spreading. *Petals* 23–34 × 7.1–11 mm, ovate, broader than the sepals, porrect, often forming a hood over the column. *Lip* 2.1–2.9 cm long when flattened; side-lobes erect, rounded, often reduced, papillose or pubescent; mid-lobe broad, 12–16 × 9–13 mm, oblong with an obtuse mucronate apex, to ovate with an acute apex, recurved, often as broad as the side-lobes when the lip is flattened, papillose or with papillose hairs, margin entire and undulating or kinked; callus in 2 ridges, converging in the apical half to form a small tube at the base of the mid-lobe. *Column* 1.2 × 1.6 cm long, arching, winged; pollinia 4, ovate, in two unequal pairs. *Capsule* about 6 cm long, fusiform, ridged, beaked, held erect and parallel to the rhachis. Pls. 35 & 36; Figs. 154: 4a–f, 174.

DISTRIBUTION. Northern India (Meghalaya: Khasia Hills), Myanmar (Burma), northern Thailand, Vietnam, central and southern China, Taiwan, Ryukyu Islands, Vietnam (Map 52); 300–2300 m (985–7545 ft). In Hong Kong, Taiwan, and eastern China, *C. sinense* occurs at elevations of 250–1000 m (820–3280 ft). In northern India, Burma, Yunnan (and probably Thailand) it occurs at higher elevations, usually between 1400–2300 m (4595–7545 ft). In China and Taiwan it grows in humus-rich soil in dense or partial shade, often near streams or water seepages (Lin, 1977; Ying, 1977; Barretto, pers. comm.). In Myanmar, it has been found in mixed deciduous and evergreen forest containing species of *Cephalotaxus*, *Podocarpus*, *Alnus* and *Shorea*. In Thailand, it occurs in semi-shaded spots in lower montane forest.

Cymbidium sinense occurs in the southernmost provinces of China, Hainan, Taiwan (Wu & Chen, 1980; Chen *et al.*, 1999) and Vietnam (Averyanov, 1994).

HABITAT. Open to dense mixed or evergreen forest, in shade or semi-shade in moist well-drained places; flowering October–March.

CONSERVATION STATUS. VU A1cd; B1ab.

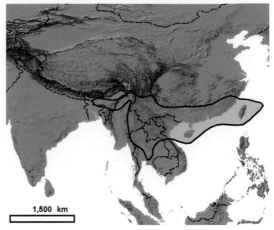

Map 52. Distribution of *C. sinense*.

50. CYMBIDIUM TORTISEPALUM

This species was originally described from Taiwan. It differs from the closely related *C. goeringii* in having longer leaves that are articulated near the base and an inflorescence of 3–7 flowers. It has minutely but sharply serrulate leaf margins, and vegetatively the plant resembles *C. goeringii*.

Although previously considered a variety of *C. goeringii* (Wu & Chen, 1980; Du Puy & Cribb, 1988), Chen & Liu (2003) argued for its recognition as a distinct species based on its having articulated longer leaves and inflorescences with 3–7 flowers. They recognise two varieties, var. *tortisepalum* and var. *longibracteatum*. We follow their treatment here.

Cymbidium tortisepalum is widespread across China, particularly in the South-west and in Taiwan. It is widely cultivated, and many selected forms of it are grown, some highly prized.

50. C. tortisepalum Fukuy. in *Jap. Bot. Mag.* 48: 304–306, t. 1 (1934); Ying in *Quart. J. Chinese Forestry* 11(2): 100–101 (1976); Lin, *Nat. Orchids Taiwan* 2: 129, t. 56–59, + fig. (1977); Mark, Ho & Fowlie in *Orchid Dig.* 50: 24, + figs. (1986). Type: Taiwan, *Fukuyama 3983* (holotype KANA).

A terrestrial *herb* with thick, fleshy roots, 5–10 mm in diameter. *Pseudobulbs* ovoid, 1–2 cm long, 0.5–1 cm in diameter, enclosed in about 6 scarious cataphylls. *Leaves* 5–7(10), 40–65 × 0.4–1.8 cm, the shortest merging with the cataphylls, linear-elliptic, acute, usually V-shaped in section, arching, usually with a serrated margin, narrowed towards the base. *Scape* erect, (2)3–7-flowered, 20–30 cm tall; peduncle up to 15(21) cm, covered in 4–8 sheaths; sheaths up to 6(9) cm long, becoming scarious, cylindrical in the basal half, expanded and cymbiform in the upper half, acute; bracts 2.5–6 cm long, linear-lanceolate, usually exceeding the ovary, scarious, cymbiform, acute, often red-tinted. *Flowers* about 4–5 cm across, porrect or nodding, sometimes scented; pedicel and ovary green to

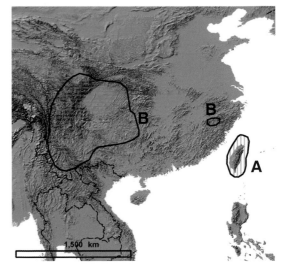

Map 53. Distribution of *C. tortisepalum*. **A.** var. *tortisepalum*; **B.** var. *longibracteatum*

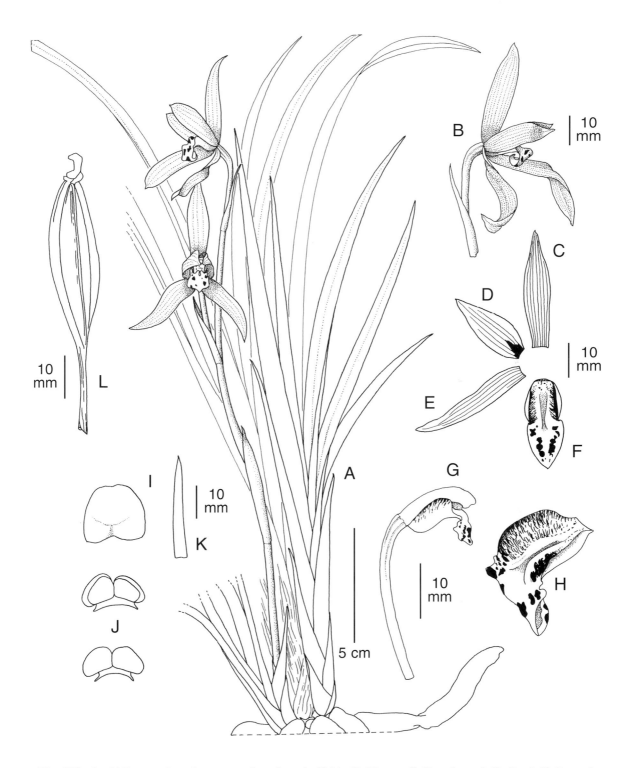

Fig. 175. *Cymbidium tortisepalum* var *tortisepalum*. **A.** Habit. **B.** Flower. **C.** Dorsal sepal. **D.** Petal. **E.** Lateral sepal. **F.** Lip, flattened. **G.** Lip, column and ovary, side view. **H.** Lip. **I.** Anther-cap. **J.** Pollinarium, two views. **K.** Bract. **L.** Capsule. Drawn from photographs and the type by Andrew Brown.

purplish; sepals and petals whitish green to yellowish green; lip cream with crimson spots and margins on the side-lobes and sparse red blotches on the mid-lobe; callus ridges cream to pale yellow; column usually pale green, cream towards the base, spotted lightly above and densely below with maroon; anther-cap cream, sometimes purple or yellow below. *Pedicel and ovary* 2.5–6 cm, curved behind the flower. *Sepals* usually oblong to oblong-lanceolate, sometimes twisted, acute, margins often incurved; dorsal sepal porrect, 3–3.9 × 0.7–0.8 cm; lateral sepals similar, spreading. *Petals* 2.5–3 × 0.8–0.9 cm, oblong-ovate to oblong-lanceolate, acute, oblique, porrect and usually closely covering the column. *Lip* 1.7–2 × 0.8–1 cm, ovate to elliptic when flattened, subentire to 3-lobed; side-lobes sometimes reduced, erect, rounded, sometimes angled at the front, papillose; mid-lobe 1–1.1 × 0.9–1 cm, broadly ovate to ovate, obtuse or rounded, recurved, papillose, the margin sometimes undulate; callus 2-ridged, convergent towards the apex and forming a short tube that extends to the base of the mid-lobe, with two large pads of callus on the side-lobes adjacent to the callus ridges. *Column* 1.4–1.5 cm long, broadening into two wings towards the apex; pollinia 4, in 2 unequal pairs. *Capsule* up to 8 cm long, fusiform, erect, and parallel to the rhachis, pedicellate, beaked.

Fig. 176. *Cymbidium tortisepalum* var. *tortisepalum*. Taiwan.

DISTRIBUTION. China: Fujian, Guizhou, Sichuan, Yunnan,Taiwan (Map 53).

Key to the varieties of *C. tortisepalum*

Leaves soft in texture, strongly arching, 4–12 mm in width; bracts as long as or shorter than the pedicel and ovary .... var. ***tortisepalum***
Leaves rigid in texture, suberect, 13–18 mm in width; bracts longer than pedicel and ovary ....
.... var. ***longibracteatum***

var. **tortisepalum** (Fukuy.) Y.S. Wu & S.C. Chen in *Acta Phytotax. Sin.* 18: 300 (1980).

C. longibracteatum Y.S. Wu & S.C. Chen var. *tortisepalum* (Fukuy.) Y.S. Wu in *Acta Phytotax. Sin.* 9: 31 (1966).

C. tsukengensis C. Cheng in *Taiwan Orchid Bull.* 8: no. 2 (1970) & *Formosan Orchids*, ed. 2.: 41 (1974); Mark *et al.* in *Orchid Dig.* 50: 27, 29, + figs. (1986); *nom. inval.*

C. tortisepalum Fukuy. var. *viridiflorum* S.S. Ying, *Coloured Ill. Ind. Orch. Taiwan* 1: 415 (1977). Type: Taiwan, Taichung Prov., Lishan, *Ying 5282* (holotype not located).

C. goeringii (Rchb.f.) Rchb.f. var. *tortisepalum* (Fukuy.) Y.S. Wu & S.C. Chen in *Acta Phytotax. Sin.* 18(3): 300 (1980).

C. goeringii var. *tortisepalum* f. *albiflorum* S.S. Ying in *Quart. J. Chinese Forestry* 20(2): 54 (1987), *nom. inval.* lacking type.

C. lianpan T. Tang & F.T. Wang ex Y.S. Wu, *nom. nud.*

C. tsukengensis C. Cheng in *Taiwan Orchid Bull.* 8: no. 2 (1970), *nom. nud.*

Leaves soft-textured, arcuate, 30–65 × 4–12 cm. Floral bracts lanceolate, longer than or about as long as the pedicel and ovary. Flowers (2)3–5(or rarely 7). Sepals and petals twisted or not. Figs. 175 & 176.

DISTRIBUTION. Taiwan, ?western Yunnan; 800–2500 m (2625–8200 ft). In Taiwan it occurs mainly on steep slopes up to about 1500 m (4920 ft).

HABITAT. It prefers an open habitat on steep slopes, often growing in full sun, in association with *Miscanthus* and other grasses in rocky meadows, also in open forests and on forest margins.

The variability of the flowers of this variety in Taiwan is well illustrated by Mark *et al.* (1986, as *C. tortisepalum* and *C. tsukengensis*), who photographed several dissimilar specimens. They also noted that it has an appealing scent.

The name *C. tsukengensis* was first used by Chow Cheng (1970), but was not validly published. Mark *et al.* (1986) recently resurrected it, but no attempt was made to publish it formally. It is applied to specimens from high elevations in Taiwan, near the highest elevations at which var. *tortisepalum* occurs. It often has a 1- or 2-flowered scape, intermediate in this character between *C. goeringii* and *C. tortisepalum*. Its flowers do not appear to be outside of the range of variation expected in var. *tortisepalum*, and it is certainly not distinct enough to be regarded as a distinct species.

It is possible that *C. tortisepalum* is a product of introgression following hybridisation with one of the other species from sect. *Jensoa* such as *C. ensifolium* or *C. kanran*, especially when the wide variation in its bract length and flower colour are taken into consideration (see also Table 19). Field studies are required to investigate this possibility. There is evidence in Taiwan of hybridisation between *C. goeringii* and several other species. Cheng (1979) illustrated *C. tortisepalum* and several other plants that appear to be intermediate between *C. goeringii* and possibly *C. ensifolium*.

var. **longibracteatum** (Y.S. Wu & S.C. Chen) S.C. Chen & Z.J. Liu in *Acta Phytotax. Sin.* 41, 1: 81 (2003). Type: China, Sichuan, Chiou-lai-shan, *Fee 2064* (holotype PE); W Sichuan, Dujiangyan, *Z.J. Liu 22318* (neotype PE).

C. longibracteatum Y.S. Wu & S.C. Chen in *Acta Phytotax. Sin.* 11, 1: 31, t. 5 (1966).

C. goeringii Rchb.f. var. *longibracteatum* (Y.S. Wu & S.C. Chen) Y.S. Wu & S.C. Chen in *Acta Phytotax. Sin.* 18(3): 300 (1980).

C. longibracteatum var. *flaccidifolium* Y.S. Wu, *Chinese Cymbidium* (ed. 2): 138 (1993), *nom. inval.*

C. longibracteatum var. *rubisepalum* Y.S. Wu, *Chinese Cymbidium* (ed. 2): 138 (1993), *nom. inval.*

Fig. 177. *Cymbidium tortisepalum* var. *longibracteatum*, cult. Sichuan.

TABLE 19

C. TORTISEPALUM VAR. *TORTISEPALUM* COMPARED WITH *C. GOERINGII* VAR. *GOERINGII* AND *C. ENSIFOLIUM*

Character	*C. goeringii*	*C. tortisepalum* var. *tortisepalum*	*C. ensifolium*
Leaf number and arrangement	5–6(8), the lowest leaves short, and merging with the cataphylls	5–8, the lowest leaves short and merging with the cataphylls	3–4, all leaves long, cataphylls and leaves well separated
Peduncle and sheaths	peduncle covered in overlapping sheaths	sheaths overlapping in the basal half of the peduncle	sheaths overlapping towards the base of the peduncle
Floral bract length	bract longer than the ovary	bract almost equalling or shorter than the ovary	bract shorter than the ovary
Ovary	ovary erect, bent towards the tip	ovary almost erect, almost parallel to the rhachis, bent towards the tip	ovary angled away from the rhachis, not bent at the tip
Flower number	1(2)	2–4(5)	3–9
Petals	petals porrect, closely covering the column	petals porrect, closely covering the column	petals forward pointing, slightly spreading, not closely covering the column

C. longibracteatum var. *tonghaiense* Y.S. Wu, *Chinese Cymbidium* (ed. 2): 138 (1993), *nom. inval.*

Leaves rigid, erect, 50–70 × 1.2–1.5 cm. Floral bracts longer than the pedicel and ovary, often enveloping the ovary. Flowers 3–5(–7). Sepals and petals not twisted. Fig. 177.

DISTRIBUTION. Southern China (northern Fujian, Guizhou, Sichuan, Yunnan); 1000–2500 m (3280–8200 ft).

HABITAT. Rocky, grassy and scrubby slopes and light woodland.

Wu & Chen (1966) described *C. longibracteatum* from Sichuan, but later reduced it to varietal rank under *C. goeringii*. Chen & Liu (2003) resurrected *C. tortisepalum* and treated this taxon as a variety of it. They distinguished var. *longibracteatum* by its more rigid suberect, broader leaves, longer floral bracts that exceed the ovary in length, the absence of the slight twist in the sepals and the differently shaped mid-lobe of the lip.

Liu and Chen described *C. teretipetiolatum* in 2002, and distinguished it by its leaves with a cylindrical tubular petiole and flower with an unlobed or obscurely lobed lip that lacks a callus. The flowers probably represent no more than a developmental or genetic abnormality. In its habit and floral structure it is close to var. *longibracteatum* and may belong here. Good photographs of the habit and flowers have been published by Liu *et al.* (2006)

SECTION PACHYRHIZANTHE

Section **Pachyrhizanthe** Schltr. in *Fedde, Repert. Sp. Nov. Regni Veg. Beih.* 4: 73 (1919). Type: *C. aberrans* (Finet) Schltr. (= *C. macrorhizon* Lindl.), lectotype chosen by Seth & Cribb (1984).
C. section *Macrorhizon* Schltr. in *Fedde, Repert. Sp. Nov. Regni Veg.* 20: 99–101 (1924). Type: *C. macrorhizon* Lindl.
C. section *Geocymbidium* Schltr. in *Fedde, Repert. Sp. Nov. Regni Veg.* 20: 101 (1924). Type: *C. lancifolium* Hook., lectotype chosen by P.F. Hunt (1970).
Pachyrhizanthe (Schltr.) Nakai in *Bot. Mag. Tokyo* 45: 109 (1931).

Cymbidium macrorhizon has flowers that closely resemble those of *C. lancifolium* and they are only separated by the heteromycotrophic habit and unusual roots of the former. Nakai (1931) justified the removal of *C. macrorhizon* to a new genus, *Pachyrhizanthe*, because of its subterranean rhizome, lacking leaves, pseudobulbs or roots. Garay & Sweet (1974) stated that these criteria did not justify Nakai's decision because autotrophic and saprophytic species are known in several other genera. The DNA sequence analyses by Yukawa *et al.* (2002) and van den Berg *et al.* (2002) indicate that this section forms a well-supported clade sister to section *Jensoa*.

51. CYMBIDIUM LANCIFOLIUM

William Hooker based his original description of *C. lancifolium* in 1823 on a cultivated plant that had been collected by Nathaniel Wallich in Nepal. A specimen in Hooker's herbarium at Kew, labelled "Dr Wallich", matches the original illustration of *C. lancifolium* published by Hooker and is considered to be the type specimen. It is characterised by its distinctive, petiolate leaves with a relatively broad,

Plate 37. *Cymbidium lancifolium* var. *aspidistrifolium*. *Andrew* s.n., cult. Kew, del. Claire Smith.

elliptic lamina, its superposed, cigar-shaped pseudobulbs, its ascending habit, and its delicate, white or greenish flowers in a short, erect spike produced from the central nodes of the pseudobulb. Its habit and flowers indicate a relationship with the saprophytic *C. macrorhizon* which, however, has an elongate rhizome, lacks leaves, pseudobulbs, well-developed roots and chlorophyll and is apparent only when its flowers emerge from the substrate. *Cymbidium lancifolium* is morphologically rather variable, as might be expected from such a widespread species, and these two factors are undoubtedly responsible for its extensive synonymy. Variation is particularly apparent in the Chinese, Taiwanese and New Guinea representatives of this species.

In 1825, Carl Blume described two species from Java, *C. cuspidatum* and *C. javanicum*, each differing slightly in leaf shape, but he compared neither with *C. lancifolium*. Their types differ little from one another or from *C. lancifolium* and most authors, following J.J. Smith (1905), have considered them to be conspecific with *C. lancifolium*. The leaf margin has been used by several authors to distinguish taxa in this group. Although not mentioned in the protologue, the leaves of the type of *C. lancifolium* have smooth margins. Maekawa (1958) examined the two sheets of *C. javanicum* and selected the left hand plant on sheet H.L.B. 902.322-1227 as lectotype, it being the only specimen with smooth margins to its leaves, the other three having serrulate margins.

Cymbidium gibsonii, described by Lindley from a cultivated specimen imported from Meghalaya in NE India, was distinguished by its 'fusiform, jointed, naked stem'. However, although the pseudobulbs in *C. lancifolium* are covered by cataphylls when they are young, as the pseudobulbs swell the cataphylls split and become scarious, soon disintegrating and often leaving the pseudobulbs exposed. It also has smooth-margined leaves.

In 1925 Rolfe based *C. kerrii* on a broad-lipped specimen with smooth-margined leaves collected by Arthur Kerr in northern Thailand. Seidenfaden (1983) compared the Thai material of *C. kerrii* and *C. lancifolium*, and concluded that the variation in lip shape was continuous and did not justify the recognition of two species.

A plant with a similarly broad and obscurely 3-lobed lip led Masamune to describe the Japanese *C. nagifolium* in 1930. It has serrulate margins to its leaves. Broad-lipped specimens have also been collected elsewhere, for example at high altitudes in Sabah (*Lamb* SAN 91582).

Cymbidium maclehoseae, described from Hong Kong by S.Y. Hu in 1972, and *C. robustum*, described from New Guinea by Gilli in 1983, are based on unusually large specimens with smooth leaf margins. It is evident that these variants fall within the range of variation encountered in *C. lancifolium* in several distant parts of its range, and they cannot be considered as specifically distinct. Maekawa's *C. lancifolium* var. *pantlingii* from Sikkim also has entire leaf margins.

In summary, only plants with leaves with smooth margins have been found in the Himalayas, Myanmar, and Indo-China. Only plants with serrulate leaves have been found in Sumatra, Japan (excluding Ryukyu islands), and the Khasia Hills. Seidenfaden (1983) reported both types of leaf margin in Thailand, while we have found both in Java, Taiwan and China.

The types of *C. caulescens* and of *C. rhizomatosum* are similar, both having long rhizomes and sterile growths of two small leaves, more or less well-separated from the longer fertile shoots. The former was collected from Koh Samoi off the coast of peninsular Thailand by H. Robinson and was described by Henry Ridley in 1915. The latter was collected in Malipo County of Yunnan by Z.J. Liu in September 2002 and was described by S.C. Chen and Liu in the same year. Both represent, in our opinion, extreme forms of *C. lancifolium*, probably the result of environmental stress possibly associated with drought in their limestone habitats. We also include *C. multiradicatum* Liu & Chen in the synonymy of *C. lancifolium*. The authors describe it as a saprophyte but its rhizome is glabrous, lacking the nodules characteristic of the rhizome of *C. macrorhizon*, whereas its roots are pubescent, suggesting that it cannot be one. Rather it is the extreme expression of the reduction of leaf-size found in the previous two taxa. Its flowers and fruits are characteristic of *C. lancifolium* and, on balance, we consider this a leafless form of that species. Whether its habit is genetically or environmentally induced is debatable and needs further observation.

The variation of the *C. lancifolium* complex in Taiwan was examined and documented by Mark *et al.* (1986). They recognised four species, all of which are considered here to be conspecific with *C. lancifolium*. The largest variant with the strongest scape was treated as *C. syunitianum* (see previously), and the medium-sized plants with greenish flowers as *C. aspidistrifolium*. They also described a new species, *C. bambusifolium*, to include the smallest variants with very weak scapes and pale green flowers. All three have entire leaf margins. A fourth, *C. nagifolium*, has serrulate leaf margins and whitish flowers. Each was further differentiated by the season of the year at which it usually flowered. There is undoubtedly great variation in plant size, flower colour and shape, and flowering time in Taiwan, but the distinctions used by Mark *et al.* to recognise their taxa appear to vary continuously from one extreme to the other, and become particularly confused when compared with specimens from outside of Taiwan. The green-flowered specimens with spreading lateral sepals are all considered here to belong to var. *aspidistrifolium* (see below). However, critical study of this species in the Far East and SE Asia is needed and may produce a more useful infraspecific classification.

51. Cymbidium lancifolium Hook., *Exot. Flora* 1: t.51 (1823). Type: Nepal, *Wallich* s.n., cult. Shepherd (holotype K!).

C. cuspidatum Blume, *Bijdr.* 8: 379 (1825); Lindley, *Gen. Sp. Orch. Pl.*: 170 (1833). Type: Java, Salak and Cereme Mts, *Blume* s.n. (holotype L!).

C. javanicum Blume, *Bijdr.* 8: 380, t. 19 (1825). Type: Java, Mt. Seribu, *Blume* s.n. (holotype L!).

C. caulescens Ridl. in *J. Fed. Mal. St. Mus.* 5: 167 (1915). Type: Thailand, Ko Samui Isl., *Robinson* s.n. (holotype SING, isotype K!).

C. gibsonii Lindl. in *Paxt. Fl. Gard.* 3: 144 (1852–3). Type: Assam, Khasia Hills, cult. Paxton (holotype K!).

C. kerrii Rolfe in *Kew Bull.* 1925: 381–382 (1925). Type: Thailand, Doi Suthep, *Kerr* 227 (holotype K!).

C. nagifolium Masam. in *Jap. Bot. Mag.* 44: 220 (1930). Type: Japan, Yakushima, *Masamune* s.n. (holotype TI!).

C. javanicum Blume var. *pantlingii* F. Maek. in *J. Jap. Bot.* 33: 320 (1958). Type: Sikkim, near Sureil, *Pantling* 75 (holotype K!).

C. maclehoseae S.Y. Hu in *Chung Chi J.* 11: 15, fig. 2 (1972) & Gen. Orchid. Hong Kong: 96 (1977). Type: Hong Kong, New Territories, *S.Y. Hu* 9369 (holotype A).

C. robustum Gilli in *Ann. Naturhist. Mus. Wien* 84: 22–3 (1983). Type: Papua New Guinea, Kompiam, *Gilli* G546 (holotype W).

C. rhizomatosum S.C. Chen & Z.J. Liu in *J. Wuhan Bot. Res.* 20, 6: 421, fig. 1 (2002), **syn. nov.** Type: China, Malipo Co., 15 Sept. 2002, *Z.J. Liu* 2559 (holotype SZWN).

C. multiradicatum Z.J. Liu & S.C. Chen in *Acta Bot. Yunn.* 26, 3: 297 (2004), **syn. nov.** Type: China, Yunnan, Malipo, *Z.J. Liu* 2614 (holotype SZWN).

A medium- to small-sized, perennial, terrestrial *herb*. *Roots* thick, fleshy, whitish, often visible as stilt-like supports keeping the rest of the plant well above the substrate. *Pseudobulbs* 3–15 × 0.4–1.5 cm, raely absent, narrowly fusiform, slightly bilaterally flattened, erect, closely spaced and often crowded towards the apex of a rhizomatous stem; the new pseudobulb is formed annually from a shoot produced slightly above the base of the mature pseudobulb, causing the plant to grow at an angle of about 45° to the horizontal. The strongly bilaterally flattened new growths are initially formed of 6–9 folded, sharply keeled, acute, distichous, overlapping cataphylls which are largest and most leaf-like towards the apex of the growth, where 2–4(5) true leaves (rarely absent) are eventually produced. The cataphylls are green at first, but became scarious and then fibrous as the pseudobulb inflates, eventually disintegrating, leaving the old pseudobulbs exposed. *Leaves* up to 9–50(60) × (1.3)1.9–5.5 cm, narrowly obovate to elliptic, the margin sometimes finely serrated towards the acute to acuminate apex, suberect to

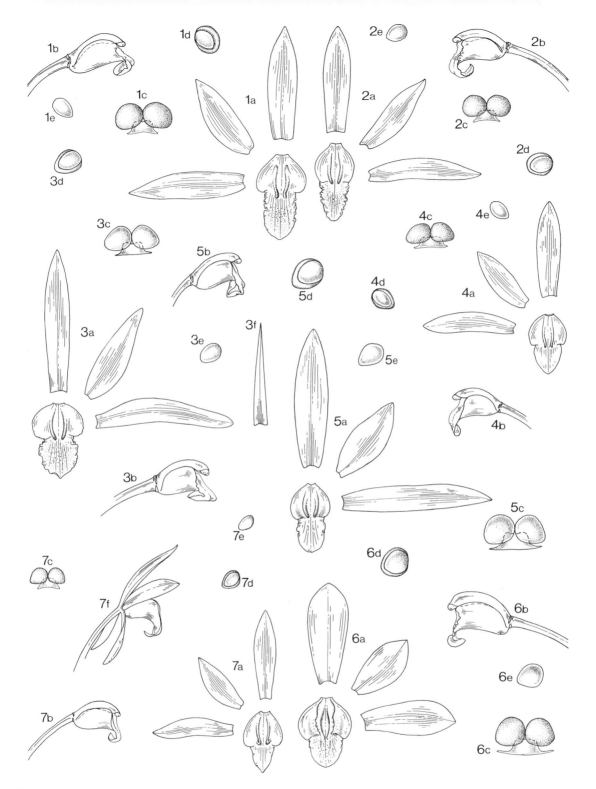

Fig. 178. *Cymbidium faberi* var. *szechuanicum* (Sichuan, *Henry* 5515). **1a.** Perianth, × 1. **1b.** Lip and column, × 1. **1c.** Pollinarium, × 4.5. **1d.** Pollinia (one pair, reverse), × 4.5. **1e.** Pollinium, × 4.5. *C. faberi* var. *faberi* (Yunnan, *Cavalerie* 2233). **2a.** Perianth, × 1. **2b.** Lip and column, × 1. **2c.** Pollinarium, × 4.5. **2d.** Pollinia (one pair, reverse), × 4.5. **2e.** Pollinium, × 4.5. *C. faberi* var. *szechuanicum* (Nepal, *Bailes* 1040; Kew spirit no. 49390). **3a.** Perianth, × 1. **3b.** Lip and column, × 1. **3c.** Pollinarium, × 4.5. **3d.** Pollinia (one pair, reverse), × 4.5. **3e.** Pollinium, × 4.5. **3f.** Bract, × 1. *C. lancifolium* var. *aspidistrifolium* (Kew spirit no. 48293). **4a.** Perianth, × 1. **4b.** Lip and column, × 1. **4c.** Pollinarium, × 4.5. **4d.** Pollinia (one pair, reverse), × 4.5. **4e.** Pollinium, × 4.5. *C. goeringii* (Kew spirit no. 48256). **5a.** Perianth, × 1. **5b.** Lip and column, × 1. **5c.** Pollinarium, × 4.5. **5d.** Pollinia (one pair, reverse), × 4.5. **5e.** Pollinium, × 4.5. *C. goeringii* (Kew spirit no. 26171). **6a.** Perianth, × 1. **6b.** Lip and column, × 1. **6c.** Pollinarium, × 4.5. **6d.** Pollinia (one pair, reverse), × 4.5. **6e.** Pollinium, × 4.5. *C. macrorhizon* (Kew spirit no. 40631). **7a.** Perianth, × 1. **7b.** Lip and column, × 1. **7c.** Pollinarium, × 4.5. **7d.** Pollinia (one pair, reverse), × 4.5. **7e.** Pollinium, × 4.5. **7f.** Flower, × 1. All drawn by Claire Smith.

horizontal, narrowing to a slender conduplicate petiole, articulated to an expanded, sheathing base 0.5–8 cm from the pseudobulb. *Scape* 7–35 cm long, with (2)4–8 flowers, erect, produced laterally on the pseudobulb from the axils of the cataphylls; peduncle covered in 5–7 overlapping sheaths up to 1.2–3.4 cm long, cymbiform, acute, with a cylindrical, sheathing base and an inflated, spreading apex; bracts 0.4–1.6(3.5) cm long, lanceolate, cymbiform, acute, becoming scarious. *Flowers* 2.5–5 cm across; not usually scented; rhachis, pedicel and ovary usually pale green; sepals and petals usually white to pale green, occasionally apple-green, with a central maroon stripe and spotting over the mid-vein which does not reach the apex, and may be weak or absent in the sepals; lip white, pale green or pale yellow, with red spots and blotches on the mid-lobe and purple-red stripes on the side-lobes which become confluent at the margin; callus white, sometimes finely speckled with red; column pale green, streaked purple-red below; anther-cap cream. *Pedicel and ovary* 1.9–4 cm long. *Dorsal sepal* 1.7–3 × 0.3–0.8 cm, narrowly oblong to obovate, acute or apiculate, erect; lateral sepals similar, oblique, spreading. *Petals* 1.5–2.3(3) × 0.4–0.8(1) cm, oblong to narrowly elliptic, oblique, acute or apiculate, porrect and tending to cover the column. *Lip* 1.4–2 × 0.8–1.6 cm, subentire to strongly 3-lobed, minutely papillose; side-lobes erect, rounded to obtuse at the apex, usually not well differentiated from the mid-lobe; mid-lobe 0.6–1.2 × 0.6–1.4 cm, broadly rounded or ovate to ligulate, usually acute to mucronate, recurved, occasionally hooded, margin entire; callus ridges 2, convergent towards the apex, forming a short tube which extends into the base of the mid-lobe, occasionally crenulate at the apex. *Column* 1–1.5 cm long, slender, arching, winged towards the apex; pollinia 4, on a broadly crescent-shaped viscidium. *Capsule* 4.5–5 × 1–1.4 cm, fusiform to club-shaped, pedicellate, held erect and parallel to the rhachis, retaining the column which forms a short apical beak.

DISTRIBUTION. N India (Himalaya, Meghalaya, Sikkim), Nepal, Bhutan, Burma, S & C China, Taiwan, Korea, Ryukyu Islands, Japan, Indo-China, W Malaysia, Java, Sumatra, Borneo, Moluccas, and New Guinea (Map 54); 300–2300 m (985–7545 ft). In its more northerly localities in Japan, Taiwan and China it is found growing at between 300 m and 1800 m (985–5905 ft) elevation. In N India, SW China, Burma and Thailand it occurs between 1000 and 2300 m (3280–7545 ft). In Malaya, the Malay Archipelago and New Guinea it has been recorded from a wide altitudinal range, from as low as 300 m (985 ft) in Sabah, up to 2000 m (6560 ft) in New Guinea.

HABITAT. It usually grows in deep shade in broad-leaved forest, usually in rich soil and deep humus and leaf litter, often in conjunction with tree roots or rotting wood on the forest floor. In China and northern Vietnam it is common in monsoonal mixed forests on limestone. In Java, it grows in montane forest in similarly shaded positions in leaf litter, often on steep slopes and ridge tops. In Thailand it has been found growing near creek beds, in very deep shade where there is little competition from other herbs. The plant is often supported above the substrate on long, stilt-like roots, with each new pseudobulb being produced slightly above the previous one, allowing the plant to survive in the accumulating leaf litter. The broad leaves also seem to be well adapted to the low light intensities of the forest floor. The flowering period is very variable, but is generally April–October in the more northern, seasonal localities, and is sporadic, throughout the year in the southern, tropical localities.

CONSERVATION STATUS. NT.

Key to the varieties of *C. lancifolium*

1. Plants with inflorescences usually as long as or shorter than the leaves; rhizome generally absent, short or obscure **2**
   Plants with inflorescences almost twice as long as the leaves; rhizome well-developed, elongate; lip with a small, concave, hook-like tip .... var. **papuanum**

2. Flowers white to off-white; lateral sepals falcate; petals more or less covering column; usually flowering in summer .... var. **lancifolium**

Flowers pale green to green; lateral sepals not falcate; petals somewhat spreading, not covering the column; usually flowering in autumn /winter .... var. **aspidistrifolium**

var. **lancifolium**

Plant large. Leaf margins entire or serrate. Inflorescence as long as the leaves or shorter. Flowers white, off-white or pale green marked with purple on the petals and lip. Sepals and petals relatively thin-textured. Figs. 180 & 183.

DISTRIBUTION. Throughout the range of the species.

var. **aspidistrifolium** (Fukuy.) S.S. Ying, *Col. Ill. Indig. Orchids Taiwan* 1: 439 (1977). Type: Taiwan, Mt. Syoagyoku-san, *Fukuyama* 4137 (holotype KANA).

Fig. 179. *Cymbidium lancifolium* var. *aspidistrifolium.* Vietnam, Vinh Phu.

Fig. 180. *C. lancifolium* var. *lancifolium.* Vietnam, Cao Bang.

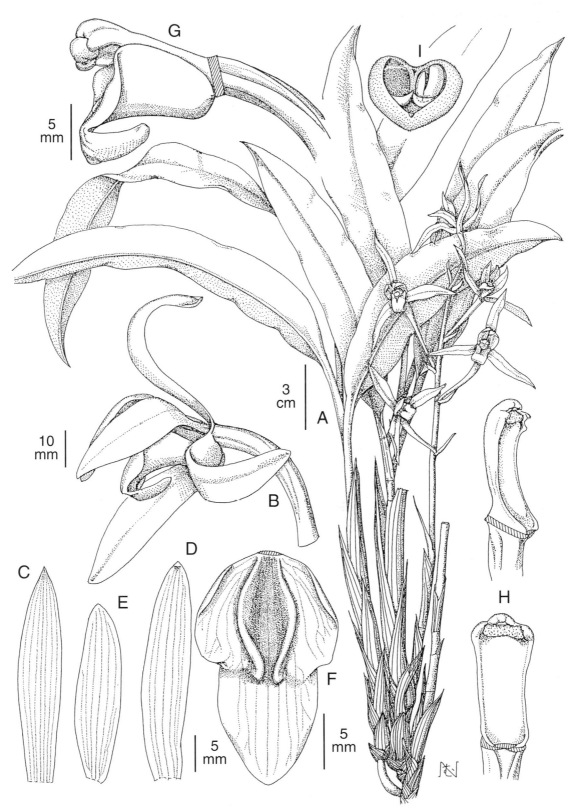

Fig. 181. *Cymbidium lancifolium* var. *aspidistrifolium*. **A.** Habit. **B.** Flower. **C.** Dorsal sepal. **D.** Petal. **E.** Lateral sepal. **F.** Lip, flattened. **G.** Column, ovary and lip, side view. **H.** Column, three views. **I.** Anther-cap. **J.** Pollinarium. Drawn from a living collection by Mutsuko Nakajima.

C. aspidistrifolium Fukuy. in *Bot. Mag. Tokyo* 48: 438, t. 213 (1934).

C. syunitianum Fukuy. in *Bot. Mag. Tokyo* 49: 757–758 (1935). Type: Taiwan, Kwarenko, Mt. Taroko-taizan, *S. Sasaki 4688* (holotype KANA).

C. javanicum Blume var. *aspidistrifolium* (Fukuy.) F. Maek. in *J. Jap. Bot.* 33: 320 (1958) & in *Wild Orchids of Japan in Colour*: 400, 479, t. 163 (1971).

C. lancifolium Hook. var. *syunitianum* (Fukuy.) S.S. Ying, *Col. Ill. Indig. Orchids Taiwan* 1: 439 (1977).

C. bambusifolium Fowlie, Mark & Ho in *Orchid Dig.* 50: 19, + figs. (1986). Type: Taiwan, nr. Taichung City, *Ho Fu-Shun s.n.*, cult. Fowlie *et al.* FMH 83 T8 (holotype UCLA).

Plant usually large. Leaf margins entire. Inflorescence as long as the leaves or shorter. Flowers pale green or green marked with purple on the petals and lip. Sepals and petals relatively thick-textured. Pl. 37; Figs. 178: 4a-e; 179 & 181.

DISTRIBUTION. Taiwan, Japan (Ryukyu Islands) and China and possibly elsewhere in SE Asia.

The name *C. aspidistrifolium* is currently widely used for plants in cultivation originating from Taiwan. Fukuyama discovered it in 1934, and distinguished it from *C. lancifolium* by its green rather than white flowers with oblong, thick-textured sepals and petals, entire leaf margins, and autumn rather than summer flowering period. However, as indicated above, *C. lancifolium* can also have smooth leaf margins. Maekawa (1958), Ohwi (1965) and Su (1975) and Liu & Su (1978) all treat the Taiwanese taxon as *C. lancifolium* var. *aspidistrifolium*, distinguished mainly by its green flower colour and its late flowering season. In Japan, China and Taiwan this variety flowers in autumn and winter, rather than in summer.

A robust variant from Taiwan was described by Fukuyama in 1935 as *C. syunitianum*. It also has entire leaf margins. The more recently described *C. bambusifolium* seems to be no more than a dwarf variant of this taxon.

var. **papuanum** (Schltr.) P.J. Cribb & Du Puy **comb. et var. nov.**

C. papuanum Schltr. in *Fedde, Repert. Sp. Nov. Regni Veg., Beih.* 1: 952 (1913) & in *op. cit.* 21: t. 336, no. 1296 (1928). Type: New Guinea, Bismack Mts, *Schlechter 18680* (holotype B†).

Distinguished by its elongated rhizome and inflorescences which greatly overtop the leaves. Fig. 182.

DISTRIBUTION. Highland New Guinea only.

HABITAT. It grows in broad-leaved forest, for example in *Castanopsis* forest (Reeve, 1984). It has often been reported that its roots are anchored on mossy tree roots or on rotting wood buried in the substrate.

Fig. 182. *Cymbidium lancifolium* var. *papuanum*, Papua New Guinea.

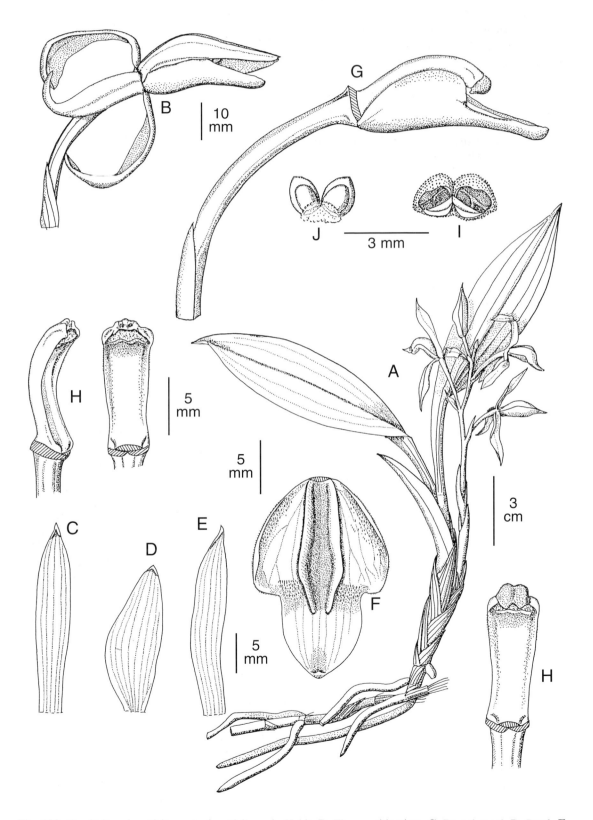

Fig. 183. *Cymbidium lancifolium* var. *lancifolium*. **A.** Habit. **B.** Flower, side view. **C.** Dorsal sepal. **D.** Petal. **E.** Lateral sepal. **F.** Lip, flattened. **G.** Column, ovary, lip and bract, side view. **H.** Column, two views. **I.** Anther-cap. **J.** Pollinarium. A: Kanagawa Pref., Japan (TI); B-J: Amani-Oshima Isl. (Spirit coll, TI). Drawn by Mutsuko Nakajima.

In 1984 Reeve collected plants agreeing with Schlechter's (1913) description of *C. papuanum* from New Guinea. Unfortunately the type specimen, in the Berlin herbarium, was destroyed during the war, leaving some doubt as to the true identity of this variant. Schlechter stated that his species differed from *C. lancifolium* by the much smaller habit of the plant, its pale yellow flowers and by the lip. Although his description of the lip is not unusual for this species, his later published drawing of the flower (Schlechter, 1928) shows the lip as having a concave, hooded apex. These three characters are found in the specimens collected by Reeve. He further differentiated this variant by its inflorescence being longer than the leaves, its slightly drooping flowers which do not open fully, and by the production of distinctive, long, creeping rhizomes which extend through the leaf litter in which the plants grow.

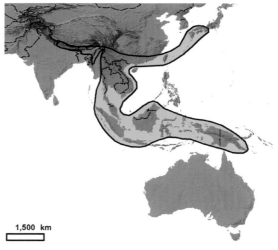

Map 54. Distribution of *C. lancifolium*.

We do not consider that these differences merit recognition of *C. papuanum* at specific rank. Plants of a similar small size have been collected from several countries, and indeed specimens from the northern extremes of the distribution, in Japan, are usually small. The inflorescence is large relative to the small size of the plant, but is similar to that of typical specimens of *C. lancifolium* collected both on New Guinea and in other parts of the range. The cream-coloured flowers are somewhat unusual, but are also known from other specimens of normal size from New Guinea, which are otherwise indistinguishable from *C. lancifolium*. Drooping, not fully open flowers have been noted in specimens from other countries, and in specimens that have flowered under poor conditions in cultivation. The creeping habit which at first seems to be so distinctive (as in *Reeve* 437 and *Stevens* LAE 58147) is also found in specimens of *C. lancifolium* from high altitudes in Burma (*Kingdon-Ward* s.n.), in Yunnan, China (*Handel-Mazzetti* 9414) and Meghalaya (*Hooker & Thompson* s.n.). Furthermore, it is also found in larger, more typical specimens of *C. lancifolium* from lower elevation in Burma (*Baldworth* 13546), Java (*Lobb* s.n.), Sabah (*Collenette* 598) and New Guinea (*Reeve* 702). It seems likely that many of the characters used to distinguish Schlechter's *C. papuanum* are high altitude adaptations of *C. lancifolium*. In the absence of further information, we are inclined to recognise this taxon at varietal rank within *C. lancifolium*.

52. CYMBIDIUM MACRORHIZON

Cymbidium macrorhizon is a small, leafless, holomycotrophic (saprophytic) orchid, only visible above ground when the flower spikes emerge from the substrate, or when the erect capsules are developing. The subterranean, branching rhizome is covered in small hairy warts which appear to be linked with the symbiotic fungal growth necessary for its survival. Early descriptions of this species suggested that it was a parasite, but there is no evidence of this, and Gustav Mann noted on one of his herbarium specimens that attempts to trace the rhizome to its host plant were unsuccessful. The flower spike is usually produced at the tip of the rhizome or of one of its branches, although developmental studies are necessary to determine whether or not it is truly terminally produced. The spike and flowers closely resemble those of other species in subgenus *Jensoa*, especially the closely related *C. lancifolium*. Further evidence of the relationship between these two species may be found in the occasional

Plate 38. *Cymbidium macrorhizon* var. *macrorhizon*. Shillong, Khasia Hills, NE India. Kew collection. del. G. Mann.

rhizomatous growth habit of *C. lancifolium*. The rhizomes of both species are segmented, each segment being subtended by a small cataphyll-like bract, although these may be difficult to observe on *C. macrorhizon* as they appear to disintegrate quickly. One specimen from Thailand (*Geesink, Phanichapol & Santisuk 5536*), at present placed in *C. macrorhizon*, appears to be somewhat intermediate between these two species. It lacks leaves, and the basal portion resembles the rhizome of *C. macrorhizon*, but the apical region is fast-growing and covered in cataphylls, and appears similar to an elongated, immature shoot of *C. lancifolium*. This specimen also resembles the type of *C. caulescens* (at present synonymised under *C. lancifolium*), except that it has no true leaves. Further investigation may show *C. caulescens* to be a distinct taxon intermediate between *C. macrorhizon* and *C. lancifolium*, living as a semi-saprophytic plant capable of the production of some green leaves and cataphylls. Investigations of the mycorrhizal relationships of *C. macrorhizon* and *C. lancifolium* may clarify their relationship.

Cymbidium macrorhizon was described by Lindley in 1833 from a specimen collected by Royle in Kashmir, at about 31°N, and at the extreme north-western edge of the distribution of the genus. It is the most westerly representative of the genus *Cymbidium*. Specimens have since been collected in China, Japan and the Ryukyus further east.

Cymbidium nipponicum (Franch. & Sav.) Rolfe is considered by Ohwi (1965) and Garay & Sweet (1974) to be the correct name for the representatives of this species from Japan and the Ryukyus. It was originally described as *Bletia nipponica* in 1879, from a specimen collected by Savatier. Rolfe transferred it to the genus *Cymbidium* in 1895 but noted its similarity to *C. macrorhizon* from northern India. Rolfe (1904) included *C. pedicellatum* Finet (1900) as a synonym of *C. nipponicum*, both being based on the same type specimen. Seidenfaden (1983) compared the types of *C. macrorhizon* and *C. nipponicum*, and concluded that there did not appear to be any justification for recognising them as distinct.

Finet also published the name *Yoania aberrans* in 1900, illustrating it alongside his *C. pedicellatum*, and noted the similarity of its flowers to those of *Cymbidium*. The illustration shows

Fig. 184 (left). *Cymbidium macrorhizon* var. *macrorhizon*. Japan. **Fig. 185** (right). Nepal. *Trudel s.n.*

325

Y. aberrans to be a smaller, fewer-flowered plant than *C. pedicellatum*, but the flowers are otherwise similar. In 1911 Schlechter transferred it to *Aphyllorchis*, and in 1919 to *Cymbidium*. Seidenfaden (1983) considered that *C. nipponicum* differed from *C. aberrans* by the larger bracts, the short ovary, the larger flowers and the shorter, broader column of the former. Maekawa (1971) also recognised them as distinct species, his illustrations demonstrating the smaller size of *C. aberrans*, its shorter bracts and fewer, more pallid, possibly albinistic flowers which do not open so widely. The latter is recognised here as var. *aberrans*.

There have been few comparisons made between the Japanese material and specimens of *C. macrorhizon* from northern India. This may be due to the disjunct distribution of these two regions. The variation in size and flower number already noted from Japan is mirrored in the Indian specimens (see also the discussion of the treatment of Nakai, 1931).

In 1919 Schlechter described *C. aphyllum*, from Sichuan in western China. His description differs from the northern Indian specimens of *C. macrorhizon* only in its obtuse apex to the mid-lobe. Wu & Chen (1980) in their revision of *Cymbidium* in China placed this species into the synonymy of *C. macrorhizon*, and their treatment is followed here.

Hu (1973) noted that the name *C. aphyllum* Schltr. was a later homonym of *C. aphyllum* (Roxb.) Sw. (1799) used for a species later transferred to the genus *Dendrobium*. She therefore published *C. szechuanense*, a new name for this taxon. This name is therefore also treated here as a synonym of *C. macrorhizon*.

The taxa now included in *C. macrorhizon* were studied by Nakai in 1931. He removed them to a new genus, *Pachyrhizanthe*, and recognised five distinct species. These were *P. macrorhizos* from northern India, *P. aphylla* from China, and *P. nipponica*, *P. aberrans* and *P. sagamiensis* from Japan. The last was distinguished by its lack of purple colouring in the flowers, and by its lip which had very poorly defined side-lobes. This is probably an albino variant, lacking the red pigment in the flower, similar to those encountered in many other species such as *C. ensifolium*, *C. sinense*, *C. goeringii*, *C. insigne*, and *C. mastersii*. Later authors (Ohwi, 1965; Maekawa, 1971; Garay & Sweet, 1974) did not recognise *P. sagamiensis* as distinct but we consider it to belong to var. *aberrans*.

Nakai (1931) keyed out the species in this group on the basis of the length of the scape, the colour of the flowers and the degree to which the lip appears to be 3-lobed. We only recognise two varieties based upon examination of material from throughout the range of *C. macrorhizon*.

52. Cymbidium macrorhizon Lindl., *Gen. Sp. Orchid. Pl.*: 162 (1833). Type: India, Kashmir, *Royle* (holotype K!).

Bletia nipponica Franch. & Sav. in *Enum. Pl. Jap.* 2: 511 (1879). Type: Japan, *Savatier* s.n. (holotype P!).
Cymbidium nipponicum (Franch. & Sav.) Rolfe in *Orchid Rev.* 3: 39 (1895).
C. pedicellatum Finet in *Bull. Soc. Bot. Fr.* 47: 268, t. 9A (1900); Rolfe in *Orchid Rev.* 12: 303 (1904). Type: as for *Bletia nipponica*.
C. aphyllum Ames & Schltr., *loc. cit.* 73, 265 (1919), *non* (Roxb.) Sw. (1799). Type: China, Sichuan, *Wilson* 4712 (holotype AMES!).
Pachyrhizanthe macrorhizos (Lindl.) Nakai, *Bot. Mag. Tokyo* 45: 109 (1931).
P. nipponica (Franch. & Sav.) Nakai, *loc.cit.* (1931).
P. aphylla (Ames & Schltr.) Nakai, *loc. cit.* (1931).
Cymbidium szechuanense S.Y. Hu in *Quart. J. Taiwan Mus.* 26: 140 (1973). Type: as for *C. aphyllum*.

A small, perennial, terrestrial *saprophyte*, without leaves or pseudobulbs. *Roots* short, often absent, occasionally produced towards the apex of the rhizome. *Rhizome* 3–8 mm in diameter, soft, fleshy, tuberculate, with nodes 2–15(19) mm apart, usually branching, whitish, subterranean. Tubercles closely spaced, each with a tuft of short hairs. Scales up to 5 mm long, scarious, often disintegrating leaving an indistinct scar; branches originating in the axils of the scales. *Scape* 7–32 cm long, erect, with (1)3–6(8) flowers, usually produced terminally on the rhizome or on its side branches; peduncle covered with 4–8, mostly overlapping sheaths, of which 1–4 are above ground; sheaths 13–27 mm long, ovate, with a cylindrical sheathing base and a loosely sheathing, somewhat spreading, cymbiform apex; bracts 2–17 mm long, oblong to lanceolate, acute, somewhat keeled, becoming scarious. *Flower* 3–4 cm across; rhachis, pedicel and ovary pale green to cream, often stained with pink or purple; sepals and petals cream to pale yellow or brownish-pink with a diffuse, central, purple-red stripe to near the apex; lip white with red spots on the mid-lobe and purple-red stripes on the side-lobes, which become confluent at the margin; callus white, sometimes stained pink; column white, stained or lightly spotted purple-pink above, with purple-red dashes below; anther-cap cream. *Pedicel and ovary* 10–37 mm long. *Dorsal sepal* (14)19–26 × (3)4–6 mm, narrowly obovate to ligulate, acute, erect; lateral sepals similar, spreading. *Petals* 14–20 × 5–7 mm, narrowly elliptic, oblique, acute, usually porrect and covering the column. *Lip* 12–17 × 9–11 mm when flattened, almost rhombic in outline, subentire to strongly 3-lobed, minutely papillose; side-lobes erect, obtuse to subacute at the apex; mid-lobe 5–7 × 4–6 mm, triangular to ligulate, acute to obtuse, recurved, the margin usually slightly undulating; callus ridges 2, convergent towards the apex, forming a short tube at the base of the mid-lobe. *Column* 9–13 mm long, slender, arching, narrowly winged towards the apex; pollinia 4, broadly ovate, in two unequal pairs; viscidium broadly crescent-shaped. *Capsule* 30–50 × 9–12 mm, fusiform to ellipsoidal, pedicellate, held erect and parallel to the rhachis, retaining the column which forms a short (5–10 mm) apical beak. 2n = 38.

DISTRIBUTION. Pakistan (N Punjab), N India (Kashmir, NW India, Sikkim, Assam, Meghalaya, Nagaland), ?Nepal, ?Bhutan, Myanmar, China (Chongqing, SW Guizhou, SW Sichuan, NE Yunnan, Taiwan), Thailand, Laos, Vietnam, Korea, Japan, Ryukyus (Map 55); 700–2500 m (2295–8200 ft).

The distribution of *C. macrorhizon* is unusual in that it is disjunct, and extends over a greater east-west range than any other *Cymbidium* species (Map 55). It occurs as far west as the hills of northern Pakistan and Kashmir. From there it extends east along the Himalaya in Nepal, Sikkim and Darjeeling although it has not been collected in Bhutan, but it occurs again in Meghalaya (Khasia Hills) and Nagaland. The distribution continues through Myanmar (including the Chin Hills) into northern Thailand, Laos and the southwestern provinces of China. From there, there is a gap to Japan, the Ryukyus and Taiwan.

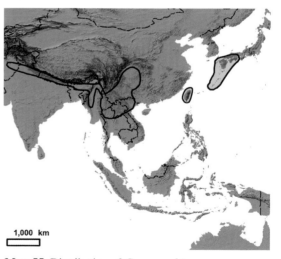

Map 55. Distribution of *C. macrorhizon*.

Its altitudinal range varies in the different regions of this distribution. In Japan and the Ryukyus it may be found almost at sea level. In south-western China, Thailand and Myanmar, it is found between 1000 and 1500 m (3280–4920 ft), in Taiwan between 1800 and 2200 m (5905–7220 ft), while in northern India the range is between 1000 and 2500 m (3280–8200 ft), It is found at lower elevations in the west of its range.

HABITAT. In riverine forest, forest margins, broad-leaved and pine forest growing in damp humus, in shade. In general, *C. macrorhizon* is a forest species, growing in shaded spots, in humus-rich soil. In Japan it has been reported as growing on hills and in fields not far from the sea, and on roadsides in China. In Thailand it has been found in open forest, growing in humus-filled rock fissures. In India, it is usually encountered in pine forest, where it thrives in the decaying pine needles, although broad-

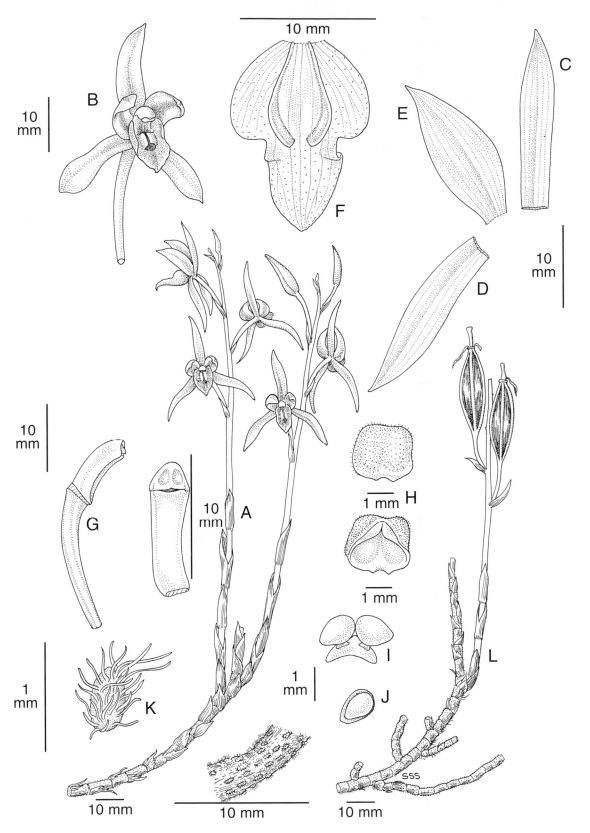

Fig. 186. *Cymbidium macrorhizon* var. *macrorhizon*. **A.** Habit. **B.** Flower. **C.** Dorsal sepal. **D.** Lateral sepal. **E.** Petal. **F.** Lip, flattened. **G.** Column, two views. **H.** Anther-cap, two views. **I.** Pollinarium. **J.** Pollinia. **K.** Root hairs. Drawn from *Pradhan* 22 by Susanna Stuart-Smith.

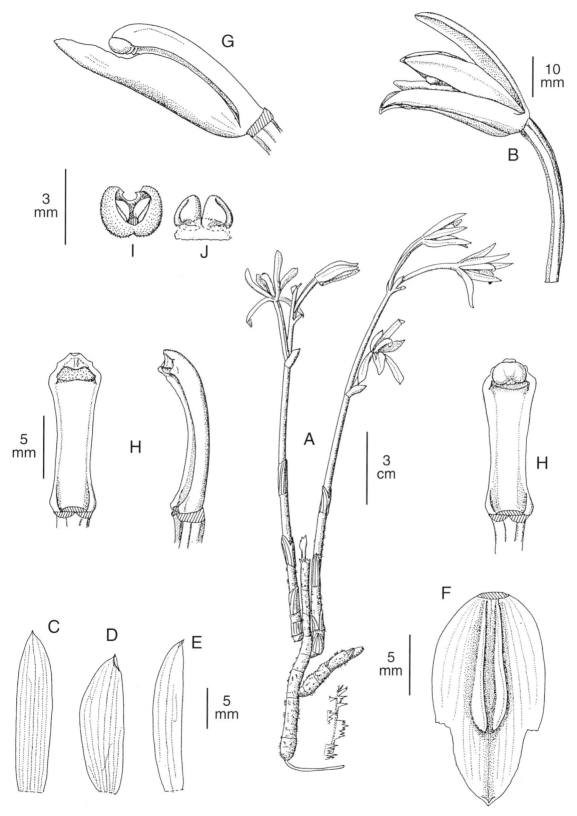

Fig. 187. *Cymbidium macrorhizon* var. *aberrans*. **A.** Habit. **B.** Flower. **C.** Dorsal sepal. **D.** Petal. **E.** Lateral sepal. **F.** Lip, flattened. **G.** Column, ovary and lip, side view. **H.** Column, three views. **I.** Anther-cap. **J.** Pollinarium. Drawn from a collection preserved in the Yokosuka Museum, Kanagawa Pref., Japan by Mutsuko Nakajima.

leaved forest of oaks and dipterocarps can also support populations of this species. Flowering time seems constant, between June and August, except in Indo-China where it flowers slightly earlier, during May and June.

CONSERVATION STATUS. VU A1cd; B1ab.

Key to the varieties of *C. macrorhizon*

Flower opening fully, white with purple markings on the sepals and lip; lip side-lobes well developed .... var. ***macrorhizon***
Flower not opening widely, white or off-white; lip side-lobes obscure .... var. ***aberrans***

var. **macrorhizon**

Flowers opening fully, white with purple markings on the sepals and lip. Lip strongly trilobed; lip side-lobes well developed. Pl. 38; Figs. 184–186.

DISTRIBUTION. Throughout the range of the species.

HABITAT. As above.

var. **aberrans** (Finet) P.J. Cribb et Du Puy **comb. et var. nov.**

Yoania aberrans Finet in *Bull. Soc. Bot. Fr.* 47: 274, t.9B (1900). Type: Japan, *sine coll.* (holotype P!).
Aphyllorchis aberrans (Finet) Schltr. in *Bot. Jahrb. Syst.* 45: 387 (1911).
Cymbidium aberrans (Finet) Schltr. in *Fedde, Repert. Sp. Nov. Regni Veg., Beih.* 4: 264 (1919).
Pachyrhizanthe aberrans (Finet) Nakai in *Bot. Mag., Tokyo* 45: 109 (1931).
P. sagamiensis Nakai, *loc. cit.:* 110 (1931). Types: Japan, Hondo, *Musashi* (florum), *Hisauchi s.n.*;
    Hondo, *Sagami* (fructum), *Hisauchi s.n.* (syntypes not located).

Flowers not opening widely, white or off-white. Lip elliptic; side-lobes obscure or absent. Fig. 187.

DISTRIBUTION. Japan and the Ryukyu Islands; 700–1000 m (2295–3280 ft). In Japan it may be found almost at sea level.

HABITAT. In Japan it has been reported as growing on hills and in fields not far from the sea. Flowering between June and August.

Yoania aberrans was described by Finet based on a specimen from Japan. Schlechter eventually identified it as a saprophytic *Cymbidium*. It differs from typical *C. macrorhizon* in lacking any purple colouration in the flowers and in having an almost entire lip. These features suggest that this variety may be possibly no more than a partially peloric, albinistic variant of *C. macrorhizon*. *Pachyrhizanthe sagamiensis*, described by Nakai in 1931 from Japan, was said to differ in its shorter dilated column but this may be no more than the effect of pollination on column shape. We do not see that it differs from var. *aberrans* enough to warrant taxonomic recognition.

LITTLE KNOWN
AND DOUBTFUL SPECIES

CYMBIDIUM AESTIVUM

Cymbidium aestivum was described by Z.J. Liu and S.C. Chen in 2004 based upon a collection from western Yunnan. They distinguished it from *C. dayanum*, with which it shares a common habit, arching inflorescence, acute sepals and petals and glandular lip callus ridges. However, they distinguished it by its larger flowers distinctively coloured whitish flowers heavily flushed with purple on the sepals and petals and marked with dark maroon on the lip mid-lobe and side lobes, longer floral bracts and mid-lobe three-fifths the length of the lip. We have examined the type, and the drawing and the photographs in Liu *et al.* (2006) and are inclined to think that this taxon has a hybrid or possibly polyploidy origin involving *C. dayanum*. If it is a hybrid, then the other parent may be one of the small, dark-flowered species that grow sympatrically with it and *C. dayanum*. Further investigation is undoubtedly necessary to ascertain its status.

Cymbidium aestivum *Z.J. Liu & S.C. Chen* in *J. Wuhan Bot. Res.* 22, 4: 323 (2004); Liu, Chen & Ru, *The Genus Cymbidium in China*: 55–58 (2006). Type: China, Yunnan, Mengla Co., Puweng, 18 July 2000, fl. Shenzhen City Wutonhgshan Nursery July 2003, *Z.J. Liu* 200234 (holo. SZWN).

CYMBIDIUM BAOSHANENSE

Cymbidium baoshanense was described in 2001 by Liu and Perner based upon a collection from the mountains of western Yunnan. Morphologically its flowers are similar to those of *C. lowianum*, which is common in the region, but they are smaller, as are its habit and leaves. Its habit, short, broad leaves, few-flowered inflorescences and flower markings are suggestive of *C. tigrinum* and we postulate that *C. baoshanense* is a natural hybrid of *C. lowianum* with *C. tigrinum* or another smaller flowered species.

Photographs of the type and other plants in the original collection show that the flowers are rather variable, some having a more spotted lip than the type (Perner, pers. comm.).

Cymbidium baoshanense *F.Y. Liu & Perner* in *Orchidee*, 52(1): 61 (2001). Type: China: Yunnan, Longling, *L.M.Zuo* 98002 (holotype KUN).

CYMBIDIUM GONGSHANENSE

This small-flowered, hard-leaved epiphytic species was described in 1989 by Li and Feng based on a plant reportedly collected by Z.H. Yang on Gongshan Xian in north-western Yunnan. The authors compared it with *C. iridioides*, to which it bears little resemblance in its flowers but more in its vegetative morphology, having a 2–4 cm long ovoid pseudobulb and leaves up to 2.5 cm broad. It is, however, almost identical to *C.* Ruby Eyes, one of the artificial hybrids of *C. floribundum* that became popular in the late 1970s and 1980s. It is possible that it is a natural hybrid of *C. floribundum* with one of the species of section *Cyperorchis*, such as *C. lowianum* or *C. iridioides*, both of which are reported from the region. It is highly unlikely that it is a distinct species.

Cymbidium gongshanense *H. Li & G.H. Feng* in *Acta Bot. Yunnanica* 11, 1: 39 (1989). Type: China, Yunnan, Gonshanxian, *Z.H.Yang* 8708 (holotype KUN).

CYMBIDIUM MICRANTHUM

This small orchid was described by Liu and Chen in 2004, based on a cultivated specimen from south-eastern Yunnan. They compared it with *C. nanulum*, distinguishing it by its small pseudobulbs, lack of an elongated rhizome, articulated leaves and the incurved tip to its lip, but nearly all of these features are found in *C. ensifolium*. The petals and sepals are less than 2 cm long whereas in allied species they exceed 2 cm in length. The drawing and photographs of the type, published by Liu *et al.* (2006), suggest that the flowers are deformed, the flowers not opening widely and the lip mid-lobe being upcurved and acute. We reserve judgement on the status of this unusual orchid and wonder if its abnormal flowers were the result of flowering in the first year after introduction into cultivation. Small flowers and an incurved lip are common aberrations in *Cymbidium* and other orchids newly introduced into culture from the wild.

Cymbidium micranthum *Z.J. Liu & S.C. Chen* in *J. Wuhan Bot. Res.* 22: 500 (2004); Liu, Chen & Ru, *The Genus Cymbidium in China*: 181–182 (2006). Type: China Yunnan, Maguan, Z.J. Liu 2705 (holotype SZWN!).

CYMBIDIUM TERETIPETIOLATUM

Cymbidium teretipetiolatum was described by Liu and Chen (2002) based upon a plant from near Jiangcheng in south-western Yunnan. The published illustration and photographs in Liu *et al.* (2006) show clearly that the type plant has semi-peloric, white or pinkish flowers, with a lip that is almost entire or only sparsely three-lobed and lacks a callus. Another distinctive feature is its elongated, somewhat coralline rhizome. Because of the lack of normal flowers, we feel that it is premature to comment on its taxonomic status. In habit, inflorescence structure and tepal shape, it appears to have affinities with *C. tortisepalum*.

Cymbidium teretipetiolatum *Z.J. Liu & S.C. Chen* in *Die Orchidee* 53 (3): 338 (2002); Liu, Chen & Ru, *The Genus Cymbidium in China*: 216–218 (2006). Type: China, S Yunnan, Simao Pref., nr. Jiangcheng, *Z.J. Liu 220210* (holotype PE, isotype SZWN!).

ADDENDUM

The collection of cymbidiums from the wild continues to the present day, and novelties are constantly being sought by growers. It is scarcely surprising that amongst the tens of thousands of wild plants collected each year, particularly in China, plants that do not fit into the descriptions of the known species will be found. Some may indeed prove to be species or infraspecific taxa new to science.

The Genus Cymbidium in China (Science Press, Beijing, ISBN 7-03-017115-2) by Z.J. Liu, S.C. Chen & Z.Z. Ru (2006) was published when the present work was in final page proof, as was an article by Liu and Chen (2006) in *Acta Phytotaxonomica Sinica* (44, 2: 179). The former contains one new species, several new combinations and descriptions, and the first published colour photographs and detailed drawings of a number of recently described Chinese species. Its publication coincided with the visit, at the kind invitation of Professors Liu and Chen, by one of the authors (PC) to the collection and herbarium (SZWN) of the National Orchid Conservation Collection at Wutongshan, Shenzhen, China. There he was able to examine the type collections of the recently described cymbidiums.

Liu *et al.* (2006) adopt a narrower species definition than that used in this work, and a narrower one than has previously been used in China. One of the authors, Chen (1999) in *Florei Reipublicae Popularis Sinicae Orchidaceae* (2: 191–227), included 29 Chinese species in his account of *Cymbidium*. The new monograph describes 49 species as occurring in China, the majority of the 20 additional species having been described since 1999, a few also being resurrected from synonymy. This is surprising when one considers that only three new Chinese species were described in the previous 25 years or so, and only five from outside of China since 1960. Natural variation is well documented in Chinese cymbidiums. Mutations of *C. goeringii* and other Chinese terrestrial species have been collected, grown and shown for over two and a half millennia. To adopt a narrow species concept is, therefore, a radical approach in the light of the known infra-specific variability of the species in the region. In such a diverse and large country it is scarcely surprising that widespread species will vary from locality to locality. Variation may take the form of distinctive local populations, mutations or even natural hybrids. The scope for natural hybridisation is immense as burgeoning human populations create habitats where species that might previously have been isolated, through different local distributions, habitat preferences, pollinators or other ecological factors. Natural hybridisation has rarely been considered in assessing whether these Chinese plants may or may not be a species.

As will already be apparent, we consider many of the newly described taxa to be variants of well-known species, or natural hybrids of them. However, our views were based upon the published descriptions and drawings of the types. With examination of the drawings and descriptions in the book by Liu *et al.* (2006), the type material, photographs and living plants of some of the species, we feel that further comments are necessary for some of the species.

The only new species described in the book is *C. sichuanicum* Z.J. Liu and S.C. Chen which they compare with *C. iridioides* D. Don. A full description and discussion of *C. sichuanicum* are provided below:

C. sichuanicum Z.J. Liu & S.C. Chen in *The Genus Cymbidium in China*: 82 (2006). Type: China, Sichuan, Mao Xian, *Z.J. Liu 3027* (holo. SZWN!).

An epiphytic or lithophytic *herb*. *Pseudobulbs* subellipsoidal, 6–10 × 2.8–3.3 cm, enclosed by persistent leaf bases. *Leaves* coriaceous, lorate, subacuminate, (30–)60–100 × 2–2.5 cm, articulated 10–15 cm above the base. *Inflorescence* arising from the base of the pseudobulb, suberect, 50–70 cm long, 10- to 15-flowered; penducle with 7–9 sheaths, each 2.5–12.5 cm long; floral bracts lanceolate, 6–20 mm long. *Flowers* 6–7 cm across, slightly scented; sepals and petals yellow-green tinged with pale red-purple, with 9–11 longitudinal red-purple stripes, the lip yellow tinged with red-brown long the margins, with purple-red longitudinal stripes and irregular dashes on both side lobes and mid-lobe, the column purple-red above, with red-purple speckles; pedicel and ovary 4–5.5 cm long, red-brown. *Sepals* elongate-elliptic, acute, 5.5–5.9 × 1.8–2 cm; lateral sepals oblique, dorsally carinate. *Petals* slightly falcate, obovate-oblong, acute, 5.2–5.5 × 1.7–1.9 cm. *Lip* 3-lobed, ovate-circular, 4.3–4.6 cm long, basally fused to the column for 3-4 mm; side-lobes hairy on adaxial veins, the innermost densely hairy, white-ciliate; mid-lobe recurved, ovate, 1.7–1.9 × 2.1–2.3 cm, adaxially white-pubescent; callus of two white-hairy lamellae, not inflated at apex. *Column* arcuate, 3.6–3.9 cm long, ventrally hairy, narrowly winged; pollinia two, cleft.

DISTRIBUTION. W Sichuan only.

HABITAT. On trees in forests or on rocks at forest margins; 1200–1600 m. Flowering in February and March.

Liu and Chen (2006) compare *Cymbidium sichuanicum* with *C. iridioides*, with which it is sympatric. They state that it differs in having broader leaves, differently coloured flowers with purple-red longitudinal stripes and irregular dashes on the mid-lobe of the lip, much wider obovate-oblong petals, and a longer column with purple-red on its upper surface. In fact, the leaves of *C. sichuanicum* fall somewhat towards the lower end of the range of those of *C. iridioides*. In addition, the column of the plant of *C. iridioides* illustrated by a colour photograph has a red-tipped column. Although it does possess some similarities, there is no doubt that *C. sichuanicum* is quite distinct from *C. iridioides* in its flower colour, petals and column pubescence. Its unusual petals are certainly different from those found in *C. iridioides* and its closely allied species such as *C. hookerianum*, *C. wilsonii* and *C. tracyanum*. They more closely resemble those of the predominantly white-flowered *C. wenshanense*, as do the purple-striped markings on its lip and the shape and attitude of the mid-lobe of its lip. Given its provenance, it seems unlikely that *C. sichuanicum* is a natural hybrid of *C. iridioides* and *C. wenshanense*, the latter occurring only well to the south, in south-east Yunnan, but it may well pay investigation to see if *C. sichuanicum* is related to *C. wenshanense* in any way.

This taxon should not be confused with *C. szechuanicum* Y.S. Wu & S.C. Chen (1966) which is ascribed by us to *C. faberi* var. *szechuanicum* (Y.S. Wu & S.C. Chen) Y.S. Wu & S.C. Chen (p. 281).

We will consider here the remaining additional taxa in the order that they appear in Liu *et al.*'s account:

C. mannii Rchb.f.
Cymbidium mannii has resurrected from subspecific status within *C. bicolor* where we have placed it as *C. bicolor* subsp. *obtusum*, Liu *et al.* (2006) argue that it occurs distant from *C. bicolor* which is found in southern India and Sri Lanka. *C. bicolor* (sensu Du Puy & Cribb, 1988), with three subspecies, is widespread throughout south and south-east Asia. Removing *C. mannii* as a distinct species means that Liu *et al.* (2006) have also raised to specific rank subsp. *pubescens* from South-east Asia and the Malay Archipelago.

C. paucifolium Z.J. Liu & S.C. Chen
This differs from *C. mannii*, according to the authors, in its darker flower colour and narrower rigid leaves. If the variation across its range is considered, we would include this, as we have with *C. mannii*, in *C. bicolor* subsp. *obtusum*.

C. aestivum Z.J. Liu & S.C. Chen
They distinguish this from *C. dayanum* by its larger and distinctively coloured flowers. The colour and size variation of *C. dayanum* across its range is considerable, and in Borneo both predominantly white-

and dark red-flowered forms occur. Having examined the colour photographs and type material of *C. aestivum*, we are less inclined to treat it as a synonym of *C. dayanum*, but it seems highly likely that it has a hybrid origin with one parent being *C. dayanum*. The identity of the other is less obvious but may be one of the other small-flowered sympatric species.

C. gaoligongense Z.J. Liu & J.Y. Zhang
We see no reason to change our opinion, after examining the type and seeing the published photographs, that this is an albinistic form of *C. iridioides* with which it is sympatric.

C. flavum Z.J. Liu & J.Y. Zhang
We see no reason to change our opinion, after examining the type and seeing the photographs, that this is an albinistic form of *C. erythraeum* which also occurs in the same area as the type of *C. flavum*. The authors state that this colour form is common in SE Yunnan, if so it probably warrants recognition as a form.

C. schroederi Rolfe
The plant illustrated with a line drawing and photographs is similar to *C. schroederi* in its vegetative and floral morphology, although its inflorescences are more erect. One of us (PC) saw this plant growing lithophytically and at the base of small trees in a karst area on limestone in northern Vietnam.

C. insigne Rolfe
The plants from Hainan illustrated by colour photographs seem closer to subsp. *seidenfadenii* from north-east Thailand than to subsp. *insigne* from southern Vietnam. The exact relationships of the Hainan plants need further investigation.

C. eburneum Lindl. var. longzhouense Z.J. Liu & S.C. Chen
This new variety was described by the authors in *Acta Phytotaxonomica Sinica* 44, 2: 179, 2006). It was described as differing from the typical variety by its heavily red-purple spotted side- and mid-lobes of its lip: an example of this colouration is provided by us on p. 183. We have seen many flowering specimens of *C. eburneum* from across its range and the degree of spotting on the lip varies considerably. Whether this variety is worth formal recognition is debatable.

C. maguanense F.Y. Liu
This species was retypified by Z. J. Liu, S.C. Chen & Z.Z. Ru (2006) in *Acta Phytotaxonomica Sinica* (44, 2: 182) because the type collection could not be traced in KUN. This taxon is closely allied to *C. mastersii* where we placed it earlier in this treatment. We have made a careful examination of the type, and of the drawings and photographs present by Liu *et al.* (2006). Our reassessment, based on the retypification, is that this is either a form of *C. mastersii* with which it is sympatric, or possibly represents a hybrid of *C. mastersii* with *C. eburneum*, with introgression to *C. mastersii*. According to Liu *et al.* (2006) *C. eburneum* occurs in all the surrounding areas but not where the type of *C. maguanense* occurs.

C. changningense (X.M. Xu) Z.J. Liu & S.C. Chen
This taxon was originally described as *C. lowianum* var. *changningense* in 2005 in the *Journal of the South China Agricultural University* (26, 3: 120, fig. 1). Liu & Chen raised it to specific rank as *C. changningense* in 2006 in *Acta Botanica Yunnanica* (27, 4: 378). The type came from western Yunnan where *C. lowianum* is common. We have no doubt that this is a natural hybrid with *C. lowianum* as one parent and possibly *C. mastersii* as the other.

C. concinnum Z.J. Liu & S.C. Chen

The excellent colour photographs of *C. concinnum* published by Liu *et al.* (2006) support our view that this is the pink-flowered variant of *C. mastersii* that occurs throughout its range.

C. quinquelobum Z.Y. Liu & S.C. Chen

Examination of the type and of the illustrative evidence confirms that this is conspecific with *C. wenshanense*, differing only in the lobing of the lip side-lobes.

C. elegans Lindl. var. **lushuiense** (Z.Y. Liu, S.C. Chen & X.C. Shi) Z.Y. Liu & S.C. Chen

This taxon was originally described in 2005 at specific rank in *Shenzhen Science and Technology* (139: 200), a journal not searched by *Index Kewensis* or other botanical nomenclators. The transfer to varietal status within *C. elegans* is made by Liu *et al.* (2006 (p. 144). According to the authors, it differs from the typical variety in details of its callus structure, the central lamellae being fused in the apical part and the presence of two short outer lamellae at the base of the lip. However, the callus structure of *C. elegans* is identical to that described for var. *lushuiense*. We do not accept that this variety warrants formal recognition.

C. baoshanense F.Y. Liu & H. Perner

Examination of the photographs and drawing presented by Liu *et al.* (2006) confirms our opinion and that of one of the authors (Perner, pers. comm.) that *C. baoshanense* is of hybrid origin. Its postulated parentage is *C. lowianum* × *C. tigrinum*. The type collection almost certainly originated to the west of Baoshan which is a market town and the base of well-known orchid dealers who trade plants from western Yunnan and adjacent Myanmar.

C. sinense (Jackson ex Andr.) Willd. var. **haematodes** (Lindl.) Z.J. Liu & S.C. Chen

The authors have transferred this variety from *C. ensifolium*, where we believe it belongs, to *C. sinense* on the basis of its "habit and floral morphology". In particular, they note that its leaves are broader than *C. ensifolium* and its inflorescence is longer than the leaves. However, its flowers are much closer to those of *C. ensifolium* and its leaf measurements, given as "2–2.5 cm", overlap those given for *C. ensifolium*. In Thailand, where *C. sinense* and *C. ensifolium* subsp. *haematodes* both occur, their flower colour is similar but *C. sinense* can always be distinguished by always having strongly porrect petals which closely cover the column, a broader more pronounced mid-lobe to the lip and broad, smoothly glossy leaves. We do not, therefore, think that their transfer to varietal status in *C. sinense* reflects its true relationship.

C. defoliatum Y.S. Wu & S.C. Chen

Examination of the line drawing and colour photographs indicates that this taxon is very close to *C. ensifolium*. Indeed, its flowers could pass as *C. ensifolium* but it differs in having fewer, deciduous leaves on only the terminal psudobulbs. Its status as a distinct species needs further examination (see our account on p. 261). *Cymbidium ensifolium* has only four leaves per pseudobulb, and they are often deciduous on the back-bulbs, or under environmental stress.

C. micranthum Z.J. Liu & S.C. Chen

The flowers shown in the photographs and drawing of *C. micranthum* by Liu *et al.* (2006) appear to us to be deformed or not fully developed. They distinguish it from *C. tortisepalum* by its articulate leaves and small flowers with the sepals and petals less than 2 cm long. Its status needs further examination (see discussion under Little known and doubtful species).

C. serratum Schltr.

Liu *et al.* (2006) have reinstated *C. goeringii* Rchb.f. var. *serratum* (Schltr.) Y.S. Wu & S.C. Chen to specific status on the basis of its narrow, non-articulated leaves and much longer, 20-30 cm long scape and non-scented, green flowers. These distinctions are, however, not as clear-cut as they state. *C. goeringii* var. *goeringii* often has long inflorescences and green flowers. We prefer to keep this taxon within *C. goeringii* in which we do not recognise varieties.

C. teretipetiolatum Z.J. Liu & S.C. Chen

The authors distinguish this on the basis of its sometimes coralline rhizome and simple or indistinctly lobed lip lacking a callus. Examination of the type, drawing and published photographs suggest that it is based on a plant with deformed, somewhat peloric flowers in which the lip is petaloid (see discussion under Little known and doubtful species). The coralline rhizome is unusual and this taxon needs further examination to determine its status (for further discussion see p. 332).

C. caulescens Ridl.

This taxon has been resurrected by Liu *et al.* (2006) at specific rank, after having been consigned to the synonymy of *C. lancifolium* by most previous authors. We have discussed the variation of *C. lancifolium* in the species account on p. 316 onwards. This falls within our concept of *C. lancifolium*.

C. rhizomatosum Z.J. Liu & S.C. Chen

The flowers of this taxon are identical with those of *C. lancifolium* where we have placed it in our treatment. Rhizomatous plants with small leaves and longer inflorescences are well-known in *C. lancifolium*, for example, the New Guinea *C. lancifolium* var. *papuanum* has them. We have also seen the same in Thai plants and this phenomenon is fully discussed in our account of *C. lancifolium*. We consider this to be probably a reaction to adverse climatic or edaphic conditions. It may, on further study, warrant recognition at infraspecific rank within *C. lancifolium*, but we are not convinced that it warrants higher recognition.

C. multiradicatum Z.J. Liu & S.C. Chen

The authors state that this is a saprophytic plant but it does not have the features of a saprophyte, having long roots typical of autotrophic species. Chlorophyll is certainly present in the plant, being illustrated in a photograph by Liu *et al.* (2006) of a plant with green fruits. Again, we consider this to be an extreme form of *C. lancifolium*, its flowers being identical. Autotrophic orchids are occasionally known to flower when leafless. The elongate rhizome shown in another of their photographs is not unlike that in *C. rhizomatosum*. Further study of these unusual plants and of the relationship between them, *C. lancifolium* and *C. macrorhizon* is necessary before their status can be fully determined.

BIBLIOGRAPHY AND
FURTHER READING

CHAPTER 1: HISTORY

Ames, O. (1925). *Orchidaceae*. In: E.D. Merrill, *Enumeration of Philippine Plants*: 252–463. Bureau of Science, Manila.

Chen, S.C. & Tang, T. (1982). A General Review of the Orchid Flora of China. In: Arditti, J., ed., *Orchid Biology, Reviews and Perspectives*, 2: 39–82. Cornell Univ. Press, Ithaca and London.

Chow, C. (1979). *Formosan Orchids*. Taichung, Taiwan.

Comber, J. & Nasution R. (1977). A new Indonesian *Cymbidium*. *Cymbidium hartinahianum*. *Bull. Kebun Raya* 3: 1–3.

Du Puy, D. & Cribb, P.J. (1988). *The genus* Cymbidium. C. Helm, London & Timber Press, Portland, Oregon.

Hooker, W.J. (1823). *Cymbidium lancifolium*. *Exotic Flora* 1: t.51.

Hu, S.Y. (1971). Orchids in the life and culture of the Chinese people. *Quart. J. Taiwan Mus.* 24: 67–103.

Lindley, J. (1833). XCVII. *Cymbidium. The Genera and Species of Orchidaceous Plants*: 161–172.

Linnaeus, C. (1753). *Species Plantarum*. L. Salvius, Stockholm.

Miyoshi, M. (1932). On the manuscripts by Matsuoka Joan. *Honzu* 6: 43–55.

Nagano, Y. (1952). Three main species of orchids in Japan. *Amer. Orchid Soc. Bull.* 21: 787–789.

Nagano, Y. (1953). History of orchid growing in Japan. *Amer. Orchid Soc. Bull.* 22: 331–333.

Nagano, Y. (1955). Miniature Cymbidiums in Japan. *Amer. Orchid Soc. Bull.* 24: 735–743.

Nagano, Y. (1960). Orchids in Japan. In: *Proc. 3rd World Orchid Conference*: 50–55. Royal Horticultural Soc., London.

Schlechter, R. (1913). *Cymbidium papuanum* Schltr. *Fedde Repert. Sp. Nov. Regnum Veg. Beih.* 1: 952.

Wood, J.J. (1983). *Cymbidium borneense*. *Kew Bull.* 38: 69–70.

Yen, T.K. (1964). *Icones Cymbidiorum Amoyensium*. Committee Sci. Tech. Amag. Fukien.

CHAPTER 2: MORPHOLOGY

Du Puy, D.J. (1986). *A taxonomic revision of the genus* Cymbidium Sw. (*Orchidaceae*). PhD thesis, University of Birmingham and Royal Botanic Gardens, Kew.

Hisanchi, K. (1958). Some instances of orchids bearing mycorhizomes. *J. Jap. Bot.* 33, 7: 220.

Seth, C.J. & Cribb, P.J. (1984). *A reassessment of the sectional limits in the genus* Cymbidium. In: Arditti, J., ed., *Orchid Biology, Reviews and Perspectives 3*: 283–322. Cornell University Press, Ithaca & New York.

Tahara, M. (2001). Artificial interspecific hybrids of Oriental *Cymbidium* species. *Proceedings of the Asia-Pacific Orchid Congress* 7: 165–167.

Winter, K., Wallace, B.J., Stocker, G.C. & Roksandic, Z. (1983). Crassulacean Acid Metabolism in Australian vascular epiphytes and some related species. *Oecologia* 57: 129–141.

Withner, C.L., Nelson, P.H. & Wejksnora, P.J. (1975). The anatomy of the orchids. In: Withner, C.L., ed., *The Orchids, Scientific Studies*: 267–347. J. Wiley & Sons, New York.

CHAPTER 3: SEED MORPHOLOGY

Ackerman, J.D. & Williams, N.H. (1980). Pollen morphology of the tribe Neottieae and its impact on the classification of the Orchidaceae. *Grana* 19: 7–18.

Ackerman, J.D. & Williams, N.H. (1981). Pollen morphology of the Chloraeinae (Orchidaceae: Diurideae) and related subtribes. *Amer. J. Bot.* 68: 1392–1402.

Arditti, J., Michaud, J.D. & Healey, P.L. (1979). Morphometry of orchid seeds. I. Paphiopedilum and native California and related species of *Cypripedium. Amer. J. Bot.* 66: 1128–1137.

Arditti, J., Michaud, J.D. & Healey, P.L. (1980). Morphometry of orchid seeds. II. Native California and related species of Calypso, Cephalanthera, Corallorhiza and *Epipactis. Amer. J. Bot.* 67: 509–518.

Barthlott, W. (1974). Morphologie der Samen. In: Senghas, K., Ehler, N., Schill, R. & Barthlott, W., Neue Untersuchungen und Methoden zur Systematik und Morphologie der Orchideen. *Orchidee* 25: 157–169.

Barthlott, W. (1976). Morphologie der Samen von Orchideen im Hinblick auf taxonomische und funtionelle Aspekte. In: *Proceedings of the 8th World Orchid Conference*: 444–455. German Orchid Soc.

Barthlott, W. & Zeigler, B. (1980). Uber ausziehbare helicale Zellwandverdickungen als Haftapparat der Samenschalen von *Chiloschista lunifera* (Orchidaceae). *Ber. Deutsch Bot. Ges.* 93: 391–403.

Barthlott, W. & Zeigler, B. (1981). Systematic applicability of seed coat micromorphology in orchids. *Ber. Deutsch Bot. Ges.* 94: 267–273.

Beer, I.G. (1863). *Biologie und Morphologie der Familie der Orchideen*. Carl Gerold, Vienna.

Bertsch, K. (1941). Fruchte und Samen. In: Reinerth, H., *Handbucher der praktischen Vorgeschichtsforschung* Band 1: 118–121. Stuttgart.

Burgeff, H. (1936). *Samenbeimung der Orchideen und Entwicklung ihrer Keimpflanzen, mit Anhang uber praktishe Orchideenanzucht*. Gustav Fischer, Jena.

Carlson, M.S. (1940). Formation of the seed of *Cypripedium parviflorum. Bot. Gaz.* 102: 295–300.

Clifford, H.T. & Smith, W.K. (1969). Seed morphology and classification of Orchidaceae. *Phytomorphology* 19: 133–139.

Davis, A. (1946). Orchid seed and seed germination. *Amer. Orch. Soc. Bull.* 15: 218–223.

Dressler, R.L. (1981). *The Orchids, Natural History and Classification*. Harvard Univ. Press, Massachusetts and London.

Dressler, R.L. & Dodson, C. (1960). Classification and phylogeny in the Orchidaceae. *Ann. Miss. Bot. Gard.* 47: 25–68.

Hoehne, F.C. (1949). *Iconographia der orchidaceas de Brasil*. Secretaria de Agricultura, Sao Paolo, Brazil.

Newton, G.D. & Williams, N.H. (1978). Pollen morphology of the Cypripedioideae and the Apostasioideae (Orchidaceae). *Selbyana* 2: 169–182.

Schill, R. (1974). Pollenmorphologie. In: Senghas, K., Ehler, N., Schill, R. & Barthlott, W., Neue Untersuchungen und Methoden zur Systematik und Morphologie der Orchideen. *Orchidee* 25: 157–169.

Schill, R. & Pfeiffer, W. (1977). Untersuchungen an Orchideen-pollinien unter besonderer Beruecksichtigung ihrer Feinskulpturen. *Pollen et Spores* 19: 5–118.

Stoutamire, W.P. (1963). Terrestrial orchid seedlings. *Australian Plants* 2: 119–122.

Thoda, H. (1983). Seed morphology in the Orchidaceae I. *Sci. Rep. Tohoku Univ. Fourth Ser. (Biol.)* 38: 253–268.

Thomale, H. (1957). *Die Orchideen*, 2nd edn. Eugen Ulmer Verlag, Stuttgart.

Williams, N.H. & Broome, C.R. (1976). Scanning electron microscope studies of orchid pollen. *Amer. Orchid Soc. Bull.* 45: 699–707.

Zeigler, B. (1981). *Micromorphologie der Orchidaceen-Samen unter beruecksichtigung taxonomisches Aspekt*. Thesis, Ruprecht Karls Universitat, Heidelberg.

CHAPTER 4: ANATOMY

Atwood, J.T. (1984). The relationships of the slipper orchids (Subfam. Cypripedioideae). *Selbyana* 7: 129–247.

Atwood, J.T. & Williams, N.H. (1978). The utility of epidermal cell features in *Phragmipedium* and *Paphiopedilum* (Orchidaceae) for determining sterile specimens. *Selbyana* 2: 356–366.

Atwood, J.T. & Williams, N.H. (1979). Surface features of the axadial epidermis in the conduplicate-leaved Cypripedioideae (Orchidaceae). *Bot. J. Linn. Soc.* 78: 141–156.

Dressler, R.L. (1981). *The Orchids, Natural History and Classification.* Harvard University Press, Massachusetts and London.

Dressler, R.L. & Dodson, C. (1960). Classification and phylogeny in the Orchidaceae. *Ann. Miss. Bot. Gard.* 47: 25–68.

Du Puy, D. (1986). *A taxonomic revision of the genus* Cymbidium *Sw.* (*Orchidaceae*). Ph.D. thesis, Univ. of Birmingham.

Kaushik, P. (1983). *Ecological and Anatomical Marvels of the Himalayan Orchids. Progress in Ecology,* 8. Today and Tomorrow's Printers, New Delhi.

Lov, L. (1926). Zur Kenntnis der Entfaltungszellen monokotylen Blatter. *Flora* 120: 283–343.

Mobius, M. (1877). Uber den anatomische Bau der Orchideen—blatter und dessen Bedeutung fur das System dieser Familie. *Jahrb. Wiss. Bot.* 18: 530–607.

Pridgeon, A.M. (1982). Diagnostic anatomical characters in the Pleurothallidinae (Orchidaceae). *Amer. J. Bot.* 69: 921–938.

Pridgeon, A.M. & Stern, W.L. (1982). Vegetative anatomy of *Myoxanthus* (Orchidaceae). *Selbyana* 7: 55–63.

Rasmussen, H. (1981a). The diversity of stomatal development in the Orchidaceae subfamily Orchidoideae. *Bot. J. Linn. Soc.* 82: 381–393.

Rasmussen, H. (1981b). Terminology and classification of stomata and stomatal development—a critical survey. *Bot. J. Linn. Soc.* 83: 199–211.

Rudall, P. (1983). Leaf anatomy and relationships of *Dietes* (Iridaceae). *Nordic J. Bot.* 3: 471–478.

Singh, H. (1981). Development and organisation of stomata in the Orchidaceae. *Acta. Bot. Indica* 9: 94–100.

Solereder, H. & Meyer, F.J. (1930). *Systematic Anatomy of the Monocotyledons—Microspsermae* vol. 6. Israel Program for Scientific Publications (I.P.S.T.) Press, Jerusalem.

Wilkinson, H.P. (1979). The plant surface (mainly leaf). Part 1: Stomata. In: Metcalfe, C.R. & Chalk, L., eds., *Anatomy of the Dicotyledons*, 2nd edn., 1: 97–117. Clarendon Press, Oxford.

Williams, N.H. (1974). The value of plant anatomy in orchid taxonomy. In: *Proc. 7th World Orchid Conf.*: 281–298.

Williams, N.H. (1979). Subsidiary cells in the Orchidaceae: their general distribution with special reference to development in the Orchidieae. *Bot. J. Linn. Soc.* 78: 41–66.

Winter, K., Wallace, B.J., Stocker, G.C. & Roksandic, Z. (1983). Crassulacean acid metabolism in Australian vascular epiphytes and some related species. *Oecologia* 57: 129–141.

Withner, C.L., Nelson, P.H. & Wejksnora, P.J. (1975). The Anatomy of Orchids. In: Withner, C.L., ed., *The Orchids, Scientific Studies*: 267–347. John Wiley & Sons, New York.

Yukawa, T. & Stern, W. L. (2002). Comparative vegetative anatomy and systematics of *Cymbidium* (Cymbidieae: Orchidaceae). *Bot. J. Linn. Soc.* 138: 383–419.

CHAPTER 5: CYTOLOGY

Aoyama, M. (1989). Karyomorphological studies in *Cymbidium* and its allied genera. *Bull. Hiroshima Bot. Gard.* 11: 1–121.

Du Puy, D.J. (1986). *A Taxonomic Revision of the genus* Cymbidium *Sw.* (*Orchidaceae*). PhD thesis, Univ. of Birmingham and Royal Botanic Gardens, Kew.

Leonhardt, K.W. (1979). Chromosome numbers and cross compatibility in the genus *Cymbidium* and related genera. *Orchid Advocate* 5: 44–51.

Tanaka, R. & Kamemoto, H. (1974). List of chromosome numbers in species of Orchidaceae. In: Withner, C.L., ed., *The Orchids: Scientific Studies*: 411–483. John Wiley & Sons, New York.

Wimber, D.E. (1957a). Cytogenetic studies in the genus *Cymbidium*. Chromosome numbers within the genus and related genera. *Amer. Orch. Soc. Bull.* 26: 636–639.

Wimber, D.E. (1957b). Cytogenetic studies in the genus *Cymbidium*. II. Pollen formation in the species. *Amer. Orch. Soc. Bull.* 26: 700–703.

Wimber, D.E. (1957c). Cytogenetic studies in the genus *Cymbidium*. III. Pollen formation in the hybrids. *Amer. Orch. Soc. Bull.* 26: 771–777.

CHAPTER 6: POLLINATION BIOLOGY AND FLORAL FRAGANCES

Ackerman, J.D. (1983a). Euglossine bee pollination of the Orchid *Cochleanthes lipsombiae*: a food source mimic. *Amer. J. Bot.* 70: 830–834.

Ackerman, J.D. (1983b). Specificity and mutual dependency of the orchid Euglossine bee interaction. *Biol. J. Linn. Soc.* 20: 301–314.

Bierzychudek, P. (1981). *Asclepias*, *Lantana* and *Epidendrum*: A floral mimicry complex? *Biotropica* 13: 54–58.

Boyden, T.C. (1980). Floral mimicry by *Epidendrum ibaguense* (Orchidaceae) in Panama. *Evolution* 34: 135–136.

Dafni, A. (1983). Pollination of *Orchis caspia*—a nectarless plant which deceives the pollinators of nectariferous species of other families. *J. Ecology* 71: 467–474.

Dafni, A. & Ivri, Y. (1981a). The flower biology of *Cephalanthera longifolia* (Orchidaceae)—Pollen imitation and facultative floral mimicry. *Plant Syst. Evol.* 137: 229–240.

Dafni, A. & Ivri, Y. (1981b). Floral Mimicry between *Orchis israelitica* Baumann and Dafni (Orchidaceae) and *Bellevalia flexuosa* Boiss. (Liliaceae). *Oecologia* 49: 229–232.

Dodson, C.H. & Hills, H.G. (1966). Gas chromatography of orchid fragrances. *Amer. Orch. Soc. Bull.* 35: 720–725.

Dodson, C.H., Dressler, R.L., Hills, H.G., Adams, R.M. & Williams, N.H. (1969). Biologically active compounds in orchid fragrances. *Science* 164: 1242–1249.

Frison, T.H. (1934). Records and descriptions of *Bremus* and *Psithymis* from Formosa and the Asiatic mainland. *Trans. Nat. Hist. Soc. Formosa* 24: 150–185.

Hills, H.G., Williams, N.H. & Dodson, C.H. (1968). Identification of some orchid fragrance components. *Amer. Orch. Soc. Bull.* 37: 967–971.

Hills, H.G., Williams, N.H. & Dodson, C.H. (1972). Floral fragrances and isolating mechanisms in the genus *Catasetum* (Orchidaceae). *Biotropica* 4: 61–76.

Kjellsson, G., Rasmussen, F.N. & Du Puy, D.J. (1985). The pollination of *Dendrobium infundibulum*, *Cymbidium insigne* (Orchidaceae) and *Rhododendron lyi* (Ericaceae) by *Bombus eximius* (Apidae) in Thailand: a possible case of floral mimicry. *J. Trop. Ecology* 1: 289–302.

Macpherson, K. & Rupp, H.M.R. (1935). The pollination of *Cymbidium iridifolium* Cunn. *North Queensland Naturalist* 3: 26.

Macpherson, K. & Rupp, H.M.R. (1936). Further notes on orchid pollination. *North Queensland Naturalist* 4: 25.

Nilsson, L.A. (1978a). Pollination ecology and adaptation in *Platanthera chlorantha* (Orchidaceae). *Bot. Notiser* 131: 35–51.

Nilsson, L.A. (1978b). Pollination ecology of *Epipactis palustris* (Orchidaceae). *Bot. Notiser* 131: 355–368.

Nilsson, L.A. (1979a). Anthecological studies on Lady's Slipper, *Cypripedium calceolus* (Orchidaceae). *Bot. Notiser* 132: 329–347.

Nilsson, L.A. (1979b). The pollination ecology of *Herminium monorchis* (Orchidaceae). *Bot. Notiser* 132: 537–549.

Nilsson, L.A. (1980). The pollination ecology of *Dactylorhiza sambucina* (Orchidaceae). *Bot. Notiser* 133: 367–385.

Nilsson, L.A. (1981). The pollination of *Listera ovata* (Orchidaceae). *Nordic J. Bot.* 1: 461–480.

Nilsson, L.A. (1983). Mimesis of bellflower (*Campanula*) by the red helleborine orchid *Cephalanthera rubra*. *Nature* 305: 799–800.

Seidenfaden, G. (1984). Orchid genera in Thailand XI. *Cymbidieae* Pfitz. *Opera Bot.* 72: 1–24.

Smythe, R. (1970). Pollination by the Common Native Bee. *Orchadian* 3: 149.

CHAPTER 7: CULTIVATION

Rittershausen, W. & B. (2000). *The Practical Encyclopedia of Orchids: A Complete Guide to Orchids and Their Cultivation*. Lorenz, London.

Rittershausen, W. & B. (2001). *The Gardener's Guide to Growing Orchids*. David & Charles, Newton Abbott.

Thorogood, E. (1992). Two views of cultivating *Cymbidium canaliculatum*. *Orchadian* 10, 8: 288–289.

Veitch, H.J. (1887–1894). *Manual of Orchidaceous Plants*. Veitch & Sons, London.

Williams, B. (1980). *Orchids for Everyone*. Salamander Books, London.

Williams, B.S. (1894). *Orchid Growers' Manual,* 7th edn. Victoria and Paradise Nurseries, London.

CHAPTER 8: HYBRIDISATION AND BREEDING

Averyanov, L. (1990). *Cymbidium ? pseudoballianum* Aver. *Bot. Zhurn.* 75, 5: 723.

Cheng, C. (1975). *Formosan Orchids: Terrestrial*. Taicheng, Taiwan.

Cheng, C. (1981). *Formosan Orchids*, 3rd edn. Taicheng, Taiwan.

Du Puy, D. & Cribb, P.J. (1988). *The Genus* Cymbidium. Christopher Helm, London.

Du Puy, D.J., Ford-Lloyd, B.V. & Cribb, P.J. (1984). A Numerical Taxonomic Analysis of *Cymbidium* section *Iridorchis* (Orchidaceae). *Kew Bulletin* 40: 421–434, + figs.

King, G. & Pantling, R. (1895). On some new orchids from Sikkim. *J. Asiatic Soc. Bengal* 64, 2: 338–339.

King, G. & Pantling, R. (1898). Orchids of the Sikkim – Himalaya. *Annals of the Royal Botanic Gardens, Calcutta* 8: 184–196, t. 247–262.

Li, H. & Feng, G.H. (1989). A new species of the sect. *Iridorchis* of the genus *Cymbidium*. *Acta Botanica Yunannica* 11, 1: 39–40.

Liu, F.Y. (1996). Two new species of *Orchidaceae* from Yunnan. *Acta Botanica Yunnanica* 18, 4: 411–414.

Liu, F.Y. & Perner, H. (2001). *Cymbidium baoshanense* und *Paphiopedilum purpuratum* var. *hainanense*, zwei neue Orchideentaxa aus China. *Orchidee* 52, 1: 61.

Maekawa, F. (1971). *The Wild Orchids of Japan in Colour*. Japan.

Masamune, G. (1984). *Cymbidium nishiuchianum* Makino. *Native Orchids of Nippon* 1: 68–69.

Rolfe, R.A. (1914). *Cymbidium cooperi*. *Orchid Rev.* 22: 94.

Sanders (1946). *Sanders' List of Orchid Hybrids*. Sanders, St. Albans.

Shaw, J.M. (2002). *Cymbidium nishiuchianum* Makino ex J.M. Shaw. *Orchid Rev.* 110 (1243): new orchid hybrids 13.

CHAPTER 9: OTHER USES

Chen, S.C. & Tang, T. (1982). *A general review of the Orchid Flora of China.* In: Arditti, J., ed., *Orchid Biology Reviews & Perspectives,* 2: 39–82. Cornell University Press, Ithaca and London.

Hu, S.Y. (1971). *The Orchidaceae of China* I. *Quart J. Taiwan Mus.* 24: 67–103.

Lawler, L.J. (1984). *Ethnobotany of the Orchidaceae.* In: Arditti, J., ed., *Orchid Biology, Reviews & Perspectives,* 3: 27–149. Cornell University Press, Ithaca and London.

CHAPTER 10: DISTRIBUTION AND BIOGEOGRAPHY

Barlow, B.A. (1981). The Australian Flora: its origin and evolution. *Flora of Australia* 1 (Introduction): 25–75. Australian Government Publishing Service, Canberra.

Burbridge, N.T. (1960). The phytogeography of the Australian Region. *Aust. J. Bot.* 8: 75–212.

Hooker, J.D. (1860). Introductory Essay, Botany of the Antarctic Voyage of H.M. Discovery ships 'Erebus' and 'Terror' in the years 1839–1843, III. *Flora Tasmaniae.* Lovell Reeve, London.

Hu, S.Y. (1971). The Orchidaceae of China 2. *Quart. J. Taiwan Mus.* 24: 38–112.

Lavarack, P.S. (1981). Origins and Affinities of the Orchid Flora of Cape York Peninsula. *Proceedings of the Orchid Symposium, 13th International Botanical Congress*: 17–26. Orchid Society of New South Wales, Sydney, Australia.

Liu, Y.F. & Nakayama, H. (2005). *Cymbidium goeringii* in Japan. *Orchids* 74, 6: 451–456.

Quisumbing, E.A. (1940). The genus *Cymbidium* in the Philippines. *Philippine J. Sci.* 72: 481–492, pl. 1–8.

Specht, R.L. (1981). Evolution of the Australian flora: some generalisations. In: A. Keast, *Ecological Biogeography of Australia*: 785–805. W. Junk, The Hague.

Van den Berg, C., Ryan, A., Cribb, P.J. & Chase, M.W. (2002). Molecular phylogenetics of *Cymbidium* (Orchidaceae: Maxillarieae): sequence data from internal transcribed spacers (ITS) of nuclear ribosomal DNA and plastid *matK. Lindleyana* 17, 2: 102–111.

Wu, Y.S., & Chen, S.C. (1980). A taxonomic review of the orchid genus *Cymbidium* in China. *Acta Phytototaxonomica Sinica* 18: 292–307.

Yukawa, T., Miyoshi, K. & Yokoyama, J. (2002). Molecular phylogeny and character evolution of *Cymbidium* (Orchidaceae). *Bull. Natn. Sci. Mus. Tokyo,* Ser. B, 28, 4: 129–139.

CHAPTER 11: PHYLOGENY

Bateman, R.M., Pridgeon, A.M. & Chase, M.W. (1997). Phylogenetics of subtribe Orchidinae (Orchidoideae, Orchidaceae) based on nuclear ITS sequences. 2. Infrageneric relationships and reclassification to achieve monophyly of *Orchis sensu stricto. Lindleyana* 12: 113–141.

Cameron, K.M., Chase, M.W., Whitten, W.M., Kores, P.J., Jarrell, D.C., Albert, V.A., Yukawa, T., Hills, H.G. & Goldman, D.H. (1999). A phylogenetic analysis of the Orchidaceae: evidence from *rbcL* nucleotide sequences. *Amer. J. Bot.* 86: 208–224.

Cox, A.V., Pridgeon, A.M., Albert, V.A. & Chase, M.W. (1997). Phylogenetics of the slipper orchids (Cypripedioideae, Orchidaceae): nuclear rDNA sequences. *Pl. Syst. Evol.* 208: 197–223.

Doyle, J.J. & J.L. (1987). A rapid DNA isolation method for small quantities of fresh tissues. *Phytochem. Bull. Bot. Soc. Amer.* 19: 11–15.

Dressler, R.L. (1993). *Phylogeny and Classification of the Orchid Family.* Dioscorides Press, Portland, Oregon.

Du Puy, D. (2005). New insights into the classification of *Cymbidium.* In: A. Raynal, A. Rougenant & D. Prat, eds., *Proceedings of the 18th World Orchid Conference*: 181–190. Dijon, France.

Du Puy, D. & Cribb, P.J. (1988). *The genus* Cymbidium. Timber Press, Portland, Oregon.

Felsenstein, J. (1985). Confidence limits on phylogenies: an approach using the bootstrap. *Evolution* 39: 783–791.

Freudenstein, J.V. & Rasmussen, F.N. (1999). What does morphology tell us about orchid relationships?—a cladistic analysis. *Amer. J. Bot.* 86: 225–248.

Johnson, L.A. & Soltis D.E. (1994). *matK* sequences and phylogenetic reconstruction in Saxifragaceae *s. s. Syst. Bot.* 19: 143–156.

Lindley, J. (1830–1840). *The genera and species of orchidaceous plants*. London.

Molvray, M., Kores, P.J. & Chase, M.W. (2000). Polyphyly of mycoheterotrophic orchids and functional influences on floral and molecular characters. In: Wilson, K.L. & Morrison D.A., eds., *Monocots: systematics and evolution*: 441–448. CSIRO Publishing, Melbourne.

Pridgeon, A.M., Solando, R. & Chase M.A. (2001). Phylogentic relationships in *Pleurothallidinae* (Orchidaceae): combined evidence from nuclear and plastid DNA sequences. *American Journal of Botany* 88 (12): 2286–2308.

Ryan, A., Whitten, W.M., Johnson, M.A.T. & Chase M.W. (2000). A phylogenetic assessment of *Lycaste* and *Anguloa* (Orchidaceae: Maxillarieae). *Lindleyana* 15: 33–45.

Schlechter, R. (1924). Die Gattungen *Cymbidium* Sw. und *Cyperorchis* Bl. *Fedde, Rep. Spec. Nov. Regni Veg.* 24: 96–110.

Seth, C.J. and Cribb, P.J. (1984). A reassessment of the sectional limits in the genus *Cymbidium* Swartz. In: Arditti, J., ed., *Orchid Biology, Reviews and Perspectives.* 3: 283–322. Cornell University Press, Ithaca.

Sun, Y., Skinner, D.Z., Liang, G.H. & Hulbert, S.H. (1994). Phylogenetic analysis of *Sorghum* and related taxa using internal transcribed spacers of nuclear ribosomal DNA. *Theor. Appl. Genet.* 89: 26–32.

Swofford, D.L. (1998). *PAUP*. Phylogenetic analysis using parsimony (*and other methods). Version 4.0.* Sinauer Associated, Sunderland, Massachusetts.

Van den Berg, C., Ryan, A., Cribb, P.J. & Chase, M.W. (2002). Molecular phylogenetics of *Cymbidium* (Orchidaceae: Maxillarieae): sequence data from interanl transcribed spacers (ITS) of nuclear ribosomal DNA and plastid *matK*. *Lindleyana* 17, 2: 102–111.

White, T.J., Bruns, T., Lee, S. & Taylor, J. (1990). Amplification and direct sequencing of fungal ribosomal RNA genes for phylogenetics. In: Innis, M., Gelfand, D., Sninsky, J. & White, T., eds., *PCR: A guide to Methods and Applications*: 315–322. Academic Press, San Diego, California.

Whitten, W.M., Williams N.H. & Chase M.W. (2000). Subtribal and generic relationships of Maxillarieae (Orchidaceae) with emphasis on Stanhopeinae: combined molecular evidence. *Amer. J. Bot.* 87: 1842–1856.

Yukawa, T., Miyoshi, K. & Yokoyama, J. (2002). Molecular phylogeny and character evolution of *Cymbidium* (Orchidaceae). *Bull. Natn. Sci. Mus. Tokyo*, Ser. B, 28, 4: 129–139.

CHAPTER 12: CONSERVATION

IUCN (2001). Red List of Threatened Species. Categories and Criteria (version 3.1). www.redlist.org.

Liu, Y.F. & Nakayama, H. (2005). *Cymbidium goeringii* in Japan. *Orchids* 74, 6: 451–456.

Luo, Y.B. (2005). The conservation and commercial development of Chinese cymbidiums. Endangered Species Scientific Newsletter CITES Commission, China 14 (www.cites.org.cn/newsletter/newsletter14-e.htm).

CHAPTER 13: THE CLASSIFICATION OF *CYMBIDIUM*

NOTE: bibliographical citations not included below may be found in the synonymy given under each species.

Ames, O. (1908). *Cymbidium. Orchidaceae* 2: 218–219.

Ames, O. (1915). *Cymbidium* Sw. *Orchidaceae* 5: 199.

Ames, O. (1925). Orchidaceae. In: Merrill, E., ed. An Enumeration of Philippines Plants. *Bureau of Science Publ.* 18: 253–458.

Ames, O. & Quisumbing, E. (1932). Genus *Cymbidium* Swartz. *Philippine J. Sci.* 49: 491–492, tt. 2 (4 & 5), 8, 21, 22.

Ames, O. & Schweinfurth, C. (1920). *Cymbidium* Sw. In: Ames, O., *Orchidaceae* 6: 212–214.

Anon. (1912). The International Horticultural Exhibition. *Orchid Rev.* 20: 163–171.

Anon. (1980). *Cymbidium kanran*. M. Ogawa, Tokyo.

Anon. (1984). *Cymbidium kanran* of Shikoku and Kishu. Seibundo-Shinkosha, Tokyo.

Averyanov, L.V. (1994). *Identification Guide to Vietnamese Orchids*. V.L. Komarov Botanical Institute, St Petersburg and Institute of Ecology and Biological Resources, Hanoi.

Averyanov, L.V. & Christenson, E. (1998). *Cymbidium schroederi. Orchids* 67: 712–713.

Averyanov, L.V., Cribb, P.J., Hiep, N.T. & Loc, P.K. (2003). *Slipper Orchids of Vietnam*. R.B.G. Kew.

Backer, C.A. & Bakhuizen, R.C. (1968). Orchidaceae. *Flora of Java* 3: 215–450.

Bailey, F.M. (1902). 15. *Cymbidium* Sw. *Queensland Fl.* 5: 1546–1548.

Bailey, F.M. (1913). *Comprehensive Catalogue of the Queensland Flora*. Govt. Printer, Brisbane.

Banerjee, M.L. & Thapa, B.B. (1978). *Orchids of Nepal*. Today & Tomorrow's Publ., New Delhi.

Barretto, G. & Youngsaye, J.L. (1980). *Hong Kong Orchids*. Urban Council, Hong Kong.

Bateman, J. (1866). *Cymbidium Hookerianum. Curtis's Bot. Mag.* 92: t. 5574.

Beaman, T.E., Wood, J.J., Beaman, R.S. & Beaman, J.T. (2001). *Orchids of Sarawak*. Natural History Publ., Kota Kinabalu, Sabah.

Benson, R. (1870). Orchid Culture. *Gardeners' Chronicle* 1870: 311, 631, 763, 796.

Bentham, G. (1873). 15. *Cymbidium* Swartz. *Fl. Australiensis* 6: 303.

Blume, C. (1825). *Cymbidium. Bijdragen* 378–380.

Blume, C.L. (1848). Orchideae. *Rumphia* 4: 38–56.

Blume, C.L. (1849). *Cyperorchis. Mus. Bot. Lugduno-Batavia* 1: 48.

Blume, C.L. (1858). *Collection des Orchidées les plus remarquables de l'archipel Indien et du Japon* 1: 90–93, t. 26.

Brown, R. (1810). *Cymbidium* Sw. *Prodromus Florae Novae Hollandiae* 331–332.

Chase, M.W., Cameron, K.M., Barrett, R.L. & Freudenstein, J.V. (2003). DNA data and Orchidaceae classification: a new phylogenetic classification. In: Dixon, K.W., Kell, S.P., Barrett, R.L. & Cribb, P.J., eds., *Orchid Classification*. Natural History Publ., Kota Kinabalu, Sabah.

Chen, S.C. (1999). *Cymbidium. Flora of China Orchidaceae* 2: 191–227.

Chen, S.C. & Liu, Z.J. (2003). Critical notes on some taxa of *Cymbidium. Acta Phytotax. Sinica* 41, 1: 83.

Chen, S.C., Tsi, Z.H. & Luo, Y.B. (1999). *Native Orchids of China in Colour*. Science Press, Beijing.

Cheng, C. (1979). *Formosan Orchids*, 2nd edn. Taicheng, Taiwan.

Cheng, C. (1981). *Formosan Orchids*, 3nd edn. Taicheng, Taiwan.

Comber, J.B. (1980). The species of *Cymbidium* in Java. *Orchid Digest* 44: 164–168, + figs.

Comber, J.B. (2001). *Orchids of Sumatra*. Royal Botanic Gardens, Kew and Natural History Publications (Borneo), Kota Kinabalu.

Comber, J.B. & Nasution, R. (1978). A new Indonesian *Cymbidium. Cymbidium hartinahianum. Orchid Dig.* 42: 55–57.

Cootes, J. (2001). *The Orchids of the Philippines*. Timber Press, Portland, Oregon.

Cribb, P.J. & Du Puy, D. (1983). *Cymbidium longifolium* and *Cymbidium elegans* (Orchidaceae). *Kew Bull.* 38: 65–67.

Cribb, P.J. & Tibbs, M. (2004). *A Very Victorian Passion. The Orchid Paintings of John Day*. Thames & Hudson, London.

Dockrill, A.W. (1966). *Cymbidium* Sw. in Australia. *Australian Pl.* 3: 293–295, + fig.

Dockrill, A.W. (1969). *Australian Indigenous Orchids* 1. The Society for Growing Australian Plants, Sydney, pp. 629–639, + figs.

Downie, S.R. (1925). Contributions to the Flora of Siam. Addit. XV. *Kew Bull.* 1925: 382–383.

Dressler, R.L. (1974). Classification of the orchid family. *Proceedings of the 7th World Orchid Conference*: 259–279.

Dressler, R.L. (1981). *The Orchids—Natural History and Classification*. Harvard University Press, Cambridge and London, 332 pp.

Dressler, R.L. (1993). *Phylogeny and Classification of the Orchid Family*. Dioscorides Press, Portland, Oregon.

Dressler, R.L. & Dodson, C.H. (1960). Classification and phylogeny in the Orchidaceae. *Annals of the Missouri Botanical Garden* 47: 25–68.

Du Puy, D.J. (1983). The Wildlife Sanctuary of Phu Luang, Thailand and its rich orchid flora. *Orchid Review* 91: 366–371.

Du Puy, D.J. (1984). Flowers of the Phu Luang Wildlife Sanctuary. *Kew Magazine* 1: 75–84.

Du Puy, D.J. (1986). *A taxonomic revision of the genus* Cymbidium *Sw. (Orchidaceae)*. PhD thesis, University of Birmingham and Royal Botanic Gardens, Kew.

Du Puy, D. (2005). New insights into the classification of *Cymbidium* (from molecular systematic analyses). In: A. Raynal-Roques, A. Rougenant & D. Prat, eds., *Proceedings of the 18th World Orchid Conference*: 181–190. Dijon, France.

Du Puy, D.J. & Cribb, P.J. (1988). *The genus* Cymbidium. Christopher Helm, London.

Du Puy, D.J., Ford-Lloyd, B.V. & Cribb, P.J. (1984). A Numerical Taxonomic Analysis of *Cymbidium* section *Iridorchis* (Orchidaceae). *Kew Bulletin* 40: 421–434, + figs.

Du Puy, D.J. & Lamb, A. (1984). The genus *Cymbidium* in Sabah. *Orchid Review* 92: 349–358, + figs.

Duthie, J.F. (1906). *Cymbidium*. In: R. Strachey, *Catalogue of the plants of Kumaon*: 176. Lovell Reeve, London.

Gagnepain, F. (1931). Trois Orchidacées nouvelles d'Indo-Chine. *Bull. Mus. Nat. Hist. Paris*, sér. 2, 3: 679–687.

Garay, L.A. & Sweet, H. (1974). *Orchids of the Southern Ryukyu Islands*. Botanical Museum, Harvard University.

Garden Life Magazine, ed. (1979). *Cymbidium goeringii* in Japan. Seibundo-Shinkosha, Tokyo.

Ghose, B.H. (1960). *Cymbidium suavissimum. Amer. Orchid Soc. Bull.* 29: 824.

Ghose, B.H. (1972). *Cymbidium tigrinum. Orchid Rev.* 80: 187.

Guillaumin, A. (1932). *Cymbidium*. In: Lecomte, H., ed., *Flore Générale de l'Indo-Chine* 6: 412–419.

Guillaumin, A. (1960). Notules sur quelques Orchidées de l'Indo-Chine XXII. *Bull. Mus. Nat. Hist. (Paris)* 2, sér. 32, 1: 115–117.

Guillaumin, A. (1961a). Notules sur quelques Orchidées de l'Indo-Chine XXV. *Bull. Mus. Nat. Hist. (Paris)* 2, sér. 32, 6: 562–565.

Guillaumin, A. (1961b). Notules sur quelques Orchidées de l'Indo-Chine XXVI. *Bull. Mus. Nat. Hist. (Paris)* 2, sér. 33, 3: 332–335.

Hara, H. (1985). On typification of David Don's names, i.e. *Cymbidium longifolium* (Orchidaceae). *Taxon* 34: 690–691.

Hara, H., Stearn, W. & Williams, L.H.J. (1978). *Cymbidium. Enumeration of the Flowering Plants of Nepal* 1: 37.

Hawkes, A.D. (1961). The correct name for *Cymbidium iridifolium. Austr. Orchid Rev.* 26: 135.

Hayata, B. (1914). Orchideae. *Icones Plantarum Formosanum* 4: 23–129, tt. 5–25.

Hayata, B. (1916). Orchideae. *Icones Plantarum Formosanum* 6: 66–94, tt. 11–14.

Herraman, R. (2005). *Cymbidium canaliculatum. Austr. Orchid Rev.* 70, 4: 53–64.

Holttum, R.E. (1953, 1957, 1964). *Orchids of Malaya*. Government Printing Office, Singapore.

Hooker, J.D. (1865). *Cymbidium pendulum* var. *atropurpureum. Curtis's Bot. Mag.* 94: t. 517.

Hooker, J.D. (1870). *Cymbidium canaliculatum. Curtis's Bot. Mag.* 96: t. 5851.

Hooker, J.D. (1891). *Cymbidium. Flora of British India* 6: 8–15.

Hooker, J.D. (1894). *Cymbidium sikkimense* Hook.f. *Icones Plantarum* 22: t. 2117.

Hooker, W.J. (1856). *Cymbidium chloranthum. Curtis's Bot. Mag.* 82: t. 4907.

Hu, S.Y. (1971a). The Orchidaceae of China, part 1. Orchids in the life and culture of the Chinese people. *Quart. Journ. Taiwan Mus.* 24: 67–103.

Hu, S.Y. (1971b). The Orchidaceae of China, part 2. The composition and distribution of orchids in China. *Quart. Journ. Taiwan Mus.* 24: 181–255.

Hu, S.Y. (1973). The Orchidaceae of China 5, *Cymbidium* and *Cyperorchis. Quart. Journ. Taiwan Mus.* 26: 134–142.

Hunt, P.F. (1970). Notes on Asiatic Orchids 5. *Kew Bulletin* 24: 93–94.

Hurraman, R. (2005). *Cymbidium canaliculatum. Australian Orchid Review* 70, 4: 53–59.

Hynniewata, T.M. (1979). Rediscovery of *Cymbidium tigrinum* Par. ex Hook. from India. *Orchid Rev.* 87: 219.

Jayaweera, D.M. (1981). 21. *Cymbidium.* In: Dassanayaka, M.D., ed., *Fl. Ceylon* 2: 182–188.

Jennings, S. (1875). *Cymbidium eburneum. Orchids and how to grow them in India*: t. 16.

Jones, H.G. (1974). *Orchidaceae novae vel minus cognitae. Reinwardtia* 9: 71–76.

King, G. & Pantling, R. (1895). On some new orchids from Sikkim. *J. Asiatic Soc. Bengal* 64, 2: 338–339.

King, G. & Pantling, R. (1898). Orchids of the Sikkim—Himalaya. *Annals of the Royal Botanic Gardens, Calcutta* 8: 184–196, t. 247–262.

Kingdon Ward, F. (1940). *Cymbidium tracyanum. Gard. Chron.* Ser. 3, 108: 155.

Kjellsson, G., Rasmussen, F.N. & Du Puy, D.J. (1985). The pollination of *Dendrobium infundibulum, Cymbidium insigne* (Orchidaceae) and *Rhododendron lyi* (Ericaceae) by *Bombus eximius* (Apidae) in Thailand: a possible case of floral mimicry. *J. Trop. Ecology* 1: 289–302.

Leaney, R.F. (1966). *Cymbidium* orchids. *Australian Pl.* 3: 291, + fig.

Lee, Y.N. (1976). *Cymbidium. Ill. Fl. Fauna Korea* 18: 771–772, 826, t. 183.

Lin, T.P. (1977). *Native Orchids of Taiwan* 2: 101–134, + figs., Chong Tao Printing Co. Ltd., Taiwan.

Lindley, J. (1833). XCVII. *Cymbidium. The Genera and Species of Orchidaceous Plants*: 161–172. Ridgways, London.

Lindley, J. (1842). *Cymbidium finlaysonianum. Bot. Reg.* 30: t. 24.

Lindley, J. (1843). *Cymbidium chloranthum. Bot. Reg.* 29: 68.

Lindley, J. (1854). *Cymbidium pendulum* var. *atropurpureum. Gard. Chron.* 1854: 287.

Lindley, J. (1858). Contributions to the orchidology of India II. *J. Linn. Soc.* 3: 1–63.

Liu, Z.J., Chen, S.C. & Ru, Z.Z. (2006). *The genus Cymbidium in China.* Science Publishing, Beijing.

Liu, T.S. & Su, H.J. (1978). *Cymbidium* Sw. *Flora of Taiwan* 5: 937–950.

Liu, Z.J. & Chen, S.C. (2000). *Cymbidium rigidum* sp. nov., a new orchid from Yunnan. *Acta Phytotax. Sinica.* 38, 6: 570–572.

Liu, Z.J. & Chen, S.C. (2002). *Cymbidium paucifolium.* A new species of Orchidaceae from China. *J. Wuhan Bot. Res.* 20, 5: 350–352.

Long, C.L., Li, H. & Dao, Z.L. (2003). A new species of *Cymbidium* Sw. (Orchidaceae) from Tibet (Xizang), China. *Novon* 13: 203.

Macpherson, K. & Rupp, H.M.R. (1936). Further notes on orchid pollination. *North Queensland Naturalist* 4: 25.

Maekawa, F. (1958). Notes on Japanese orchids I. *J. Jap. Bot.* 33: 320.

Maekawa, F. (1971). *The Wild Orchids of Japan in Colour.* Japan.

Makino, T. (1902). Observations on the Flora of Japan. *Bot. Mag. Tokyo* 16: 10–11.

Makino, T. (1912). *Cymbidium. Iinuma, Somoku-Dzusetsu* 18: 1179–1185.

Mark, F., Ho, H.S. & Fowlie, J.A. (1986). An artificial key to the *Cymbidium* species in Taiwan. *Orchid Digest* 50: 13–24.

Menninger, E. (1961). Catalogue of *Cymbidium* species with synonyms and excluded names. *Amer. Orchid Soc. Bull.* 30: 865–876.

Menninger, E. (1965). Parishii regained. *Amer. Orchid Soc. Bull.* 34: 892–897.

Micholitz, W. (1904). Two new Cymbidiums. *The Garden* 66: 141.

Mueller, F. (1859). *Cymbidium albuciflorum. Fragm. Phyt. Austr.* 1: 188.

Mueller, F. (1879). 4) *Cymbidium hillii* Ferd. Muell. *Gartenflora* 28: 138–139.

Nagano, Y. (1955). Miniature Cymbidiums in Japan. *Amer. Orchid Soc. Bull.* 24: 735–743.

Nagano Y. & Nagano Mrs (1957). *Miniature Orchids in Japan and China.* Nagano & Nagano, Tokyo.

Nakai, T. (1931). *Notulae ad plantas Japoniae et Korae. Bot. Mag. Tokyo* 45: 91–137.

Ohwi, J. (1965). *Flora of Japan* (in English). Smithsonian Inst., Washington.

Parish, C. (1883). Order Orchideae. In: Mason, F., ed. *Burma, its People and Productions* 2: 148–202.

Paxton, J. (1843). *Cymbidium devonianum. Mag. Bot.* 10: 97–98, + fig.

Perner, H. (2002). *Cymbidium goeringii* (Rchb.f.) Rchb.f. 1852 in China. *Die Orchidee* 53, 6: 722–733.

Pfitzer, E. (1887). *Entwurf einer naturlichen Anordnung der Orchideen.* Heidelberg.

Pradhan, U.C. (1979). *Indian Orchids: Guide to Identification and Culture* 2: 465–80. Thomas Press Ltd., India.

Pridgeon, A.M., Cribb, P.J., Chase, M.W. & Rasmussen, F., eds. (1999, 2001, 2003, 2005). *Genera Orchidacearum* I–IV. Oxford University Press.

Quisumbing, E. (1940). The genus *Cymbidium* in the Philippines. *Philippine J. Sci.* 72: 481–483, t. 1 (1 & 2), t. 4.

Raizada, M.B., NaithariNaithani, H.B. & Saxena, H.O. (1981). *Orchids of Mussoorie.* Bishen Singh Malendra Pal Singh, Dehra Dun.

Reeve, T. (1984). The native *Cymbidium* species of Papua New Guinea. *Orchadian* 8: 33–35.

Reichenbach, H.G. (1852). Orchidaceae (*Cymbidium* and *Cyperorchis*). *Walpers Ann. Bot.* 3: 547–548.

Reichenbach, H.G. (1856). 41. *Cymbidium variciferum* Rchb.f.. *Bonplandia* 4: 324.

Reichenbach, H.G. (1864). 1427. *Cymbidium. Walpers Ann. Bot.* 6: 622–627.

Reichenbach, H.G. (1872). *Cymbidium mannii. Flora* 55: 274.

Reichenbach, H.G. (1875). *Cymbidium elegans* Lindl. *Gard. Chron.* n.s. 13: 429.

Reichenbach, H.G. (1878). *Cymbidium leachianum. Gard. Chron.* n.s. 10: 106.

Reichenbach, H.G. (1880). *Cymbidium elegans* (Lindl.) var. *obcordatum var. nov. Gard. Chron.* n.s. 13: 41.

Rendle, A. (1898). Two new Queensland Cymbidiums. *J. Bot.* 36: 221.

Rendle, A. (1901). Queensland orchids. *J. Bot.* 39: 197.

Rolfe, R.A. (1903). *Cymbidium atropurpureum. Orchid Rev.* 11: 190–191.

Rolfe, R.A. (1904). *Cymbidium nipponicum. Orchid Rev.* 12: 303.

Rolfe, R.A. (1914). *Cymbidium cooperi. Orchid Rev.* 22: 94.

Rolfe, R.A. (1917). *Cymbidium aloifolium* and its allies. *Orchid Rev.* 25: 173–175.

Rolfe, R.A. (1919). *Cymbidium chloranthum. Orchid Rev.* 27: 128–129.

Roxburgh, W. (1814). *Cymbidium* Swartz. *Hort. Bengal.*: 63.

Roxburgh, W. (1832). *Cymbidium* Swartz. *Fl. Indica* 3: 458.

Rupp, H.M. (1930). X. *Cymbidium* Sw. *Guide to the Orchids of New South Wales*: 48.

Rupp, H.M. (1934). Review of the genus *Cymbidium* in Australia 1. *Proc. Linn. Soc. New South Wales* 59: 94–100, f. 1–3.

Rupp, H.M. (1937). Notes on Australian orchids III. A review of the genus *Cymbidium* in Australia II. *Proc. Linn. Soc. New South Wales* 62: 299–302.

Rupp, H.M. (1939). Fitzgerald's "Australian Orchids". *Australian Orchid Rev.* 4: 65–66, + fig.

Rupp, H.M. (1943). 36. *Cymbidium* Sw. *The Orchids of New South Wales*: 128.

Schlechter, R. (1910). XIII. *Orchidaceae novae et criticae. Fedde, Repert. Sp. Nov. Regni Veg.* 8: 560–574.

Schlechter, R. (1911). Zur Kenntnis der Orchideen van Celebes. *Fedde, Repert. Sp. Nov. Regni Veg.* 10: 177–213.

Schlechter, R. (1913). XX. *Orchidaceae novae et criticae. Fedde, Repert. Sp. Nov. Regni Veg.* 12: 109.

Schlechter, R. (1918). Ueber einige neue Cymbidien. *Orchis* 12: 45–48.

Schlechter, R. (1919). *Orchideologiae Sino-Japonicae Prodromus. Fedde, Repert. Sp. Nov. Regni Veg. Beih.* 4: 1–319.

Schlechter, R. (1924). Die Gattungen *Cymbidium* Sw. und *Cyperorchis* Bl. *Fedde, Repert. Sp. Nov. Regni Veg.* 20: 96–110.

Schlechter, R. (1925). Die Orchidaceen der Insel Celebes. *Fedde, Repert. Sp. Nov. Regni Veg.* 21: 197.

Schlechter, R. (1928). *Cymbidium papuanum. Fedde, Repert. Sp. Nov. Regni Veg., Beih.* 21: t. 336, no. 1296.

Seidenfaden, G. (1975). *Contribution towards a revision of the orchid flora of Cambodia, Laos & Vietnam.* Olsen & Olsen, Friedenborg, Denmark.

Seidenfaden, G. (1983). Orchid Genera in Thailand 11, *Cymbidieae* Pfitz. *Opera Bot.* 72: 65–93.

Seidenfaden, G. (1992). The orchids of Indochina. *Opera Bot.* 114: 5–502.

Seth, C.J. (1982). *Cymbidium aloifolium* (Orchidaceae) and its allies. *Kew Bull.* 37: 397–402.

Seth, C.J. & Cribb, P.J. (1984). A reassessment of the sectional limits in the genus *Cymbidium* Swartz. In: Arditti, J., ed., *Orchid Biology, Reviews and Perspectives* 3: 283–322. Cornell University Press, Ithaca and London.

Shaw, J.M. (2002). *Cymbidium nishiuchianum* Makino ex J.M. Shaw. *Orchid Rev.* 110 (1243): new orchid hybrids 13.

Smith, J.J. (1905). Die Orchideen von Java. *Flora von Buitenzorg* 6: 475–484.

Smith, J.J. (1910). *Cymbidium atropurpureum* Rolfe var. *olivaceum* J.J.S. n. var. *Bull. Dép. Agric. Ind. Néerland.* 43: 60–62.

Smith, J.J. (1911). *Die Orchidaceen von Java. Figuren-Atlas:* tt. 363–368.

Smith, J.J. (1933). Enumeration of the Orchidaceae of Sumatra. *Fedde, Repert. Sp. Nov. Regni Veg.* 32: 129–386.

Su, H.J. (1975). *Native Orchids of Taiwan,* 2nd edn. Harvest Farm Magazine, Taipei.

Summerhayes, V. (1942). *Cymbidium schroederi. Curtis's Botanical Magazine* 163: t. 9637.

Summerhayes, V. (1963). *Cymbidium goeringii. Curtis's Botanical Magazine* 174: t. 413.

Swartz, O. (1799). *Cymbidium grandiflorum. Nov. Act. Soc. Sc. Upsal.* 6: 76.

Tanaka, R. & Kamemoto, H. (1974). List of chromosome numbers in species of Orchidaceae. In: Withner, C.L., ed., *The Orchids, Scientific Studies:* 411–483. John Wiley & Sons, New York.

Taylor, P. & Woods, P. (1976). *C. giganteum* cv. Wilsonii. *Curtis's Bot. Mag.* 181: n.s. t. 704.

Teijsmann, J.E. & Binnendijk, S. (1866). *Cymbidium sanguineum. Cat. Hort. Bog.:* 51.

Thunberg, C.P. (1794). *Epidendrum ensatum. Icon. Fl. Japan* 1: t. 8.

Tierney, W.F. (1957). Australian *Cymbidium* species. *Amer. Orchid Soc. Bull.* 26: 168.

Trimen, R. (1885). *Cymbidium. A systematic catalogue of the Flowering Plants and Ferns indigenous to or growing wild in Ceylon:* 89.

Trimen, R. (1898). *Cymbidium* Sw. *Handb. Fl. Ceylon* 4: 179–180, t. 90.

Trudel, N. (1983). *Cymbidium whiteae* King & Pantl. Ein wiederfundenes *Cymbidium. Die Orchidee* 34, 3: 100–102, + figs.

Tso, C.L. (1933). Orchid Flora of Canton. *Sunyatsenia* 1: 131–156.

Tuyama, T. (1941). In: Nakai, T., *Icon. Pl. As. Orient.* 4: 363–365, t.118.

Van den Berg, C., Ryan, A., Cribb, P.J. & Chase, M.W. (2002). Molecular phylogenetics of *Cymbidium* (Orchidaceae: Maxillarieae): sequence data from internal transcribed spacers (ITS) of nuclear ribosomal DNA and plastid *matK. Lindleyana* 17, 2: 102–111.

Wight, R. (1852). *Cymbidium. Icones Plantarum Indiae Orientalis* 5: 11, 21, tt. 1687–1688, 1753.

Willdenow, C. (1805). 1604. *Cymbidium* Swartz. *Species Plantarum* ed. 4, 4: 106.

Wimber, D.E. & D.R. (1968). Floral characteristics of diploid and neo-tetrapoid cymbidiums. *Amer. Orchid Soc. Bull.* 37: 572–576.

Winter, K., Wallace, B.J., Stocker, G.C. & Roksandic, Z. (1983). Crassulacean acid metabolism in Australian vascular epiphytes and some related species. *Oecologia* 57: 129–141.

Wood, J.J. (1983). A new species of *Cymbidium* from Borneo. *Kew Bull.* 38: 69–70.

Wood, J.J. & Du Puy, D. (1984). A recently described *Cymbidium* from Borneo. *Cymbidium borneense* J.J. Wood. *Orchid Dig.* 48: 115–116, + figs.

Wu, Y.S. & Chen, S.C. (1966). *Tres species novae generic Cymbidii e Provincial Szechuan. Acta Phytotaxonomica Sinica* 11: 31–34.

Wu, Y.S. & Chen, S.C. (1980). A taxonomic review of the orchid genus *Cymbidium* in China. *Acta Phytotaxonomica Sinica* 18: 292–307.

Yen, T.C. (1964). *Cymbidium ensifolium* var. *susin. Icon. Cymbid. Amoyens.* D.b. 1.

Ying, S.S. (1976). *Cymbidium tortisepalum. Quart. J. Chinese Forestry* 11(2): 100–101.

Ying, S.S. (1977). *Coloured Illustrations of the Indigenous Orchids of Taiwan* 1: 415.

Yukawa, T., Miyoshi, K. & Yokoyama, J. (2002). Molecular phylogeny and character evolution of *Cymbidium* (Orchidaceae). *Bull. Natn. Sci. Mus. Tokyo,* Ser. B, 28, 4: 129–139.

Yukawa, T. & Stern, W. (2002). Comparative vegetative anatomy and systematics of *Cymbidium* (Cymbidieae: Orchidaceae). *Bot. J. Linn. Soc.* 138, 4: 383–419.

CYMBIDIUM SPECIMENS EXAMINED

C. aliciae

Philippines: Luzon: *Ramos & Edaòo* 75 (AMES, K). **Hort.:** cult. Walker s.n. (K).

C. aloifolium

Bangladesh: Wallich 7352 (K); *J.D. Hooker & T. Thompson* 596 (K). **Cambodia:** *Regnier* s.n. (P); *Thorel* 1236 (P). **China:** Guangdong: *McClure* 13419 (K), *Metcalfe* 17070 (K); Guangxi: *Morse* 102 (K); *Qin Hai-ning et al.* 891124 (K, PE); Guizhou: *Cavalerie* 3424 (K, P), Yunnan: *Cavalerie* 8185 (K, P); *Henry* 12968 (K), 12968A (K); *C.W. Wang* s.n. (AMES); Without exact loc.: Hort. Soc. London (K-LINDL). **India:** Andaman Islands: *Kurz* s.n. (K); Assam: *Clarke* 40722 (K); Herb. Hookerianum s.n. (K); *Keenan* s.n. (K); *Koelz* 29670 (K); *Parry* 714 (K); *Tessier-Yandell* T-7.15 (K); *Thakur Rup Chand* 7055 (K); Bengal: *Gamble* 9433 (K); Madras: *Baulur* 1200 (K), 1201 (K); *Fischer* 4692 (K); *Matthew* 45400 (C, K); *Matthew & Paramasivan* 24280 (C, K); *G. Thomson* s.n. (K); *Rottler* s.n. (K); Meghalaya: *J.D. Hooker & Thompson* s.n. (K); Orissa: *Mooney* 471 (K); Salem Distr.: *Perumal & Manoharam* RHT22018 (c); *Venugopal* RHT15822 (c); Sikkim, *Clarke* 11811 (K); *J.D. Hooker* 228 p.p. (K-LINDL), 236 (K), s.n. (K); *Pantling* 268 (K, P, iso. of *C. simulans*); *White* s.n. (K); N India: *Barnwell* s.n. (K); Malabar, *Abraham* 3136 (K); *Barnes* 2150 (K), H39/1944 (K); *Christenson & Amy* 929 (CONN, K, MSU), 990 (CONN, K, MSU); *Ritchie* 1418 (K), 1419 (K); *Stocks & Law* s.n. (AMES); *Wallich* 7357 (K-LINDL). **Indonesia:** Java: *Comber* 1367 (K). **Laos:** *d'Orleans* s.n. (P); *Pierre* 1354 (P); *Spire* 926, 1092 & 1489 (P); *Vidal* 4177 (P). **Myanmar:** Shan States: *Manders* 92 (K); *Samuel* 13548 (K). **Philippines:** *Cuming* 2121 (P). Sri Lanka: *Thwaites* 3379 (P). **Sri Lanka:** *Comber* 1605 (K); *Cramer* 4330 (K); Herb. Hookerianum s.n. (K); *Thwaites* 754 (K), 3379 (K); *Tirvengadam, Cramer & Waas* 488 (K). **Thailand:** *Bare* s.n. (K), *Clulow* s.n. (K); *Collins* 142 (K); *Cumberledge* 751 (K); *Kerr* 42 (K), 0345 (K), 0519 (C, K), 0724 (K), 0804 (C, K), 01014 (K); *Merrill King* 5443 (AMES, C, K); *Put* 1530; *Schomburgk* s.n. (K-LINDL). **Vietnam:** *Averyanov* 2313 (253) (C, LE) & s.n. (LE); *Averyanov, Binh & Tam* CBL 958 (LE); *Averyanov & Klaravueva* 843 (C, LE); *L. Averyanov, N.T. Ban, A. & L. Budantsev, N.T. Hiep, D.D. Huyen, P.K. Loc & G. Yakovlev* VH 1043 (LE, P); *Averyanov* 1985 (130) (LE); *Averyanov & Hiep* VH 4888 (LE); *Balansa* 2011 (P) & 2012 (K, P); *Bon* 1794, 1936, 2053 & 2112 (P); *Demange* s.n. (P); *Dournes* s.n. (P); *Evrard* 868 (P); *Fleury* 32417 & 32429 (P); *N.T. Hiep* 365 (LE) & 708 (P); *Hiep & Averyanov* NTH 2626 (LE); *Hiep, Loc & Averyanov* NTH 3875 (LE); *Kovetsko-Vyethamska* 3265 (LE); *Krempf* 1446 (P); *Loc, Hoang & Averyanov* CBL 1591 (LE); *Poilane* 32432 (P); *Vu Tu Tien & H.A. Arpopov* 12195b (LE); *W.T. Tsang* 29345 (AMES, C, K, P). **Hort.:** cult. Shepherd s.n. (K, type of *C. limbatum*).

C. atropurpureum

Indonesia: Banka: *van Leeuwen-Reijnvaan* s.n. (NSW). Java: *J.J. Smith* 48 (K); hort. Bogor 54 (NSW); hort. Rollisson (K, type of *C. atropurpureum*). **Malaysia:** Sabah: *Lamb* 339/85 (K, SAN); *Mason* s.n. (K). **Philippines:** Leyte: *Wenzel* 367 (AMES, K); Mindanao: *Copeland* 648 (K); *Sulit* 10161 (K). **Vietnam:** *Averyanov* 2267 (C, LE); *Kovetsko-Vyethamska* 4262 (LE).

C. × ballianum

Hort.: cult. Holford s.n. (K); cult. Kew s.n. (K); cult. Mansell & Hatcher s.n. (K); cult. Sander s.n. (K, type of *C. ballianum*).

C. banaense
Vietnam: *Poilane* 29022 (P, type of *C. banaense*).

C. bicolor subsp. bicolor
India: *J.D. Hooker* 234 (K-LINDL); *Matthew & Rajendram* RHT 48878 (C); *Stocks & Law* s.n. (K). **Sri Lanka:** *Codrington* 38 (K); *Hooker* s.n. (E); *Macrae* 54 (K-LINDL, type of *C. bicolor*); cult. Peradeniya B.G. 819 (K), s.n. (K); *Roelfsema* 913074 (K, L); *Walker* s.n. (K); *Wilhelm* 133 (K). **Hort.:** cult. Bull s.n. (K); cult. Sander s.n. (K); cult. van Imschoot s.n. (K).

C. bicolor subsp. obtusum
Bangladesh: *Wallich* 7357 (K). **China:** Hainan: *S.K. Lau* 478 (K); *C.L. Lei* 479 (K); *Metcalf* 17069 (K); *Tsang Wai-Tak* 17065 (K). **India:** Sikkim: *J.D. Hooker* 228 (K); *Pantling* 441 (K); W. Himalayas: *Inayat* 24097 (K). **Myanmar:** *Chin* 5754 (K); *Daun* 66 (K); *Rule* 5388 (K), s.n. (K); *Venning* 48 (K). **Thailand:** *Bare* s.n. (K); *Cumberledge* 712 (K), s.n. (K); *Hennipman* 3967 (C, K, L); *Kerr* 126 (K), 0145 (K), 0929 (K); *Menzies & Du Puy* 120 (K, type of subsp. *obtusum*); *Seidenfaden* N3.66 (C). **Vietnam:** *Tixier* 27 (P).

C. bicolor subsp. pubescens
Indonesia: Java: *Comber* 1281 (K); Sulawesi: *Koorders* 29514 (K, L), 29515 (K, L), 29517 (K, L); Sumatra: *W.N. & C.M. Bingham* 618 (AMES); *Schlechter* 15857 (K). **Malaysia:** Sabah: *Beaman* 10506 ((K, MSC, UKMS); *J. Clemens* s.n. (AMES); *Darnton* 314 (BM); *Giles* 611 (K); *Giles & Woolliams* PB 134 (K), PB 137 (K); *Lamb* SAN 87483 (K, SAN); *Lugas* 1744 (K, Sabah Parks); *Mason* s.n. (K); *Wood* 719 (K); Sarawak: *Haviland* s.n. (K). **Philippines:** Palawan: *Birk* s.n. (K). **Singapore:** *Cuming* s.n. (K-LINDL, type of *C. pubescens*). **Thailand:** *Schomburgk* s.n. (K-LINDL).

C. borneense
Malaysia: Sabah: *Carr* SFN27789 (AMES); *Lamb* 7 (K), 108/83 (K, SAN), C 18 (K, SAN), MAL 18 (K, SAN), SAN 93357 (K, SAN); *Lomudin Tandong* 349 (K, Sabah Parks); Sarawak: *Chai* S39461 (K, L, SAR, UA); *Vallack, Hollis & Lewis* in *Lewis* 314 (K, type of *C. borneense*).

C. canaliculatum
Australia: *Abell* 10027 (NSW) (var. *sparkesii*), 10028 (NSW) (var. *sparkesii*); *Barrett* s.n. (NSW); *Brown* sn. (K), 5503 (BM, K, type of *C. canaliculatum*); *G. Burrow* s.n. (NSW); *Cady* s.n. (NSW); *Campbell* s.n. (var. *sparkesii*); *Chippendale* 7/1961 (NSW); *Clarkson* 3582 (BRI, K, NSW); *Coe* s.n. (NSW); *Coveny & Wilson* 11728 (K, NSW); *Cunningham* s.n. (K); *Dwyer* s.n. (NSW); *Froggatt* 5550/18 (NSW); *Hann* 257 (K); *Hind* 630 (NSW); *Johnson & Blaxell* 1049 (NSW); *Kehoe* 6632/11 (NSW) (var. *sparkesii*); *Lazarides* 6606 (CAN, K); *MacConachie* NT 28124 (NSW); *Maiden* s.n. (NSW); *Mitchell* 148 (K), 454 & 469 (K-LINDL); *Mueller* s.n. (K); *Musgrave* s.n (NSW); *Nicholls* s.n. (NSW); *Oldfield* s.n. (K); *Paddison & Baker* 246 (NSW); *S. Parker* 128 (NSW); *Pedley* 705 (BRI, K); *Perry* 1160 (CAN, K, NSW) (var. *sparkesii*); *Rupp* 7 (NSW), 387d (NSW), s.n. (K, NSW); *Sparks* s.n. (K, type of *C. sparkesii*); *Swain* s.n. (NSW); *Taylor* 2056 (CAN, K); *Tierney* in *Rupp* 387f (NSW); *van Royen* 10036 (BISH, K); *Veitch* s.n. (K); *K.Wilson* 1410 (NSW); *Yeomans* s.n. (NSW); *Yeoward* 6371/11 (NSW) (var. *sparkesii*). **Hort.:** cult. O'Brien (K).

C. chloranthum
Malaysia: Sabah: *Carr* 26596 (AMES); *J. Clemens* 50 (AMES, K), 51 (AMES); *J. & M.S. Clemens* 40456 (BM, K); *Darnton* 310 (K); *Lamb* 104/83 (K, SAN). **Without exact loc.:** cult. Loddiges (K-LINDL, ?type of *C. variciferum* Rchb.f.).

C. cochleare

India: Meghalaya: *Mann* s.n. (K); Sikkim: *J.D. Hooker* 235 (K-LINDL, type of *C. cochleare*), s.n. (K, K-LINDL); *Pantling* 352 (K). **Thailand:** *Roebelen* s.n., cult. Sander (K). **Hort.** Cult. Charlesworth s.n. (K); cult. R.B.G. Kew s.n. (K).

C. cyperifolium

Bhutan: *Griffith* 454 (K-LINDL, type of *C. viridiflorum*). **Cambodia:** *Godefroy-Lebeuf* 427 (K). **China:** Fukien, *de Latouche* s.n. (P); Guangdong, *W.T. Tsaang* 26729 (P). **India:** Assam: *Kingdon-Ward* 8087 (K); *Koelz* 32173 (K), 32266 (K); *Parry* 573 (K); Thakur Rup Chand 7313 (K); Manipur: *Watt* 5988 (K); Meghalaya: *Bor* 20894 (K); *Griffith* 5264 (K), s.n. (K); *J.D. Hooker & Thompson* s.n. (K); Naga Hills: *Clarke* 41850A (K); *Kingdon-Ward* 7770 (K); Sikkim: *Edgeworth* 27 (K); *J.D. Hooker & Thompson* 267 (K-LINDL); *Pantling* 306 (K); *Wallich* 7353 (K, K-LINDL, type of *C. cyperifolium*). **Myanmar:** *Swinhoe* 133 (K). **Thailand:** *Garrett* 769 (K); *Kerr* 3999 (K, type of subsp. *indochinense*); *Put* 3972 (K, type of subsp. *indochinense*). **Vietnam:** *Averyanov, Loc & Tam* CBL 740 (LE); *Averyanov & Binh* VH 4006 (LE); *Poilane* 23969 (P).

C. dayanum

China: Guangdong, *T.W. Tak* 16356 (AMES); *W.T. Tsang* 26729 (C, K); Hainan, *F.C. Chow* 73770 (AMES); *N.K. Chun & C.L. Tso* 44117 (K, NY); *W.T. Tsang* 15750 (K). Taiwan: *Henry* 1352 (K). **India:** Sikkim: cult. Day s.n. (K, W, type of *C. dayanum*); *Pantling* 51 (K, type of *C. simonsianum*); *T. Lobb* s.n. (K). **Indonesia:** Sumatra: *Comber* 1433 (K). **Laos:** cult. Kew s.n. (K). **Malaysia:** Sabah: *Amin & Francis* SAN 123419 (K, SAN); *Amin & Jarius* SAN 114341 (K, SAN); *Carr* 26799 (AMES, K); *Chew, Corner & Stainton* 1380 (K); *J. Clemens* 39 (AMES, K), 74 (AMES, type of *C. angustifolium*), 82 (AMES); *J. & M.S. Clemens* 26124 (K), 26771 (K); *Lamb* 108/83 (K); **Myanmar:** *Rule* s.n. (K). **Philippines:** Luzon: *Curtis* 35 (K); *Loher* 565 (K). **Thailand:** *Kerr* 113 (K, type of *C. sutepense*), s.n. (K); *Koyama* T61486 (AMES).**Vietnam:** *Averyanov* 1950, 2100, 2354 (184) & s.n. (LE); *Averyanov & Klaravueva* 191 (LE); DPB 83 N 157 (LE); *COBETCKO-BbETHAMCKAR* 3652 & 4128 (LE). **Hort.:** cult. Glasnevin B.G. (K); cult. R.B.G. Kew (K).

C. devonianum

India: Sikkim: *Gamble* 10333 (K); *Griffith* 128/228 (K-LINDL, type of *C. syringodorum*); *J.D. Hooker* 228 p.p. (K-LINDL), s.n. (K, type of *C. sikkimense*); *J.D. Hooker & T. Thompson* 225 (K, K-LINDL); *Pantling* 265 (K); *Subrahmanyam* s.n. (K). **Thailand:** *Williams* s.n. (K). **Vietnam:** *Hoch* s.n. (P); *Sepaldi* 211 (P); *Tixier* s.n. (P). Hort.: cult. Jackson s.n. (K); cult. R.B.G. Kew s.n. (K); cult. Summers s.n. (K). **Hort.:** cult. Sander s.n. (K).

C. eburneum

China: Yunnan: *Forrest* 26146 (E, K), 26650 (E); *H.T. Tsai* 58739 (AMES); *T.T. Yu* 17912 (E). **India:** *Griffith* s.n. (K-LINDL); cult. Loddiges (K-LINDL, type of *C. eburneum*); Assam: *Bourne* s.n. (K); Bengal: *Prakash* s.n. (K); Meghalaya: *Assam Forest Herbarium* 329 (K); Sikkim: *Clarke* 13971 (K); *Pantling* 108 (K); **S. Vietnam**, *Averyanov, Binh, Hiep, Loc & Lowry* VH 4460 (LE); *Averyanov, Ban, Binh, A. & L. Budantzev, Huyen, Loc, Tam & Yakovlev* VH 399 (LE). **Hort.:** cult. Aldworth s.n. (K); cult. Castle s.n. (K); cult. Rothschild s.n. (K); cult. R.B.G. Kew s.n. (K); cult. Veitch s.n. (K).

C. elegans

China: Yunnan, *d'Orleans* s.n. (P); *Forrest* 26164 (P). **India:** Assam, *Clarke* 41760A (K); *Griffith* s.n. (K); *Koelz* 24082 (K); Megalhaya: *J.D. Hooker & Thompson* s.n. (K); *Griffith* 73 (K), 229 (K-LINDL, type of *C. densifolium*), 5262 (K), 5263 (K), 5266 (K), s.n. (K-LINDL); *J.D. Hooker &*

Thompson 231 & 232 p.p. (K-LINDL); *Mann* 42 (K), s.n. (K); *Parry* 562 (K); *Prain* 23 (K); Sikkim: *Clarke* 9408 (K), 9594 (K); *Gamble* 2056A (K), 25144B (K); *J.D. Hooker* 49 (K), 229p.p. (K), 232 (K), s.n. (K); *Pantling* 14 (K); *Treutler* s.n. (K); Without exact loc.: *Junghuhn* 314 (K-LINDL). **Myanmar:** cult. Braine s.n. (K); *Daun* 60 (K); *Forrest* 27694 (E, K); cult. Low (K); cult. J.W. Moore s.n. (K); *Venning* 55 (K). **Nepal:** *Sharma & White* 175 (K); *Wallich* 7354 (K-LINDL, type of *C. elegans*).. **Hort.:** cult. Glasnevin s.n, (K); cult. Godman s.n. (K).

C. elongatum

Malaysia: Sabah: *Bailes & Cribb* s.n. (K); *J. & M.S. Clemens* s.n. (BM), 34331 (BM); *Collenette* A33 (BM), A46 (BM), A47 (BM, K, type of *C. elongatum*); *Lamb* 43/83 (K), 807/87 (K, SAN).

C. ensifolium subsp. ensifolium

Cambodia: *Hamand* 419 & 427 (P); *Thorel* s.n. (P). **China:** Guangdong: *Y.W. Taam* 200 (K); Hainan: *N.K. Chun & C.L. Tso* 44117 (AMES); Hong Kong: *Champion* 525 (K, K-LINDL); *Fowlie* s.n. (K); *S.Y. Hu* 8212 (K). Shanghai: *Levaille* s.n. (E). Sichuan; *Fang* 2302 (K), 2303 (AMES, K); *Kingdon Ward* 4289 (E); **Taiwan:** *Egerod* 62.713-1 (K); *Price* s.n. (K). **Yunnan:** *Bons d'Anty* s.n. (P); *Maire* s.n. (K, P); without exact loc., *Schauer* s.n. (K-LINDL, type of *C. micans*); H.H.J. s.n. (K-LINDL); hort. (K, K-LINDL, type of *C. xiphiifolium*). **India:** *J.D. Hooker & Thompson* 2469 (K); *CL* 3694 (K). **Indonesia:** Java: *Lobb* s.n. (K-LINDL), *Lobb* 302 (K-LINDL); **Japan:** *Maximowicz* s.n. (K); *Kaempfer* drawing (K-LINDL, type of *C. ensatum*). **Laos:** *Vidal* 4398 (P). **Thailand:** *Kerr* 145 (K), 0333 (K), 428 (K), s.n. (K); *van Beusekom & Santisuk* 2843 (P). **Vietnam:** *Bon* 1694 (P); *Chevalier* 29078 (P); *d'Alleizette* s.n. (P); *Hack* 5(P); *Hayak* s.n. (P); *Pierre* 114 (P); *Poilane* 12646 (P); *Simond* s.n. (P).

C. ensifolium subsp. haematodes

India: Mysore: *Ramamoorthy & Ghandi* HPF 2677 (E). **Indonesia:** Java: *Schiffner* 1764 (AMES, BO, K); Sumatra: *Bünnemeyer* 4302 (BO, K): *de Wilde & de Wilde Duyfies* 15719 (K, L); *Lörzing* 12045 (BO, K, L). **Malaysia:** Sabah: *Bailes & Cribb* 746 (K); *Fowlie & Ross* 8281 (K); *Lamb* 877/87 (K), SAN 91507 (K, SAN), 91553 (K, SAN). **Sri Lanka:** *Cramer* s.n. (K); *Macrae* 12 (K, K-LINDL, type of *C. haematodes*). **Thailand:** *Kerr* 0136 (C, K), 242 (K, type of *C. siamense*), s.n. (K); *Menzies & Du Puy* 326 (K); *Satisuk* 2843 (E); *van Beusekom & Santisuk* 2843 (C).

C. ensifolium subsp. acuminatum

Sulawesi: *O'Byrne* s.n. (SING). **Papua New Guinea:** *Cruttwell* 367 (K); *Reeve* 1174 (CBG, K, LAE, L, E, NSW) *Sands* 1850 (K); *Streimann & Kairo* NGF 39369 (K, LAE).

C. erythraeum

Bhutan: *Ludlow, Sherriff & Hicks* 21249 (E). **China:** Sichuan: *Pratt* 771 (K); Yunnan, *Delavay* s.n. (P); *Ducloux* 2983 (P); *Henry* 11100 (K), 11100A (K), 13531 (K), 11371 (K), 11371A (K); *Maire* 927 (E), 2470 (E), 2622 (AMES, K); *Wang* 66960 (AMES) & s.n. (AMES); *T.T. Yu* 17984 (AMES, E). **India:** Assam: *Kingdon-Ward* 6773 (K), 8704 (K); *Stonor* s.n. (E); Bengal: *Clarke* 9546 (K); *Griffith* 5268 (K); *Gamble* 1982 (K), 1982A (K), 1982B (K), 9996 (K), 10008 (K),s.n. (K); *J.D. Hooker* 229 (K-LINDL, type of *C. erythraeum*), 232 p.p. (K-LINDL), s.n. (K); *Wallich* 7356 (K-LINDL); *Pantling* 8 (K); Meghalaya: *Clarke* 44480B. **Myanmar:** cult. *Kirk* s.n. (K); *Venning* 46 (K). **Nepal:** *Buchanan* s.n. (K); *Wallich* s.n. (K). **N. Vietnam:** *Takhtajan* 0733 (LE). **Hort.:** cult. James s.n. (K); cult. Mitten s.n. (K); cult. Sander s.n. (K); cult. Seeger & Tropp s.n. (K); cult. Tate s.n. (K); cult. Veitch s.n. (K).

C. erythrostylum

Vietnam: *J. & M.S. Clemens* 4328 (K, P); *Everard* 2249 (P); *Micholitz* s.n. (K); *Poilane* 3465, 3563, 3660, 4407 (P); *Tixier* s.n. (P). **Hort.:** cult. Easton s.n. (K); cult. Glasnevin B.G. (K, type of *C. erythrostylum*).

C. faberi

China: Guizhou, *Bodinier* 1503 (E, P); *Cavalerie* 3960 (P), & s.n. (P); *Tsi, Luo, Cribb & McGough* ASBK 106 (K, PE); Hupeh: *Silvestri* 276 (FI, K); *Wilson* 605 (P), 606 (K), 1900 (K); Sichuan, *Biet* 1434 (P); *K.L. Chu* 2550 (E); *Farges* 948 (K, PE); *Henry* 5515 (E, K, syntype of *C. faberi*); *Soulié* 1434 (P), s.n. (AMES, K); Taiwan: *Hayata* s.n. (AMES, type of *C. oiwakensis*); *Du Puy* s.n. (K); Yunnan, *Cavalerie* 2233 (K), 2763 (E), 4055 (K), 7561 (K); *Delavay* s.n. (P); *Ducloux* 3558 & 7612 (K, P); *K.M. Feng* 1087 (AMES); *Forrest* 4856 (E); *Henry* 11125 (K, type of *C. yunnanense*); *Maire* 3734 (AMES); *Monbeig* s.n. (K), 92 (K), 243 (E); Zhekiang, *Armacost* 40 (AMES), *Faber* 94 (K, syntype of *C. faberi*). **India:** Himalpradesh: *Bailes* 1064 (K); *Mackinnon* 21743 (K), 22719 (K). **Myanmar:** *Cooper* 6017 (E). **Vietnam:** *Balansa* 315 (P). Cult.: *Camapanini* (K).

C. finlaysonianum

Brunei: *Ariffin* BRUN 16888 (BRUN, K); *Suhaili, Zinin, Ariffin, Sharbini, Ibrahim & Ham* 15023 (BRUN, K). **Cambodia:** *Couderc* s.n. (P); *Godefroy-Lebeuf* 249 (P), 254 (K). **Indonesia:** Java: *Kuhl & van Hasselt* s.n. (L, NSW); *Zollinger* 679 (K, L); Sulawesi: *Grimes* 1214 (K); *Schlechter* 20685 (K); *van Balgooy* 3604 (K, L); Sumatra: *Lörzing* 12104 (K,L). **Malaysia:** Malay peninsula, *Curtis* s.n. (P); *Samsuri Ahmad* SA1007 (C); *Wallich* 7352 (K-LINDL, type of *C. wallichii*); Sabah, *Beaman* 9681 (K, MSC, UKMS); *Carr* SFN27414 (AMES); *Creagh* s.n. (K); *Giles* 600 (K); *Kamarudin Mat Salleh* s.n. (K); *Mason* s.n. (K); *Talip* SAN 87649 (K, SAN); Sarawak: *Haviland* s.n. (K); *Hose* 565 (K); *Jacobs* 5478 (K, L); *Purseglove* P5577 (K); *Synge* 40 (K). **Philippines:** Bohol: *McGregor* 1689 (K); *Cuming* 2049 (K), 2082 (K), 2121 (K), s.n. (K, K-LINDL); *Cuming* s.n. (K-LINDL, type of var. *brevilabre*); Dinagata: *Ramos & Pascasio* 35238 (K); Luzon: *Loher* 566 (K), 5371 (K, 15112 (AMES); *Lyon* 101 (AMES, K); *Merrill* 4111 (K); *Ramos & Edaòo* 28530 (K); *Whitford* 87 (K); Polillo; *McGregor* 10460 (AMES); Rizal: *Loher*; Mindanao: *Hutchinson* 4829 (K); *Taylor* s.n. (AMES); Palawan: *Foxworthy* 635 (K). **Singapore:** cult. Loddiges (K-LINDL). **Thailand:** *Geesinck, Hattinck & Phengklai* 6507 (C, L); *Kerr* 0168 (K), 0641 (K), 0701 (K), s.n. (K); *Schmidt* 143 (C). **Vietnam:** *J.B.* 3351 (P); *J. & M.S. Clemens* 4071 (P); *Everard* 2691 (P); *Finlayson in Wallich* 7358 (K-LINDL, type of *C. finlaysonianum*); *Poilane* 3793, 11059 & 11211 (P); *Regnier* 8 & 320 (P); *Thorel* 1236 (P). **Without exact loc.,** without coll. (K-LINDL). **Hort.:** cult. Van Imschoot s.n. (K).

C. floribundum

China: Fukien: *Dunn* 7257 (AMES). Guangxi, *A.N. Steward & Chow* 1192 (P); *Y.W. Taam* 554 & 678 (P); Guizhou: *Bodinier* 2209 (P); *Cavalerie* s.n. (K); *Cavalerie* 3988 (K, P); *Esquirol* 6697 (P). Guangdong, *A.N. Steward & H.C. Chow* 129 (AMES, P); *Y.W. Taam* 554 (K), 678 (K); *W.T. Tsang* 20491 (K, P), 20702 (K); Shanghai, *Debeaux* s.n. (P); Sichuan, *W.P. Fang* 798 (K, P) & 2303 (P); *Monbeig* s.n. (P); *Soulié* 1452 & s.n. (P); Yunnan, *Beauvais* 310 (P); *Cavalerie* 2233 p.p. (K); 7493 (K), 8114 (K); *Forrest* 16529 (K), 19076 (K); *Hancock* 580 (K); *Handel-Mazzetti* 13059 (AMES); *Henry* 12971 (K); *Maire* 6415 (AMES); *Monbeig* s.n. (type of *C. pumilum*, 5 sheets); *Rock* 8765 (AMES); *Soulie* s.n. (K); without exact loc., Chinese drawing ex Hort. Soc. (K-LINDL, type of *C. floribundum*); Zhekiang: *Hu* 196 (K). **Japan:** cult. Kew 200-1900 (K, paratype of *C. pumilum*). **Vietnam:** *Pételot* 2278 (AMES).

C. gammieanum

India: *Du Puy* 120 (K); *Gamble* 9991 (K); *Gammie* s.n. (K); *Pantling* 299 (K, type of *C. gammieanum*). **Hort.:** cult. R.B.G. Kew (K); cult. Sander (K).

C. goeringii

China: Guangdong, *Bowring* 10180 (K), 10246 (NSW); *Y.W. Taam* 554 (AMES); Guizhou: *Bodinier* 2073 (P); *Cavalerie* 7493 (E), s.n. (P); *Esquirol* s.n. (photo. from B, type of *C. serratum*); Shanghai, *Poli* s.n. (P); Sichuan, *H.H.Chung* 30 (AMES) & 31 (AMES); *Farges* 944 (K, P), s.n. (K); *Z.Y. Liu* 15238 (K, PE); *Soulié* 1529 (K, P) & s.n. (P).Yunnan, *R.C. Ching* 20563 (AMES); *Forrest* 328 (K), 11485 (K), 12283 (E), 27747 (K, P); *Henry* 360 (K); *Maire* 29902 (P); *Monbeig* s.n. (K). Zhejiang: *Faber* s.n. p.p. (K). **India:** Himapradesh: *Mackinnon* 22709 (K), 24152 (K, type of *C. mackinnoni*). **Japan:** *Buerger* s.n. (K); *Clohan* 577 (K); *Dickens* s.n. (P); *Faurie* 36 (K,P), 157 (P), 248 (P) & 11996 (P); *Furuse* 10721 (K), 47358 (K); *Goering* 365 (P); *Greatrex* H2320/51 (K); *Hong Kong Bot. & Forestry Dept.* s.n. (K); *Hotta* 11781 (C, K, NSW, P); *Kanai* 9417 (K); *Kii* s.n. (P); *Krebs* s.n. (C); *Maximowicz* s.n. (C, K, P); *Musashi* s.n. (P); *Okamoto* TNS1391 (C, K, NSW, P); *Oldham* s.n. (C); *Savatier* 1329 (K, P); *Togasi* TNS1282 (C, K, NSW, P); *Tsuchida* 90 (K); *Watanabe* s.n. (K); *Zollinger* s.n. (P); cult. Rollissons (K-LINDL, type of *C. virescens*). **Korea:** *Wilford* 806 (K, K-LINDL).

C. hartinahianum

Indonesia: Sumatra: *de Wilde & de Wilde-Duyfies* 13265 (K, L), 15331 (K, L), 16405 (K, L), 16081 (K, L).

C. hookerianum

Bhutan: *Griffith* 5270 (K); *Kingdon-Ward* 6427 (K); *Ludlow, Sherriff & Hicks* 20511 (E). **China:** Guizhou, *Cavalerie* 4609 (P); Sichuan, *Monbeig* s.n. (K, P); *Soulié* 1530 & s.n. (P); Yunnan, *d'Orleans* s.n. (P); *Ducloux* 2981 & 3165 (P); *Rock* 7120 (AMES); S. Xizang: *Ludlow & Sherriff* 1245 (E). **India:** Assam: *Kingdon-Ward* 8018 (K); Sikkim, *Clarke* 27043 (K), *Gamble* 4007A (K); *Griffith* 5265 (K), s.n. (K-LINDL); *J.D. Hooker* 227 (K-LINDL), 227 p.p. (K-LINDL), 233 (K-LINDL) & s.n. (AMES, K, P); *Pantling* 63 (K). **Myanmar:** *Naw Mu Pa* 15503 (K); Parish 3 (K).

C. insigne subsp. insigne

China: Hainan, *McClure* 1844 (AMES), 1974 (AMES) & s.n. (AMES). **Vietnam:** *Averyanov, Ban, Binh, A. & L. Budantzev, Huyen, Loc, Tam & Yakovlev* VH 130 (LE), VH 566 (LE), VH 2291a (LE, P) &VH 2291b (LE); *Bronckart* 43 (K, type of *C. insigne*); *Chevalier* 30789 (P); *de Sigaldi* s.n. (P); *Everard* 22167 (P); *Jacquet* 631 (P); *Micholitz* s.n. (K). **Hort.:** cult. Gurney-Fowler s.n. (K); cult. Kew s.n. (K).

C. insigne subsp. seidenfadenii

Thailand: *Menzies & Du Puy* 500 (K), 501 (K); *Nuyomdham & Vidal* 447 (C, P); *Sorenson, Larsen & Hansen* 6301 (C).

C. iridioides

China: Yunnan: *Henry* 12965 (K), 12965A (K); *Rock* 2873 (AMES). **India:** Assam: *Bor* 20927 (K); *Hinde* s.n. (K); *Parry* 402 (K); Bengal: *Clarke* 9482 (K); *Gamble* 9914 (K); *Griffith* 5269 (K); Himalpradesh: *Strachey & Winterbottom* 26 (K); Megalhaya: *J.D. Hooker & Thompson* s.n. (K); *Mann* 4844 (K); Manipur: *Meerbold* 6918 (K), 10808 (K); *Prain* 22 (K); Sikkim: *Griffith* s.n. (K-LINDL); *J.D. Hooker* 227 p.p. (K-LINDL), 227 (K); *Pantling* 12 (K). **Myanmar:** *Micholitz* s.n. (K). **Nepal:** J.D. *Hooker* s.n. (K); *Sharma & White* 176 (K); *Wallich* 7355 (K). **Hort.:** cult. Glasgow B.G. (K); cult. RBG Kew s.n. (K); cult. Lady Sainsbury (K).

C. lancifolium var. lancifolium

China: Guangxi: *H.N. Qin et al.* 896120 (K, PE); Jiangxi: *S.K. Lau* 4828 (AMES); Taiwan: *Strickland* s.n. (K); SE Xizang: *Rock* 10184 (AMES); Yunnan: *Delavay* s.n. (P). **India:** *Gibson* s.n. (K, K-LINDL); *Griffith* s.n. (K-LINDL); *Griffith* 17 (K-LINDL), 5390 (K); *J.D. Hooker & T. Thompson* 139 (K- LINDL); *Pantling* 75 (K); *Prain* 12 (K); Meghalaya: *Clarke* 44445 (K); *J.D. Hooker & Thompson* 1539 (K), s.n. (K); *Koelz* 23414 (K); Naga Hills: *Sepoy orderly* 24 (K). **Indonesia:** Java: *Comber* 1145 (K), 1267 (K); *Kostermans* 6310 (BO, K, L); *Kuhl & van Hasselt* s.n. (NSW); *Lobb* 187 (K, K-LINDL), s.n. (K); *L. Linden* s.n. (K); *J.J. Smith* s.n. (BO, K); Sumatra, *Comber* 1432 (K); *Ericsson* s.n. (K); *de Wilde & de Wilde-Duyfies* 16431 (BO, K, L); *Hagerup* s.n. (C); *Rahmat Si Boeea* 6026 (AMES), 10222 (AMES, BO) & 10449 (AMES). **Japan:** *Greatrex* H2320/51 (K); *Maximowicz* s.n. (K). **Laos:** *Kerr* 10013 (K). **Malaysia:** Sabah, *Carr* 3691 (AMES); *Collenette* 598 (K); *Lamb* SAN 91582 (K, SAN), SAN 91570 (K, SAN), SAN 93468 (K, SAN), T22 (K); *Lim, Shahril & Soinin* SAN 143704 (K, SAN); *Wood* 720 (K); Sarawak: *Christensen & Apu* 97 (AAU, K); *Lewis* 299 (K). **Nepal:** *Wallich* 7351 (K-LINDL, type of *C. lancifolium*). **Myanmar:** *Baldworth* 13546 (K); *Kingdon-Ward* 8491 (K); *Venning* 20 (K). **Nepal:** *Schilling* 4 (K); *Wallich* s.n. (K, type of *C. lancifolium*). **Papua New Guinea:** *Rees & Reeve* 391 (K); *Reeve* 702 (CBG, E, K, L, LAE, NSW). **Philippines:** Palawan: *Edaòo* 390 (AMES). **Thailand:** *Garrett* 571 (K); *Kerr* 227 (K, type of *C. kerrii*); *Smitinand* 986 (P). **Vietnam:** *Averyanov* 1826 (69) & 2290 (LE), 2376(296) (LE); *Averyanov & Binh* VH 3703 & 3777 (LE); *Averyanov, Binh, Duy & Loc* VH 2593 (LE); *Averyanov, Binh & Loc* VH 3177 (LE) & 3529 (LE); *Averyanov & Hiep* VH 4964 (LE); *Averyanov, Hiep & Huyen* HG 081 (LE); *Averyanov, Hiep & Loc* VH 4603 (LE); *Averyanov, Ban, Binh, A. & L. Budantzev, Huyen, Loc, Tam & Yakovlev* VH 330 (LE, P), 622 (LE), 838c (LE) & 1108 (LE); *de Sigaldi* 106/TS (P); *Everard* 2085 (P); *Hack* s.n. (P); *Hiep, Loc & Averyanov* NTH 3719 (LE) & 3805 (LE); *Kovetsko-Vyemthanska* 1577 (LE) & s.n. (LE); *Loc, Hoang & Averyanov* CBL 1276 (LE), CBL 1749 (LE) & CBL 1942 (LE); *Takhtajan* 37 (LE) & 63 (LE). **Without exact loc.:** hort. (K-LINDL).

C. lancifolium var. aspidistrifolium

China: Hong Kong: *Barretto* 264 (K); Taiwan: *Strickland* s.n. (K); Yunnan: *Henry* 12723 (K), 12975 (K).

C. lancifolium var. papuanum

Papua New Guinea: *Reeve* 437 (E, K, L, LAE, NSW); *Reeve* 3436 (NSW); *Reeve* 4821 (NSW); *Stevens* LAE 58147 (K, LAE).

C. lowianum

Burma: *Dickason* 9220 (AMES) & 9710 (AMES). **China:** Yunnan, *Beauvais* s.n. (P); *R.C. Ching* 20452 (AMES); *Ducloux* 2554 (P); *K.M. Feng* 12651 (AMES); *Forrest* 4859 (E); *Henry* 12211 (K), 12211A (K); *Maire* 470 (E, K); *H.T. Tsai* 55692 (AMES, K), 55861 (AMES, K); *C.W. Wang* s.n. (AMES*); T.T. Yu* 23024 (AMES). N. Vietnam, *Hiep, Hoang & Averyanov* s.n. (LE). **Myanmar:** *Bogg* s.n. (E); *Boxall* s.n. (K); *Kingdon-Ward* 206 (E). **Thailand:** *Garrett* 345 (K), 657 (K); *Kerr* 409 (K), s.n. (K); *Put* 3760 (K). **Hort.:** cult. Eastwood s.n. (K, type of var. *concolor*); cult. Grose-Smith s.n. (K); cult. R.B. G. Kew s.n. (K); cult. Lawrence s.n. (K); cult. Seth s.n. (K); cult. Veitch s.n. (K).

C. lowianum var. iansonii

Hort.: cult. Glasnevin B.G. (K); cult. Kew (K); cult. Low s.n (K, type of *C. iansonii*); cult. Wigan s.n. (K).

C. macrorhizon

China: Guizhou, *Cavalerie & Fortunat* 1217 (P); Yunnan, *Ducloux* 3562 & 7072 (P); *C.W. Wang* s.n. (AMES). **India:** *Clarke* 38319 (K); *Coventry* 1039 (K); *Inayat* s.n. (K); *Kurzweil* 144 (K); *Mackinnon* 25403 (K), s.n. (K, NSW); *Mann* s.n. (K); *Parker* s.n. (K); *Pradhan* 22 (K), s.n. (K); *Rich* 300 (K); Royle sketch (K-LINDL); *Sander's coll.* s.n. (K); *Saunders* s.n. (K). **Japan:** *S. Arimoto* s.n. (AMES); Imperial University s.n. (K). **Thailand:** *Garrett* 787 (K); *Geesinck, Phanichapol & Santisuk* 5534 (C); *Kerr* 230 (K).

C. madidum

Australia: *Cecil* 37 (K); *Dallachy* s.n. (K); *Domin* s.n. (K); *Longman* s.n. (K). **Hort.:** cult. Bennett-Poe s.n. (K0; cult. Bull s.n. (K); cult. Chcarnley s.n. (K); cult. Glasnevin B.G. (K); cult. Kew (K); cult. Raphael s.n. (K); cult. Petrop. (LE, type of *C. queenianum*); cult. Rollissons (K-LINDL, type of *C. madidum*); cult. Veitch s.n. (K).

C. mastersii

India: Assam: *Hinde* s.n. (K); Manipur: *Koelz* 27020 (K); Meghalaya: *Clarke* 15392 (K); *Griffith* 5267 (K), s.n. (K, K-LINDL, type of *C. micromeson*); *J.D. Hooker & Thomson* s.n. (K); *P.L.* 359 (K); *Mann* s.n. (K); Bengal: *Griffith* s.n. (K, K-LINDL, type of *C. affine*); *J.D. Hooker* 230 (K-LINDL) & 232 p.p. (K-LINDL); *Pantling* 101 (K). **Myanmar:** *Vernon* s.n. (K). **Thailand:** *Garrett* 907 (K); *Kerr* 204 (K), 320 (K); *Menzies & Du Puy* 399 (K). **Without exact loc.:** cult. Glasnevin s.n. (K); cult. Loddiges (K-LINDL, type of *C. mastersii*); cult. J.W. Moore s.n. (K); cult. T. Moore s.n. (K0; cult. Veitch s.n. (K); cult B.S. Williams s.n. (K).

C. munronianum

India: *Cowan* s.n. (K); *Haines* 595 (K); *Pantling* 256 (K).

C. omeiense

China: Sichuan: *Farges* 944 (K, P).

C. parishii

Myanmar: *Parish* 56 (K, W, type of *C. parishii*).

C. rectum

Indonesia: Kalimantan: *de Vogel* s.n., cult. Leiden 913298 (K, L). **Malaysia:** Malay Peninsula: *Ridley* 11370 (K); Sabah: *Chan* 1 (K), s.n. (K); *Lamb* 76/83 (K).

C. roseum

Indonesia: Java: *Micholitz* s.n. (K); Sumatra, *Micholitz* s.n. (AMES).

C. sanderae

Myanmar: cult. Menniger s.n. (K). **S Vietnam:** *Micholitz* s.n. (K, type of *C. parishii* var. *sanderae*); cult. Sander s.n. (K). **Hort.:** cult. Menninger s.n. (K); cult. Rothschild s.n. (K).

C. schroederi

Vietnam: *Averyanov, Ban, Binh, A. & L. Budantzev, Huyen, Loc, Tam & Yakovlev* VH 108 (LE, P) & 295 (LE, P); *Averyanov, Binh, Duy & Loc* VH 2699 (LE); *Averyanov, Binh, Hiep, Loc & Lowry* VH 4437a (LE); *Averyanov, Binh & Loc* VH 4245a (LE); *Chevalier* 30789 (P); *Dalat Inst.* 302 (LE); *Delacour* s.n. (P); *de Sigaldi* 96 & 181 (P); *Gaurdon* 213 (P); *Poilane* 24519 & 32079 (P). **Hort.:** cult. Glasnevin B.G. (K); cult. R.B.G. Kew s.n. (K, type of *C. schroederae*); cult. Low s.n. (K).

C. sigmoideum

Indonesia: Java: *J.J. Smith* 150 (BO, K); Sumatra: *Micholitz* s.n. (K). **Malaysia:** Sabah: *Chan & Nais* s.n. (SAN, SNP, type of *C. kinabaluense*).

C. sinense

China: Guangdong; *Metcalfe* s.n. (K); Guangxi, *W.T. Tsaang* 23316 (P); without exact loc., cult. (K-LINDL); Hong Kong: *Barretto* s.n. (K); *Herklots* s.n. (K); Yunnan: *Henry* 12722 (K); Taiwan: *Egerod* 62.708.2 (K). **India:** cult. Chandra Nursery s.n. (K); *Clarke* 44473A (K); *J.D. Hooker & T. Thompson* 226 (K, K-LINDL), 2496 (K); *Mann* s.n. (K). **Thailand:** *Bare* s.n. (K); *Kurzweil* 634 (K). **S. Vietnam:** 17 March 1995, *L. Averyanov, N.T.Ban, A. & L. Budanttsev, N.T. Hiep, D.D.Huyen, P.K. Loc & G. Yakovlev* VH 838b (LE); *Averyanov* 2016 (LE); *Harder, Loc, Hiep & P.K. Long* DKH 4637 (LE); *Averyanov* 136-CPB-85 (LE).

C. suave

Australia: *Brown* 5504 (BM, K); *Coveny, Hancock & Hind* 8460 (K, NSW); *Coveny & Hind* 9106 (K, NSW); *Mueller* s.n. (K); *Rodway* 1208 (K), 2390 (K); *Stuart* s.n. (K); *Verreaux* 155 (K-LINDL).

C. suavissimum

Myanmar: cult. R. Brown s.n. (K). **Vietnam:** *L. Averyanov, P.K. Loc, N.X Tam* CBL 400 (LE) & CBL 401 (LE). **Hort.:** cult. Brown s.n. (K).

C. tigrinum

Myanmar: *Parkinson* 5 (K); *Parish* 149 (K, W, type of *C. tigrinum*); cult. R.B.G. Kew s.n. (K); cult. Shuttleworth s.n. (K); cult. Shuttelworth & Carder s.n. (K). **Thailand:** *Berkeley* s.n. (K).

C. tortisepalum var. tortisepalum

China: Taiwan, *Fukuyama* 3983 (holo. KANA).

C. tortisepalum var. longibracteatum

China: Sichuan, *Soulié* 1565 (P); Yunnan, *Ducloux* 602 & 3972 (P); *Maire* s.n. (P).

C. tracyanum

China: Yunnan: *Ducloux* 362 (K), s.n. (K); *Forrest* 19227 (K); *Hancock* 583 (K); *Henry* 11097 (K), 11858 (K); *H.T. Tsai* 56700 (K), 59687 (K); *T.T. Yu* 23024 (E). **Myanmar:** cult. R.B.G. Kew 000.69.16904 (K). **Thailand:** *Bare* s.n. (K); *Cumberlege* 1033 (K); *Garrett* 609 (K); *Put* 4502 (K); *Kurzweil* 566 (K, NBG).

C. wenshanense

China: Yunnan: cult. Kunming B.G. (K).

C. whiteae

India: Sikkim, *Pantling* 425 (K, P, type of *C. whiteae*).

C. wilsonii

China: Yunnan: *R.A.* 840 (E); *Wilson* s.n. (K), type of *C. wilsonii*); cult. Edinburgh B.G. (E, K). **Myanmar:** *Prazer* s.n. (K). **S Vietnam:** *Averyanov, Ban, Binh, A. & L. Budantzev, Huyen, Loc, Tam & Yakovlev* VH 2291c (LE). **Hort.:** cult. Kew (K).

GLOSSARY

abaxial – the side of an organ away from the axis.

abscission layer – zone of detachment of leaf or other organ.

acuminate – having a gradually diminishing point.

amplexicaul – clasping the stem.

anther – the part of the stamen containing the pollen or pollinia.

anticlinal – perpendicular to the surface.

apiculate – with a short sharp point.

arcuate – curved.

articulated – jointed.

auricles – small ear-like flaps.

autotrophic – applied to plants which produce their own food by photosynthesis.

axil – angle formed by leaf- or bract-base and stem.

bract – leaf-like organ subtending a flower.

callus – structure on the upper surface of the lip, usually comprising two or three ridges in *Cymbidium*.

canaliculate – with a longitudinal groove.

capsule – a dry dehiscent seed-vessel.

cataphyll – the early leaf forms of a shoot.

caudicle – the cartilaginous strap that joins the pollinia to the viscidium.

chloroplast – cell-organelle that contains the chlorophyll.

ciliate – bearing hairs along the margin.

clade – a monophyletic group.

cladistic analysis – a method of systematics used to construct phylogenies of organisms and to construct classifications which are based on the identification of clades and their inter-relationships.

cladogram – a branching diagram of taxa showing relationships as defined by synapomorphies (derived or novel states).

clavate – club-shaped.

column – the combination of the stamens and styles into the central organ of the orchid flower.

column-foot – the basal extension of the column.

conduplicate – folded together lengthwise.

convolute – rolled longitudinally with the margins overlapping.

cordate – heart-shaped.

coriaceous – leathery.

corolla – petals.

crenulate – with small teeth.

cucullate – hooded.

cultivar – a cultivated variety; a taxonomic rank used for varieties maintained in cultivation.

cuneate – wedge-shaped.

cuspidate – tipped with a rigid point.

cuticle – the outermost layer.

cymbiform – boat-shaped.

decurved – curved down.

dimorphic – occurring in two forms.

diploid – an organism or cell with twice the haploid number of chromosomes in its nuclei.

disc – the area in the basal part of the lip between the sidelobes and the midlobe.

distichous – borne in two ranks.

duplicate – folded.

emarginate – notched.

epiphyte – growing on a plant.

erose – gnawed.

exserted – protruding beyond.

falcate – sickle-shaped.

filiform – thread-shaped.

fimbriate – with a border of long slender processes.

fractiflex – zig-zag.

fusiform – spindle-shaped.

glabrous – lacking hairs.

haploid – a nucleus or individual containing only one representative of each chromosome of the chromosome complement.

holotype – the one specimen forming the basis for the original description of a new species and designated as such.

homonym – a name rejected because of an earlier application to another taxon.

hyaline – translucent.

hypochile – the basal part of a lip.

indumentum – any covering.

inflorescence – the flowers on the floral axis.

involute – with the edges rolled inwards.

isolectotype – duplicate specimen of the holotype.

lamina – the blade as of a leaf or petal.

lanceolate – narrow and tapering to each end.

lectotype – a specimen chosen from among syntypes.

lignified – converted into wood.

ligulate – tongue-shaped.

lip – the modified third petal of the orchid flower.

lithophyte – plant growing on a rock.

meiotic – applied to the reduction division of chromosomes.

mentum – the chin-like structure formed by a column-foot and the enclosing bases of the lateral sepals.

mesophyll – the interior parenchyma of the leaf.

mitotic – referring to normal cell division.

monopodial – a stem with a single axis.

motile – moveable.

mucilaginous – slimy.

mucro – sharp terminal point.

obcordate – inversely heart-shaped.

obovate – inversely ovate.

ovary – the part of the flower that contains the ovules.

ovate – shaped like the longitudinal section of an egg.

palisade cells – perpendicular elongated cells on the surface of most leaves.

paniculate – furnished with a branched raceme.

papillae – soft superficial protuberances.

pedicel – flower stalk.

peduncle – the inflorescence axis below the lowermost flower.

peloric – relating to an irregular flower becoming regular, e.g. when the lip of an orchid becomes petaloid.

perianth – the calyx and corolla.

periclinal – curved in the same direction as the surface.

petiole – the leaf stalk.

phloem – vascular tissue transporting nutrients around the plant.

plicate – pleated.

pollinarium – the pollinia, stalk and viscidium in an orchid.

pollinium (pl. pollinia) – the pollen masses in orchids.

polyad – structure in pollinium of many pollen grains.

polyploid – of organisms or cells with three or more complete sets of chromosomes in their nuclei.

porrect – pointing forwards.

pubescence – hairiness.

pyriform – pear-shaped.

raceme – a simple inflorescence bearing pedicellate flowers.

reclinate – turned or bent downwards.

reniform – kidney-shaped.

rhachis – the part of the inflorescence axis that bears flowers.

rhizome – a horizontal stem.

rhombic – an equilateral oblique-angled figure.

rostellum – a narrow extension of the upper edge of the stigma in orchids.

saccate – pouched.

saprophyte – a plant living upon dead organic matter.

scape – floral axis.

scarious – thin, dry and membranous.

schlerenchyma – thick-walled cells.

sepal – segment of calyx.

serrulate – bearing small saw-like teeth.

sessile – without a stalk.

sigmoid – S-shaped.

sinus – a recess.

s.n. – sine numero (without number).

spathulate – shaped like a spatula.

stigma – the pollen receiving part of the gynoecium.

stipe – stalk joining pollinium and viscidium.

stoma (pl. stomata) – pore on the leaf surface allowing gaseous exchange.

subulate – with a fine sharp point.

sulcate – grooved.

superposed – placed on top of.

sympodium – stem made up of successive growths.

synapormorphy – shared derived character states (term used in cladistics).

syntype – one of two or more specimens cited with the description of a new taxon when none is designated the holotype.

terete – circular in cross-section.

testa – outer covering of seed.

tetrad – a body of four cells as in the formation of pollen.

tetraploid – cell or individual with four times the haploid chromosome number.

triploid – cell or individual with three times the haploid chromosome number.

truncate – cut off.

t.s. – transverse section.

vascular tissue – the cells which transport water (xylem) and nutrients (phloem) around the plant.

velamen – the layer of dead cells that sheaths the root in orchids.

viscidium – the sticky disc to which the pollinia are attached in orchids.

xeromorphic – with adaptations to dry conditions.

zygomorphic – bilaterally symmetrical, of flowers.

INDEX OF SCIENTIFIC NAMES